Sprinkler Fitting

Level Two

Trainee Guide
Fourth Edition

nccer

AFSA
American Fire Sprinkler Association

Pearson

NCCER

President and Chief Executive Officer: Boyd Worsham
Vice President – Innovation and Advancement: Jennifer Wilkerson
Chief Learning Officer: Lisa Strite
Sprinkler Fitting Project Managers: Lauren Corley, Veronica Westfall
Senior Manager of Projects: Chris Wilson
Senior Production Manager: Erin O'Nora
Testing/Assessment Project Manager: Elizabeth Schlaupitz
Testing and Assessment Project Coordinator: Chelsi Santana
Lead Technical Writer: Veronica Westfall
Art Manager: Carrie Pelzer
Permissions Specialists: Kay Lewis, Amanda Smith
Desktop Publishing Manager: James McKay
Production Specialists: Gene Page, Eric Caraballoso, Daphney Milian
Digital Content Coordinator: Rachael Downs
Managing Editor: Graham Hack
Editors: Jordan Hutchinson, Karina Kuchta, Hannah Murray, Zi Meng
Production Artist: Judd Ivines

Pearson

Manager, Program Management: Vanessa Price
Employability Solutions Coordinator: Monica Perez
Senior Content Producer: Alexandrina B. Wolf
Content Producers: Alma Dabral, Aurodeep Mukherjee

Composition: NCCER
Text Fonts: Palatino and Univers
Content Technologies: Gnostyx

Cover Image

Cover image of activated sprinkler courtesy of Tyco/Johnson Controls, Inc.

Credits and acknowledgments for content borrowed from other sources and reproduced, with permission, in this textbook appear at the end of each module.

AFSA
American Fire Sprinkler Association

Pearson

ISBN-13: 978-0-13-747571-1

Preface

To the Trainee

Sprinkler fitting is dedicated to the preservation of life and property, and carries with it a tremendous amount of responsibility. The work of a sprinkler fitter may save lives and thousands of dollars in property. Therefore, it is essential that sprinkler fitters demand excellence and precision from themselves and coworkers.

Sprinkler fitters may install new systems, retrofit old buildings with updated sprinkler systems, or repair existing systems. Accordingly, the demand for sprinkler fitters grows with new construction and code changes. All of this results in higher demand for trained workers, which generates higher wages. Stay on top of industry developments, code changes, and emerging technologies, to broaden opportunities and earning potential.

NCCER wishes you success as you embark on your second year of training in the sprinkler fitting trade, and we hope that you'll continue your training outside of this series. By taking advantage of training opportunities as they arise, you'll demonstrate initiative and a desire to learn—qualities that are present in the industry's best professionals.

We invite you to visit the NCCER website at **https://www.nccer.org** for information on the latest product releases and training.

Your feedback is welcome. You may email your comments to **curriculum@nccer.org** or send general comments and inquiries to **info@nccer.org**.

New with *Sprinkler Fitting Level Two*

NCCER and Pearson are pleased to present *Sprinkler Fitting Level Two*. In partnership with the American Fire Sprinkler Association (AFSA), this book has been updated and enhanced to provide the most comprehensive apprenticeship training in the sprinkler industry. AFSA is the voice of sprinkler fitting professionals and the leading authority in the sprinkler industry for education, training, consulting, information, representation, and advocacy.

Some of the features of this edition include the following:

- Revised by subject matter experts who sit on the code-making panels of the National Fire Protection Association (NFPA) to reflect the latest edition of NFPA 13, *Standard for the Installation of Sprinkler Systems*, as well as all other applicable codes and standards
- Updated with hundreds of new images to represent the latest in tools, technology, and materials used in the sprinkler industry, with input and review by major manufacturers, including Johnson Controls, Inc.; Reliable Automatic Sprinkler Co., Inc.; Victaulic; The Viking Corporation; and more
- Instructional material revised by leading educators and contractors in the sprinkler industry to reflect the most pertinent information and practical examples for sprinkler apprentices that can be applied immediately, both in the classroom and on the job
- New shop drawings and 3D images from NICET-certified sprinkler system layout technicians to reflect actual project drawings

NCCER Standardized Curricula

NCCER is a not-for-profit 501(c)(3) education foundation established in 1996 by the world's largest and most progressive construction companies and national construction associations. It was founded to address the severe workforce shortage facing the industry and to develop a standardized training process and curricula. Today, NCCER is supported by hundreds of leading construction and maintenance companies, manufacturers, and national associations. The NCCER Standardized Curricula was developed by NCCER in partnership with Pearson, the world's largest educational publisher.

Some features of the NCCER Standardized Curricula are as follows:

- An industry-proven record of success
- Curricula developed by the industry, for the industry
- National standardization providing portability of learned job skills and educational credits
- Compliance with the Office of Apprenticeship requirements for related classroom training (*CFR 29:29*)
- Well-illustrated, up-to-date, and practical information

NCCER also maintains the NCCER Registry, which provides transcripts, certificates, and wallet cards to individuals who have successfully completed a level of training within a craft in NCCER's Curricula. *Training programs must be delivered by an NCCER Accredited Training Sponsor in order to receive these credentials.*

For information on NCCER's credentials and the NCCER Registry, contact NCCER Customer Service at 1-888-622-3720 or visit **https://www.nccer.org**.

Special Features

In an effort to provide a comprehensive and user-friendly training resource, this curriculum showcases several informative features. Whether you are a visual or hands-on learner, these features are intended to enhance your knowledge of the construction industry as you progress in your training. Some of the features you may find in the curriculum are explained below.

Introduction

This introductory page, found at the beginning of each module, lists the module Objectives, Performance Tasks, and Trade Terms. The Objectives list the knowledge you will acquire after successfully completing the module. The Performance Tasks give you an opportunity to apply your knowledge to real-world tasks. The Trade Terms are industry-specific vocabulary that you will learn as you study the module.

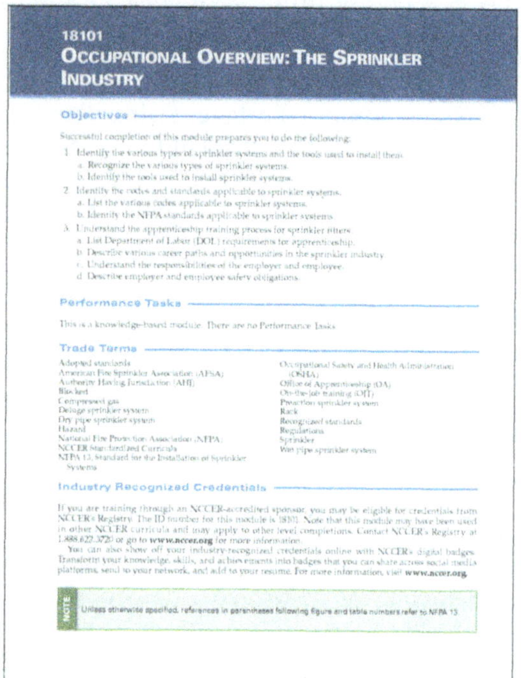

Trade Features

Trade features present technical tips and professional practices based on real-life scenarios similar to those you might encounter on the jobsite.

Bowline Trivia

Some people use this saying to help them remember how to tie a bowline: "The rabbit comes out of his hole, around a tree, and back into the hole."

Figures and Tables

Photographs, drawings, diagrams, and tables are used throughout each module to illustrate important concepts and provide clarity for complex instructions. Text references to figures and tables are emphasized with *italic* type.

Figure 31 Marking a cutting line.

Notes, Cautions, and Warnings

Safety features are set off from the main text in highlighted boxes and categorized according to the potential danger involved. Notes simply provide additional information. Cautions flag a hazardous issue that could cause damage to materials or equipment. Warnings stress a potentially dangerous situation that could result in injury or death to workers.

NOTE

Nameplates must be posted on each material handling device. The nameplate must indicate

CAUTION

It is essential to note the revision designation on a construction drawing and to use only the latest

WARNING!

Saw teeth are very sharp. Use gloves and do not handle the saw teeth with bare hands. When cutting with a saw, ensure that your fingers remain clear of the teeth at all times.

Going Green

Going Green looks at ways to preserve the environment, save energy, and make good choices regarding the health of the planet. Through the introduction of new construction practices and products, you will see how the greening of America has already taken root.

> **Going Green**
>
> **Big Horn Wind Farm**
>
> The Big Horn wind farm uses 133 GE turbines, each rising 119 meters (389 feet) from the ground to the top of the blade tips. The wind farm uses only a

Did You Know?

The *Did You Know?* features offer hints, tips, and other helpful bits of information.

> **Did You Know?**
>
> **How Blueprints Started**
>
> The process for making blueprints was developed in 1842 by an English astronomer named Sir John F. Herschel. The method involved coating a paper with a special chemical. After the coating dried, an original hand drawing was placed on top of the paper. Both papers were then covered with a piece of glass and set in the sunlight for about an hour. The coated paper was developed much like a photograph. After a cold-water wash, the coated paper turned blue, and the lines of the drawing remained white.

Step-by-Step Instructions

Step-by-step instructions are used throughout to guide you through technical procedures and tasks from start to finish. These steps show you not only how to perform a task but also how to do it safely and efficiently.

> Perform the following steps to erect this system area scaffold:
>
> *Step 1* Gather and inspect all scaffold equipment for the scaffold arrangement.
>
> *Step 2* Place appropriate mudsills in their approximate locations.
>
> *Step 3* Attach the screw jacks to the mudsills.
>
> *Step 4* Adjust the screw jacks to near their lowest position.
>
> *Step 5* Determine the location of the highest base.

Trade Terms

Each module presents a list of Trade Terms that are discussed within the text and defined in the Glossary at the end of the module. These terms are denoted in the text with **bold, blue type** upon their first occurance. To make searches for key information easier, a comprehensive Glossary of Trade Terms from all modules is located at the back of this book.

> During a rigging operation, the **load** being lifted or moved must be connected to the apparatus, such as a crane, that will provide the power for movement. The connector—the link between the load and the apparatus—is often a sling made of synthetic, chain, or **wire rope** materials. This section focuses on three types of slings:
>
> - Synthetic slings
> - Alloy steel chain slings
> - Wire rope slings

Section Review

Section Review questions can be found at the end of each section to test your knowledge of the content.

> **1.0.0 Section Review**
>
> 1. The type of sprinkler system that might be found in an aircraft hangar is a _____.
>   a. dry pipe sprinkler system
>   b. wet pipe sprinkler system
>   c. preaction sprinkler system
>   d. deluge sprinkler system
> 2. A type of ladder that provides a platform for standing is a(n) _____.
>   a. A-frame
>   b. extension trestle ladder
>   c. podium ladder
>   d. movable leg ladder

Review Questions

Review Questions are provided to reinforce the knowledge you have gained. This makes them a useful tool for measuring what you have learned.

> **Review Questions**
>
> 1. Identification tags for slings must include the _____.
>   a. type of protective pads to use
>   b. type of damage sustained during use
>   c. color of the tattle-tail
>   d. manufacturer's name or trademark
> 2. The type of wire rope core that is susceptible to heat damage at relatively low temperatures is the _____.
>   a. fiber core
>   b. strand core
>   c. independent wire rope core
>   d. metallic link supporting core
> 3. Synthetic slings must be inspected _____.
>   a. once every month
>   b. visually at the start of each work week
>   c. before every use
>   d. once wear or damage becomes apparent
> 4. An alloy steel chain sling must be removed from service if there is evidence that _____
> 7. Chain hoists are able to lift heavy loads by utilizing a _____.
>   a. rope and pulley system
>   b. rigger's strength
>   c. stationary counterweight
>   d. gear system
> 8. Before attempting to lift a load with a chain hoist, make sure that the _____.
>   a. hoist is secured to a come-along
>   b. load is properly balanced
>   c. tag lines are properly anchored
>   d. tackle is connected to its power source
> 9. A hitch configuration that allows slings to be connected to the same load without using a spreader beam is a _____.
>   a. double-wrap hitch
>   b. choker hitch
>   c. bridle hitch
>   d. basket hitch
> 10. To make the emergency stop signal that is used by riggers, extend both

NCCER Standardized Curricula

NCCER's training programs comprise more than 80 construction, maintenance, pipeline, and utility areas and include skills assessments, safety training, and management education.

Construction and Maintenance

Boilermaking
Cabinetmaking
Carpentry
Concrete Construction
Construction Craft Laborer
Construction Technology
Core: Introduction to Basic Construction Skills
Drywall
Electrical
Electronic Systems Technician
Heating, Ventilating, and Air Conditioning
Heavy Equipment Operations
Heavy Highway Construction
Hydroblasting
Industrial Coating and Lining Application Specialist
Industrial Maintenance Electrical and Instrumentation Technician
Industrial Maintenance Mechanic
Instrumentation
Ironworking
Manufactured Construction Technology
Masonry
Mechanical Insulating
Millwright
Mobile Crane Operations
Painting
Painting, Industrial
Pipefitting
Pipelayer
Plumbing
Reinforcing Ironwork
Rigging
Roofing
Scaffolding
Sheet Metal
Signal Person
Site Layout
Sprinkler Fitting
Tower Crane Operator
Welding

Maritime

Maritime Industry Fundamentals
Maritime Electrical
Maritime Pipefitting
Maritime Structural Fitter
Maritime Welding
Maritime Aluminum Welding

Green/Sustainable Construction

Building Auditor
Fundamentals of Weatherization
Introduction to Weatherization
Sustainable Construction Supervisor
Weatherization Crew Chief
Weatherization Technician
Your Role in the Green Environment

Energy

Alternative Energy
Introduction to the Power Industry
Introduction to Solar Photovoltaics
Power Generation Maintenance Electrician
Power Generation I&C Maintenance Technician
Power Generation Maintenance Mechanic
Power Line Worker
Power Line Worker: Distribution
Power Line Worker: Substation
Power Line Worker: Transmission
Solar Photovoltaic Systems Installer
Wind Energy
Wind Turbine Maintenance Technician

Pipeline

Abnormal Operating Conditions, Control Center
Abnormal Operating Conditions, Field and Gas
Corrosion Control
Electrical and Instrumentation
Field and Control Center Operations
Introduction to the Pipeline Industry
Maintenance
Mechanical

Safety

Fall Protection Orientation
Field Safety
Safety Orientation
Safety Technology

Supplemental Titles

Applied Construction Math
Tools for Success

Management

Construction Workforce Development Professional
Fundamentals of Crew Leadership
Mentoring for Craft Professionals
Project Management
Project Supervision

Spanish Titles

Acabado de concreto: nivel uno (*Concrete Finishing Level One*)
Aislamiento: nivel uno (*Insulating Level One*)
Albañilería: nivel uno (*Masonry Level One*)
Andamios (*Scaffolding*)
Carpintería: Formas para carpintería, nivel tres (*Carpentry: Carpentry Forms, Level Three*)
Currículo básico: habilidades introductorias del oficio (*Core Curriculum: Introductory Craft Skills*)
Electricidad: nivel uno (*Electrical Level One*)
Herrería: nivel uno (*Ironworking Level One*)
Herrería de refuerzo: nivel uno (*Reinforcing Ironwork Level One*)
Instalación de rociadores: nivel uno (*Sprinkler Fitting Level One*)
Instalación de tuberías: nivel uno (*Pipefitting Level One*)
Instrumentación: nivel uno, nivel dos, nivel tres, nivel cuatro (*Instrumentation Levels One through Four*)
Orientación de seguridad (*Safety Orientation*)
Paneles de yeso: nivel uno (*Drywall Level One*)
Seguridad de campo (*Field Safety*)

Acknowledgments

This curriculum was revised as a result of the vision and leadership of the following sponsors:

Cen-Cal Fire Systems, Inc.
Fire & Life Safety America
Johnson Controls, Inc.
North East Fire Protection Systems, Inc.
Piper Fire Protection
Reliable Automatic Sprinkler Co., Inc.
Security Fire Sprinkler
Strickland Fire Protection
Victaulic
The Viking Corporation
Wolfe Fire Protection, Inc.

This curriculum would not exist were it not for the dedication and unselfish energy of those volunteers who served on the Authoring Team. A sincere thanks is extended to the following:

Charles Bamford
Leslie Clounts
John Denhardt
Robert Dunn
Kevin Kelly
Mac McGaffagan
Steven Rasch
Matt Rouleau
Taylor Schumacher
Byron Weisz

NCCER Partners

NCCER partnering organizations are national associations and organizations that share a common interest in the goals and objectives of NCCER. To learn more about NCCER business partners, go to **https://www.nccer.org/about-us/partners**.

You can also scan this code using the camera on your phone or mobile device to view these partnering organizations.

Contents

Module Seven

Dry Pipe Sprinkler Systems

This module describes the operation and installation of dry pipe sprinkler systems. It covers differential and mechanical dry pipe valves and trim, including switches, gauges, air sources, accelerators, and drains. It also provides an overview of basic system troubleshooting and acceptance testing procedures, including the completion of a Contractor's Material and Test Certificate for Aboveground Piping for a dry pipe sprinkler system. (Module ID 18207; 25 Hours)

Glossary

This course map shows all of the modules in *Sprinkler Fitting Level Two*. The suggested training order begins at the bottom and proceeds up. Skill levels increase as you advance on the course map. The local Training Program Sponsor may adjust the training order.

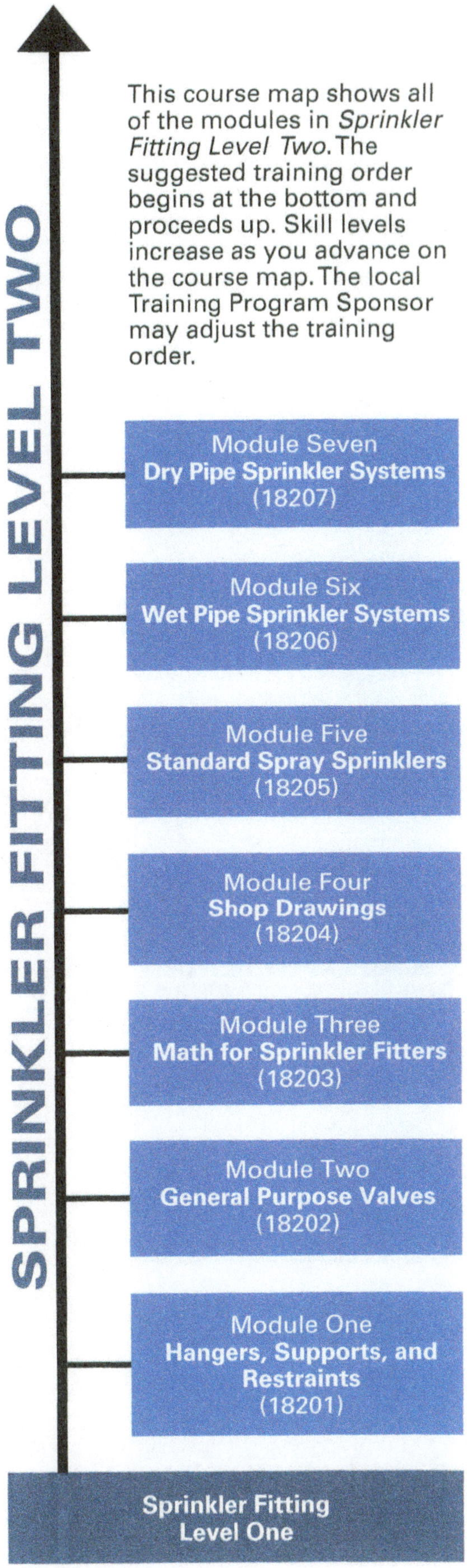

Hangers, Supports, and Restraints

Overview

Proper support is essential to the safe installation of a sprinkler piping system. This module identifies strength/spacing requirements, types, and installation of pipe hangers, supports, and restraints. It also covers the installation of firestopping.

Module 18201

Trainees with successful module completions may be eligible for credentialing through the NCCER Registry. To learn more, go to **www.nccer.org** or contact us at 1.888.622.3720. Our website, **www.nccer.org**, has information on the latest product releases and training.

Your feedback is welcome. You may email your comments to **curriculum@nccer.org**, send general comments and inquiries to **info@nccer.org**, or fill in the User Update form at the back of this module.

This information is general in nature and intended for training purposes only. Actual performance of activities described in this manual requires compliance with all applicable operating, service, maintenance, and safety procedures under the direction of qualified personnel. References in this manual to patented or proprietary devices do not constitute a recommendation of their use.

18201 V4.0

13 2024

18201

Hangers, Supports, and Restraints

Objectives

Successful completion of this module prepares you to do the following:

1. Select and install pipe hanger assemblies.
   a. List general hanger installation guidelines.
   b. Identify and install various pipe hangers.
   c. Size and install all-thread rods (ATRs).
   d. Identify and install attachments in wood, steel, and concrete structures.
   e. Identify and install pipe supports and other attachments.
2. Install seismic bracing.
   a. Identify NFPA 13 requirements for seismic bracing.
   b. Identify the components in sway bracing systems.
3. Identify firestopping assemblies.
   a. Identify various firestopping products and their ratings.
   b. Install firestopping in floor and wall penetrations.

Performance Tasks

Under supervision, you should be able to do the following:

1. Install a hanger on a wood joist.
2. Install a post-installed anchor.
3. Cut a hanger rod to a specified length.
4. Make up and install a seismic brace.

Trade Terms

All-thread rod (ATR)
Anchor
Endothermic
Firestopping
Hanger
Intumescent
Load rating
Pipe riser
Restraint
Section modulus
Seismic bracing
Seismicity
Support
Sway brace
Takeout
Water hammer

Industry Recognized Credentials

If you are training through an NCCER-accredited sponsor, you may be eligible for credentials from NCCER's Registry. The ID number for this module is 18201. Note that this module may have been used in other NCCER curricula and may apply to other level completions. Contact NCCER's Registry at 1.888.622.3720 or go to **www.nccer.org** for more information.

You can also show off your industry-recognized credentials online with NCCER's digital badges. Transform your knowledge, skills, and achievements into badges that you can share across social media platforms, send to your network, and add to your resume. For more information, visit **www.nccer.org**.

NOTE

Unless otherwise specified, references in parentheses following figure and table numbers refer to NFPA 13.

Contents

Figures and Tables

SECTION ONE

1.0.0 SELECTING AND INSTALLING PIPE HANGER ASSEMBLIES

Objective

Select and install pipe hanger assemblies.

a. List general hanger installation guidelines.
b. Identify and install various pipe hangers.
c. Size and install all-thread rods (ATRs).
d. Identify and install attachments in wood, steel, and concrete structures.
e. Identify and install pipe supports and other attachments.

Performance Tasks

1. Install a hanger on a wood joist.
2. Install a post-installed anchor.
3. Cut a hanger rod to a specified length.

Trade Terms

All-thread rod (ATR): Rod that is fully threaded end to end and often cut to size and used to suspend hangers and fasteners.

Anchor: A device used to attach a hanger to a structure.

Hanger: A device from which a pipe is suspended that is used to carry the weight of the pipe from above.

Load rating: The usable weight capacity of a hanger or support.

Pipe riser: A vertical pipe extending from an opening in a floor or ceiling.

Restraint: A device used to prevent pipe from shifting in any direction other than downward.

Section modulus: A measurement of the load capacity of a trapeze hanger or other support.

Seismic bracing: Bracing installed on a piping system to protect it from damage caused by seismic activity, such as an earthquake.

Support: A structure or saddle on which a pipe rests that is designed to support the weight of the pipe from below.

Takeout: The distance from the center of a pipe to the bottom of the threaded rod in a hanger assembly.

Water hammer: The pressure or shock wave within piping caused by a sudden change in water flow, such as when starting a pump or quickly closing a valve.

There are many types of pipe hangers, supports, and restraints. The devices selected for a given system depend on many factors, including the weight to be supported, the intended movement of the pipe within the device, and the surrounding structure.

Hangers and supports are used to manage the forces of gravity acting on a sprinkler system. A **hanger** is a device used to carry the weight of the pipe from above, while a **support** is a structure upon which a pipe rests and the support is provided from below.

Movement within sprinkler systems is usually caused by vibration and **water hammer**. Water hammer is the pressure or shock wave within piping caused by a sudden change in water flow such as that which occurs while starting a pump or quickly closing a valve. Water hammer can cause significant piping movement.

A **restraint** is used to prevent the movement of pipe other than in a downward direction. Restraints are installed at established intervals along a pipe to control this movement.

Pipe movement can also be caused by seismic activity. **Seismic bracing** is installed on a sprinkler system to protect it from damage caused by seismic activity, such as an earthquake.

Hangers, supports, and restraints are installed according to approved sprinkler shop drawings. Correctly installed hangers, supports, and restraints allow pipes to drain, expand, contract, and vibrate without damaging the sprinkler system.

> **NOTE**
> Thermal expansion in steel pipe is rarely a factor in sprinkler systems since the normal temperature range of sprinkler piping is between 40°F and 120°F. However, for nonmetallic pipe systems, such as CPVC, thermal expansion is a consideration.

1.1.0 General Hanger Installation Guidelines

Every pipe hanger and support has a maximum **load rating** or strength requirement. This load rating includes a safety load factor of five; that is, the hanger or support must be capable of handling a load five times greater than the weight of the water-filled pipe, plus an additional 250 pounds. For example, a hanger rated at 500 pounds must support at least 2,750 pounds before failing. Supports must retain adequate strength after being exposed to rust, vibration, water hammer, corrosion, seismic activity, and other factors. Manufacturers supply information about the maximum load rating and/or maximum pipe sizes of their hanger and assembly attachments. Never exceed the load rating of any hanger or other support device.

> **NOTE**
> Supports must be Listed for the application, installed in accordance with NFPA 13, or certified by a registered professional engineer. NFPA 13 and 13R standards require that devices directly attached to a pipe or structure in a sprinkler system be Listed. The Listing process for these devices requires them to be corrosion tested and pull tested. NFPA 13D refers to local plumbing codes for proper hangers.

> **CAUTION**
> Refer to the manufacturer's instructions for proper spacing of hangers and supports. Never exceed the maximum specified distance.

The following rules for hangers and supports are in accordance with NFPA 13 requirements and must be observed when installing sprinkler piping:

- The pipe must run straight without sagging. Always check the shop drawings when determining pipe hanger spacing. Examples of locations where additional hangers may be necessary include lightweight steel joists, Z purlins, and lightweight engineered wood joists.
- The hanger system must support the pipe at regular intervals. Even spacing ensures that no individual hanger is overloaded. Spacing limitations of hangers for various types of pipe are based on the flexibility and nominal size of the pipe. Closer spacing is sometimes used to overcome the strength limitations of some hangers and/or structural members.
- If the shop drawings do not match what is in the field, consult the layout technician before any field changes are made. Any modifications to the system may affect approvals.
- All hanger and support materials must be made of ferrous materials (iron and steel).
- Branch lines should run perpendicular across the joists or purlins to minimize the cost of supporting devices.
- The hanging of any equipment, devices, conduit, ceiling, wiring, or other pipe from sprinkler piping is prohibited by NFPA 13, except for equipment that serves sprinkler systems.
- The specifying engineer may require greater hanger support capability than the minimums in NFPA 13. Check the shop drawings for support capabilities of the system to be installed.

1.2.0 Pipe Hangers

There are three basic components in a hanger assembly: the fastener that attaches the hanger assembly to the structure, the rod between the fastener and the hanger, and the hanger itself. Different hanger assemblies are used based on the pipe size and the structure to which the assemblies are attached (wood, concrete, or steel). *Figure 1* shows a hanger assembly attached to slotted strut channel, which is in turn attached to structural steel.

Common types of pipe hangers include adjustable swivel ring hangers, adjustable clevis hangers, and trapeze hangers. Maximum spacing may vary depending on the job conditions. *Table 1* shows the maximum distances between hangers for various types of pipe per NFPA 13.

> **NOTE**
> Hanger spacing will be indicated in the approved shop drawings. Always follow the shop drawings.

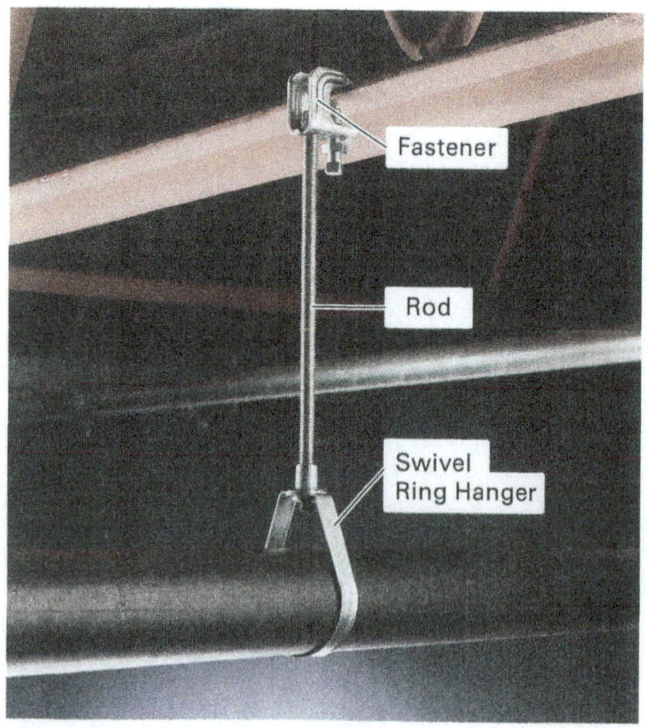

Figure 1 Pipe hanger assembly.

Table 1 Maximum Distance Between Hangers (ft-in) [Data from NFPA 13, *Table 17.4.2.1(a)*]

Pipe Material	Nominal Pipe Size										
	¾"	1"	1¼"	1½"	2"	2½"	3"	4"	5"	6"	8"
Steel pipe except threaded lightwall	N/A	12-0	12-0	15-0	15-0	15-0	15-0	15-0	15-0	15-0	15-0
Threaded lightwall steel pipe	N/A	12-0	12-0	12-0	12-0	12-0	12-0	N/A	N/A	N/A	N/A
Copper tube	8-0	8-0	10-0	10-0	12-0	12-0	12-0	15-0	15-0	15-0	15-0
CPVC	5-6	6-0	6-6	7-0	8-0	9-0	10-0	N/A	N/A	N/A	N/A
Ductile-iron pipe	N/A	N/A	N/A	N/A	N/A	N/A	15-0	15-0	N/A	15-0	15-0

N/A: Not applicable.

1.2.1 Adjustable Swivel Ring Hangers

Adjustable swivel rings are recommended for supporting stationary pipelines and are available in sizes to fit ½" to 8" pipe. A thumb nut is used to secure the pipe to the threaded rod and allows for vertical adjustment once the hanger is in place. *Figure 2* shows an adjustable swivel ring hanger. The pipe can either be slid into the hanger or the thumb nut removed, and the hanger opened to insert the pipe.

Figure 2 Adjustable swivel ring hanger.

1.2.2 Adjustable Clevis Hangers

Adjustable clevis hangers consist of a U-shaped clevis that holds the pipe, a clevis with a flat end that attaches to the rod, and a through bolt that holds them together. To install a clevis, the bolt is removed, then the flat end is secured to a threaded rod attached to the structure. The pipe is installed, then the bolt is reattached. Alternatively, the hanger can be installed as a unit and the pipe slid into place.

Adjustable clevis hangers are available for pipe up to 30" in diameter. The hanger is attached to threaded rod using a nut that is tightened to provide vertical adjustment and serve as a restraint. *Figure 3* shows an adjustable clevis hanger.

Hinged clevis hangers are also available. These hangers eliminate the use of a through bolt. Instead, they have a hinged U-shaped clevis that swings down for pipe installation and then clips back into place (*Figure 4*).

1.2.3 Trapeze Hangers

Trapeze hangers are available commercially and can also be shop- or field-fabricated. They are suspended by two rods and designed for top-loading of pipe only. The trapeze hanger is designed to distribute the load of water-filled pipe over at least two joists or purlins. Vertical adjustments are made by tightening or loosening the bolts at the bottom of the hanger. *Figure 5* shows a trapeze pipe hanger. Trapeze hangers can also be fabricated using angle iron. When angle iron is used, the long end must be installed vertically.

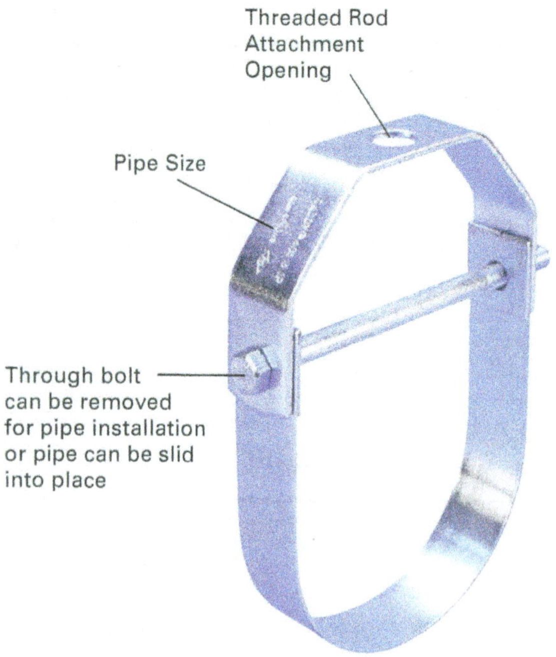

Figure 3 Adjustable clevis hanger.

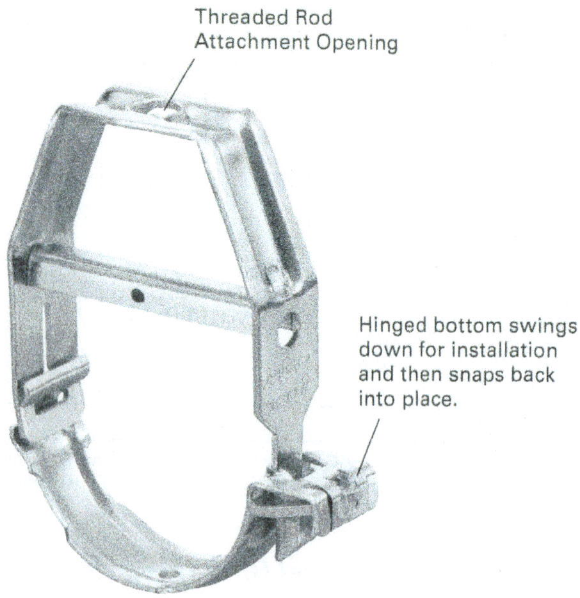

Figure 4 Hinged clevis hanger.

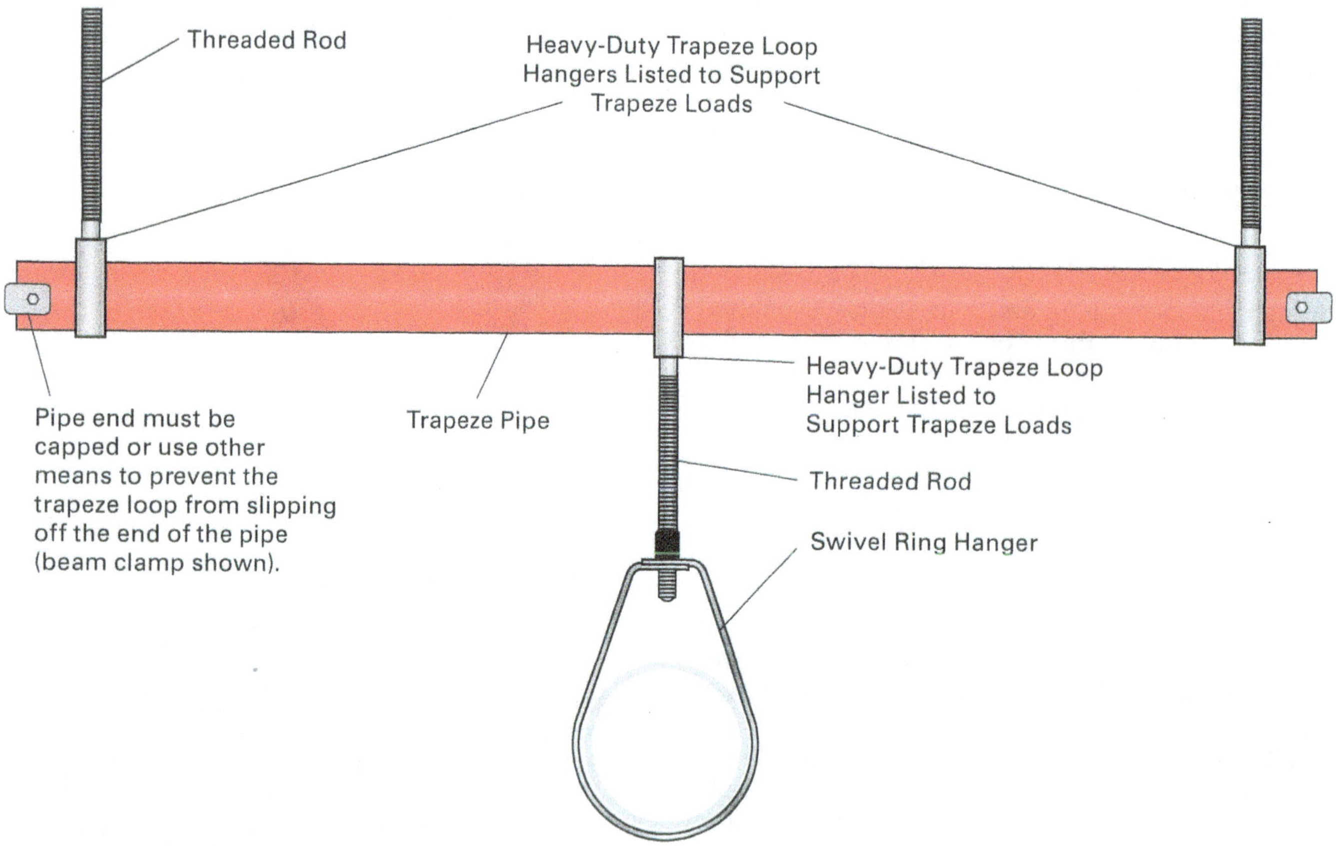

Figure 5 Trapeze pipe hanger.

> **CAUTION**
>
> Make sure the pipe hangers used on a trapeze unit can support the sprinkler pipe load in accordance with NFPA 13. Trapeze hangers may require the use of heavy-duty hangers and fasteners. Refer to the shop drawings for component details.

The minimum required strength of a trapeze hanger is known as its section modulus. Section modulus is a measure of load handling capability and varies depending on three factors: the distance between the supports (span), pipe size, and pipe schedule.

The section modulus is greater for larger or heavier pipe and for longer spans between supports. For example, the minimum section modulus for 4" pipe is less than that for 8" pipe of the same type. Similarly, the minimum section modulus for Schedule 10 pipe is less than that for Schedule 40 pipe of the same diameter. In addition, the longer the span between supports, the greater the minimum section modulus.

> **CAUTION**
>
> Sprinkler system layout technicians use the tables in NFPA 13 to calculate trapeze hanger requirements. Always follow the shop drawings for hanger size and type. The size of the hangers cannot be reduced. For example, if a shop drawing specifies a trapeze hanger/support of Schedule 40 pipe, this support requirement must not be reduced to Schedule 10 pipe. However, if the trapeze design is based on Schedule 10 pipe, it is acceptable to make that trapeze out of Schedule 40.

NFPA 13 lists minimum allowable section modulus values for three types of trapezes: Schedule 10 pipe, Schedule 40 pipe, and angle iron. Consult the manufacturers' literature for the section modulus of Unistrut or similar material.

The ends of trapeze hangers must be secured to prevent the trapeze loops from slipping off. Solutions include the use of rods, U-bolts, and beam clamps.

> **NOTE**
>
> Trapeze hangers are more expensive to fabricate and install than standard hangers. They are not normally used if there is an approved alternate method of support.

1.3.0 All-Thread Rods and Couplings

All-thread rod (ATR) is used to connect a pipe hanger to an anchor or rod attachment. ATR can be cut to the desired length using a hacksaw, battery-powered or electric reciprocating saw or bandsaw, or a special rod-cutting tool such as that shown in *Figure 6*. ATR is available in several diameters to match the hanger and rod attachment. NFPA 13 requires that ATR support a load of five times the weight of the water-filled pipe plus 250 pounds. *Table 2* shows hanger rod sizes.

ATR must not be bent. If a change in direction is required, ATR cannot be used unless it is attached to a U-joint rod coupling (*Figure 7*). Where offsets are needed, or eye rods are required, unthreaded mild steel rod must be formed, then threaded on each end before use.

With hangers, the term takeout refers to the distance from the center of the pipe to the bottom of the threaded rod in the hanger assembly. Hanger rods require adjustments in length so that the rods do not go to the center of the pipe. Penetration of the rod into the arrangement that attaches to the structure must also be considered when determining proper rod length. Both ATR and end-threaded rods are cut to an accuracy of $\frac{1}{2}$". If the dimension is not an even $\frac{1}{2}$", the rod length is rounded up.

Figure 6 ATR cutter.

Table 2 Hanger Rod Sizes (Data from NFPA 13, *Table 17.2.1.1*)

Pipe Size	Diameter of Rod
Up to and including 4"	3/8"
5" to 8"	1/2"
10"	5/8"
12"	3/4"

Figure 7 U-joint rod coupling.

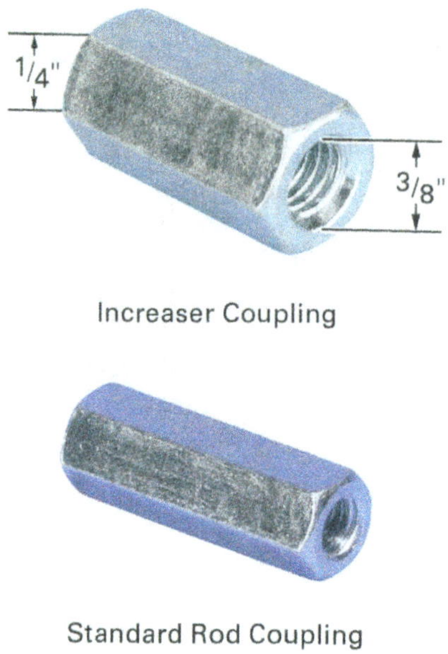

Figure 8 Rod couplings.

There are two configurations for takeouts. The first configuration is a normal rod location, and the second configuration prevents upward thrust of the pipe. In the second configuration, the hanger rod becomes a restraint. The threaded rod is adjusted within 1/8" of the pipe.

CAUTION

Be careful when adjusting the rod downward. Maintain sufficient rod thread engagement to the attachment or structure above.

Rod couplings are used to link two rods together or to link rods to other threaded anchors (*Figure 8*). Rod couplings are useful when a hanger needs to be longer or when eye rods are extended with ATR. Increaser couplings are required when 1/4" powder-driven studs are used and 3/8" mild steel hanger rods are connected to standard adjustable ring hangers. These couplings are easy to install and allow for some vertical adjustment.

1.4.0 Structural Attachments

Pipe hanger assemblies can be attached to wood joists, structural steel beams or plates, or concrete. Pipe hanger fasteners attach the pipe hanger to the supporting structure and must be able to withstand the same load that is exerted on the hanger.

1.4.1 Fasteners for Wood

Common wood fasteners and attachments include the following:

- Threaded anchor fasteners
- Lag bolts and coach screw rods
- Eye rods
- Drive screws
- U-hooks

WARNING!

Do not use common nails to attach sprinkler system components. They do not have the holding power required to ensure a safe and secure installation.

Think About It

Determining the Length of the Hanger Rod

You must be able to size both the hanger and rod to hang pipe at the proper height. This figure shows a 6" pipe in a hanger located 16'-3" above the floor and extending 1" into the anchor.

1. What is the distance from the bottom of the top deck to the bottom of the pipe hanger?
2. In the figure, the pipe hanger takeout measurement is 10". What is the distance between the bottom of the top deck and the top of the hanger?
3. Knowing that the hanger rod will extend into the anchor 1", how long must the hanger rod be to extend into the hanger $2\frac{1}{2}$"?

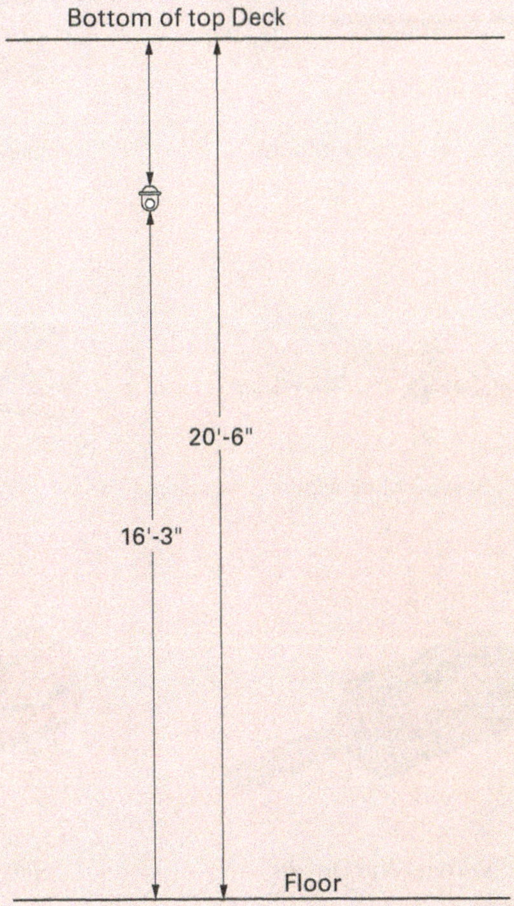

Threaded Anchor Fasteners

Threaded anchor fasteners are a simple and convenient means of rod anchoring. They provide good holding strength and can be mounted from the bottom or side of the work. These fasteners are available in self-drilling, self-tapping styles for use in steel, wood, or concrete, as well as a toggle style for use in hollow-core concrete, metal lathe, and plaster. Side-mount and swivel styles are also available. *Figure 9* shows some examples of these threaded fasteners, along with the installation tools required for each type.

Lag Bolts and Coach Screw Rods

Lag bolts and coach screw rods are types of wood screws used to provide proper holding power for sprinkler systems. A pilot hole that is $\frac{1}{8}$" less in diameter than the root diameter of the thread is predrilled into the wood supporting member. The preferred use of these rods is in the side of a beam where the load is in shear on the lag bolt. *Figure 10* shows a lag bolt and coach screw rod plus shear and tension load directions. *Table 3* shows minimum bolt or lag screw sizes. *Table 4* shows minimum coach screw rod sizes. *Table 5* shows minimum plank thicknesses and beam or joist widths.

Eye Rods

A basic type of connecting unit is the eye rod. Eye rods are made with an eye, or loop, at one end and threads on the other end. There are many types of eye rods. Some have a welded eye and others have a formed eye (*Figure 11*).

CAUTION

NFPA 13 does not allow eye rods to be made with ATR.

Table 6 lists minimum eye rod sizes. Eye rods must be secured with lock washers to prevent lateral (side-to-side) motion. When secured to wood, the eye rod must be backed with a large flat washer bearing directly against the structural member, in addition to the lock washer. Eye rods are specified for mild steel and do not have to be Listed per NFPA 13.

Figure 9 Threaded anchor fasteners and installation tools.

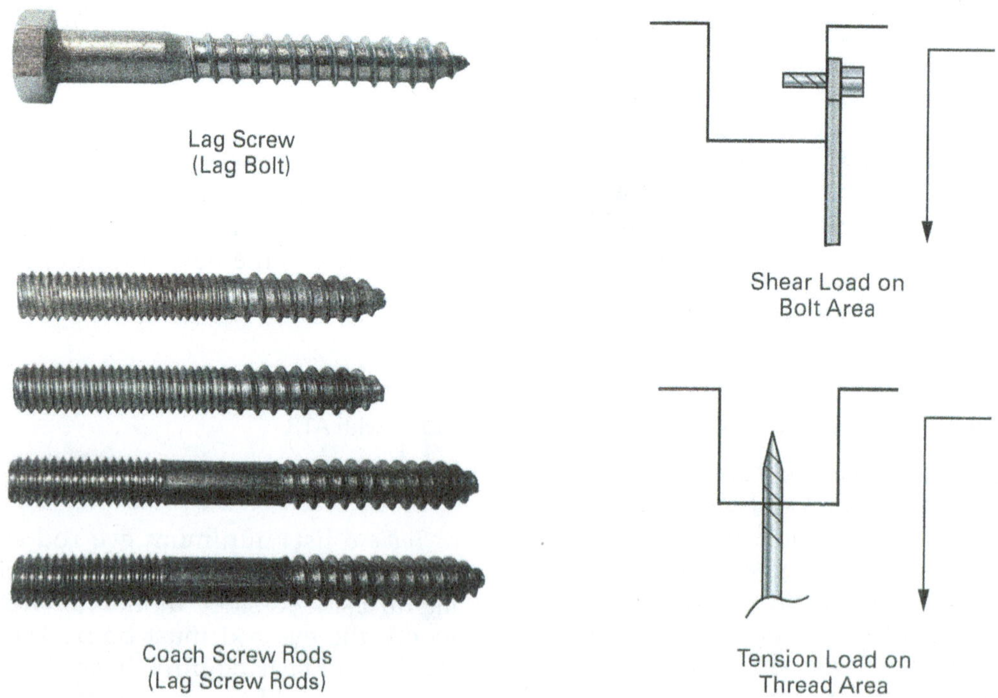

Figure 10 Lag bolt, coach screw rods, and shear load directions.

Table 3 Minimum Bolt, Rod, or Lag Screw Sizes for Side of Beam Installation (Data from NFPA 13, *Table 17.2.4.3.1*)

Pipe Size	Size of Bolt, Rod, or Lag Screw	Length of Lag Screw Used with Wood Beams
Up to and including 2"	⅜"	2½"
2½" to 6" (inclusive)	½"	3"
8"	⅝"	3"

Note: This training material is not affiliated with nor has it been reviewed or approved by the NFPA. Reproduced with permission of NFPA from NFPA 13, *Standard for the Installation of Sprinkler Systems*, 2019 edition. Copyright© 2018, National Fire Protection Association. For a full copy of NFPA 13, please go to www.nfpa.org.

Table 4 Minimum Coach Screw Rod Size (Data from NFPA 13, *Table 17.2.4.7.1*)

Pipe Size	Diameter of Rod	Minimum Penetration
Up to and including 4"	⅜"	3"
Larger than 4"	Not permitted	Not permitted

Note: This training material is not affiliated with nor has it been reviewed or approved by the NFPA. Reproduced with permission of NFPA from NFPA 13, *Standard for the Installation of Sprinkler Systems*, 2019 edition. Copyright© 2018, National Fire Protection Association. For a full copy of NFPA 13, please go to www.nfpa.org.

Table 5 Minimum Plank Thicknesses and Beam or Joist Widths (Data from NFPA 13, *Table 17.2.4.7.2*)

Pipe Size	Nominal Plank Thickness	Nominal Width of Beam or Joist Face
Up to and including 2"	3"	2"
2½"	4"	2"
3", 4"	4"	3"

Note: This training material is not affiliated with nor has it been reviewed or approved by the NFPA. Reproduced with permission of NFPA from NFPA 13, *Standard for the Installation of Sprinkler Systems*, 2019 edition. Copyright© 2018, National Fire Protection Association. For a full copy of NFPA 13, please go to www.nfpa.org.

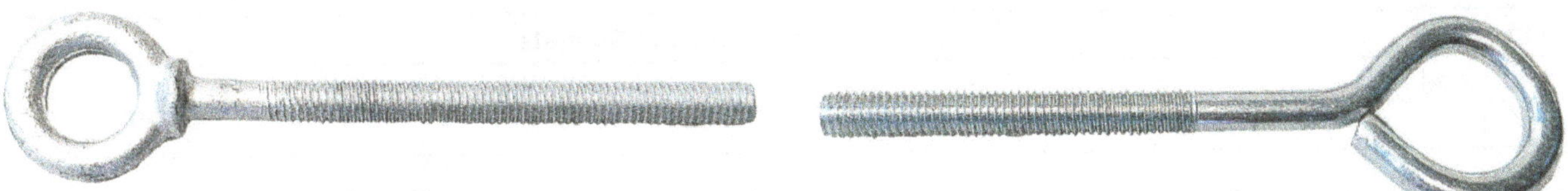

Figure 11 Eye rods.

Drive Screws

Drive screws have limited holding power. These are only used in the sides of wood structural members with U-hooks or other Listed attachment devices where the load is in shear across the diameter of the drive screw. They must be used in the horizontal position. Drive screws are not allowed with any hanger other than U-hooks and only with pipe diameters of 2" and smaller, unless specifically Listed for another application. U-hooks require two points of attachment (two screws). *Figure 12* shows a drive screw.

U-Hooks

U-hooks are used as hangers and restraints. Where sloping joists are required for support, an unequal leg U-hook is useful. The wraparound type serves as a restraint and is used to prevent upward thrust. *Figure 13* shows two types of U-hooks. *Table 7* shows U-hook rod sizes.

1.4.2 Fasteners for Steel

In addition to the steel threaded anchors discussed previously (*Figure 14*), there are various fasteners that clamp directly to steel structural members. The most common connecting units for steel are beam clamps, C-clamps, and purlin clamps. The type of fastener selected depends on the structure of the steel.

> **NOTE**
> Most metal building manufacturers have specific requirements for attachments to their structures. Check with the general contractor or building supplier for specific information.

Table 6 Eye Rod Sizes (Data from NFPA 13, *Table 17.2.1.5.1*)

Pipe Size	Diameter of Rod	
	With Bent Eye	With Welded Eye
Up to and including 4"	3/8"	3/8"
5", 6"	1/2"	1/2"
8"	3/4"	1/2"

Note: This training material is not affiliated with nor has it been reviewed or approved by the NFPA. Reproduced with permission of NFPA from NFPA 13, *Standard for the Installation of Sprinkler Systems*, 2019 edition. Copyright© 2018, National Fire Protection Association. For a full copy of NFPA 13, please go to **www.nfpa.org**.

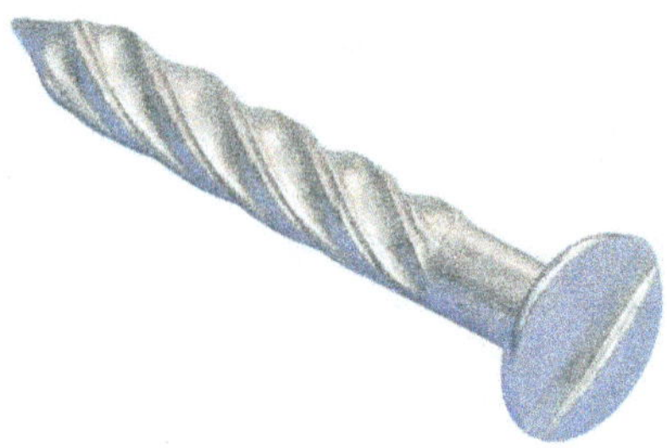

Figure 12 Drive screw.

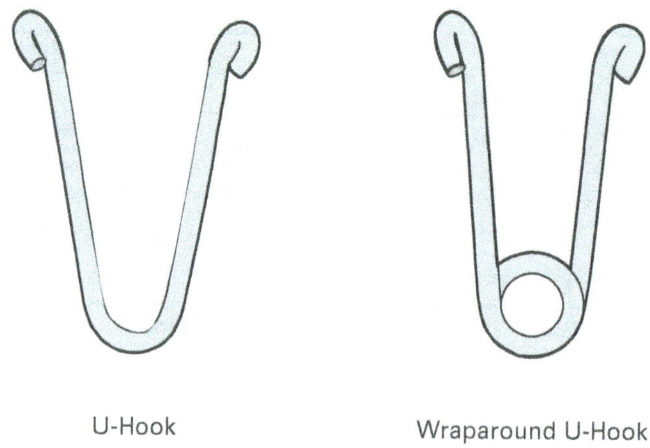

Figure 13 U-hooks.

Table 7 U-Hook Rod Sizes (Data from NFPA 13, *Table 17.2.1.4*)

Pipe Size	Hook Material Diameter
Up to and including 2"	5/16"
2½" to 6"	3/8"
8"	1/2"

Note: This training material is not affiliated with nor has it been reviewed or approved by the NFPA. Reproduced with permission of NFPA from NFPA 13, *Standard for the Installation of Sprinkler Systems*, 2019 edition. Copyright© 2018, National Fire Protection Association. For a full copy of NFPA 13, please go to **www.nfpa.org**.

Beam Clamps

A beam clamp is a universal, reversible clamp that can be attached to the top or bottom of a steel I-beam (*Figure 15*). Check the shop drawings to make sure hangers are attached to the correct point. The threaded hole for the hanger rod is offset from the setscrew that clamps to the beam. The threaded rod can be adjusted up through the clamp as far as the structure overhead will allow. Beam clamps come in a variety of types and are available to support up to 5,000 pounds. In areas of seismic activity, beam clamps are used in conjunction with retaining straps.

CAUTION

Beam clamps may fracture or deform if overly tightened. Refer to the manufacturer's instructions regarding the appropriate torque.

Purlin Clamps

Purlin clamps are designed to fit around steel structures with a 45-degree or 90-degree lip. Typical purlin clamps are shown in *Figure 16*.

Did You Know?

NFPA 13 requires that hangers be designed to support five times the weight of the water-filled pipe plus 250 pounds. But why the extra 250 pounds? Because 250 pounds is the estimated weight of a sprinkler fitter with a typical equipment load who might fall and need to grab the nearest pipe.

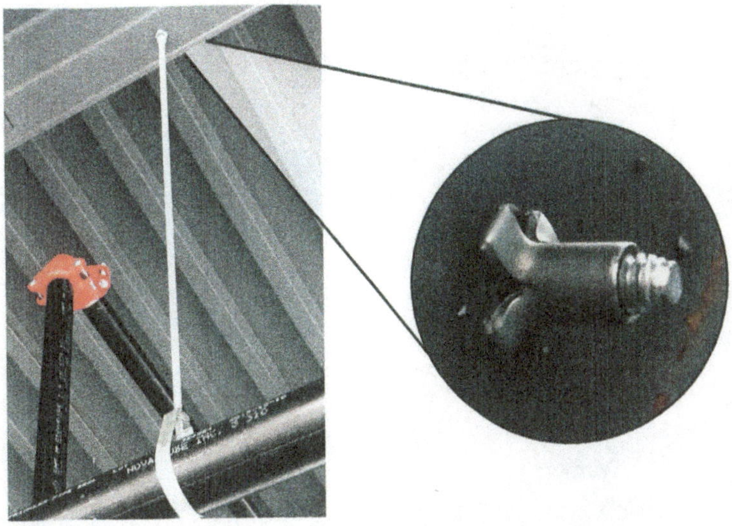

Figure 14 Expanding anchor installed in steel.

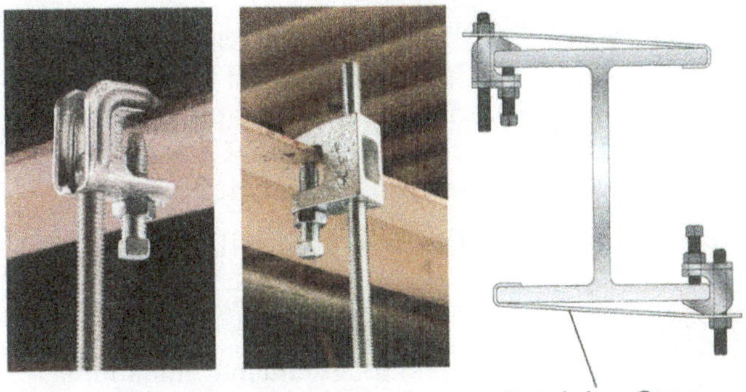

Figure 15 Beam clamps.

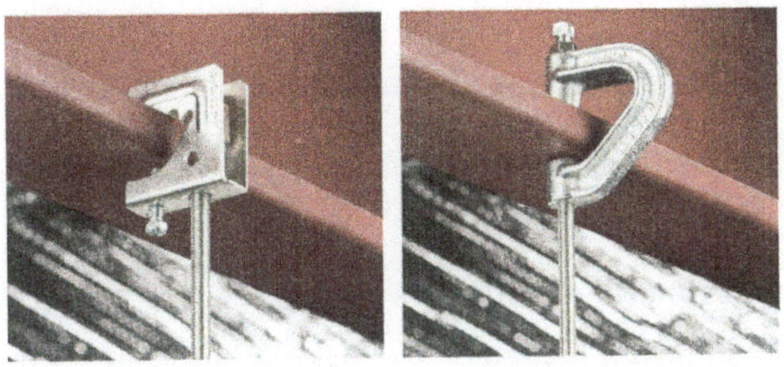

Figure 16 Purlin clamps.

C-Clamps

C-clamps attach to the top of steel beams. Typical C-clamps are shown in *Figure 17*.

> **NOTE**
> NFPA 13 requires clamps used in sprinkler systems to be Listed by a recognized testing laboratory.

1.4.3 Fasteners for Concrete

There are many types of concrete fasteners. In addition to the concrete threaded anchors discussed previously, the following fasteners are used to attach piping components to concrete:

- Concrete pre-installed inserts (anchors)
- Post-installed anchors
- Powder-driven studs

Concrete Pre-Installed Inserts

Concrete pre-installed inserts are placed into the forms before the concrete is poured (*Figure 18*). To install the insert, the top is struck with a hammer to drive the nails into the deck. After the concrete has cured and the form has been removed, the breakaway nails are knocked off and the lower face of the insert is exposed to reveal internal threads for the attachment of threaded rod.

Pre-installed inserts have some disadvantages. Special trips must be made to the jobsite to place inserts in the forms at the appropriate time, and the inserts must be positioned carefully to be useful. Sprinkler fitters must have a dimensioned plan to properly place concrete inserts. On fast-track projects, the sprinkler piping layout is not complete when the concrete is being poured and rerouting of the piping after the placement of inserts renders them useless. Small adjustments can be made if the long dimension of the insert is placed at right angles to the line of pipe travel.

Figure 17 C-clamps.

Another problem with pre-poured concrete inserts is that other trades use the sprinkler inserts to support their own material or cover them up during construction.

Post-Installed Anchors

Expanding fasteners are installed after the concrete has been poured and hardened. They are made in a variety of sizes and heads. The size of an expanding fastener is based on the size of the bolt in use. These fasteners may not grip if the drilled holes are oversized. Always follow the manufacturer's instructions concerning hole size and maximum allowable loads and check the Listing requirements. All post-installed fasteners must be installed per approved shop drawings.

Special hammer drills are available for drilling holes in concrete. The combination of rotary and hammering action produces a hole much faster than with either motion alone. Drill bits specially designed for working in concrete must be used. These are usually carbide tipped for longer life. *Figure 19* shows various concrete anchors.

Expansion anchors can be used with ³⁄₈", ½", ⁵⁄₈", and ¾" threaded rods. Follow these steps to install an expansion anchor:

Step 1 Identify the type of hanger to be used.

Step 2 Locate the area where the support is to be fastened or attached.

Step 3 Lay out and mark hole locations.

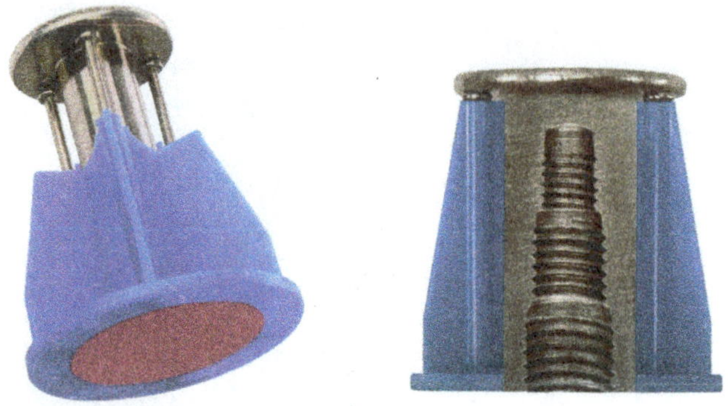

Figure 18 Concrete pre-installed insert.

Step 4 Identify the type and size of post-installed anchor, as indicated on the shop drawings.

Step 5 Select a drill bit to match the anchor chosen. The drill bit selected for an expansion shield is typically larger than the threaded rod being used. Check with the manufacturer for proper drill bit size.

Step 6 Assemble the drill and drill bit.

> **WARNING!** Never use standard rotary drill bits in hammer drills.

Step 7 Drill the holes for the fasteners.

> **WARNING!** Drilling in concrete generates noise, dust, and flying debris. Always use all appropriate personal protective equipment (PPE), including respiratory, hearing, eye, and fall protection. Maintain a firm grip on the drill and ensure solid footing because drilling in concrete produces a great amount of torque.

Step 8 Clean out the holes when drilling is complete.

Step 9 Insert the expansion shield into the hole.

Step 10 Screw the threaded rod into the expansion shield. The sides of the shield will spread and grip the inside of the hole. The fastener is now ready for a hanger.

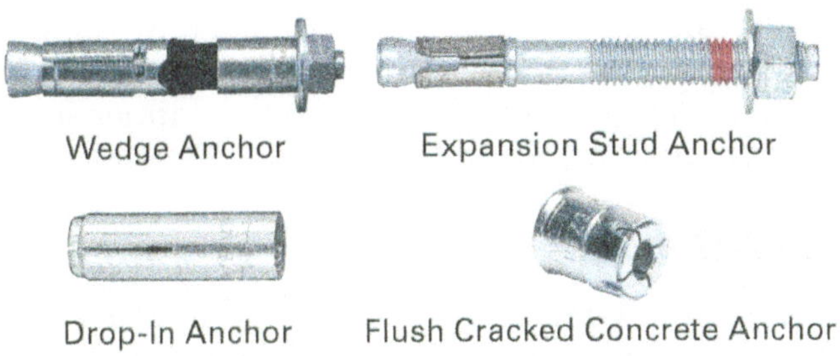

Figure 19 Concrete anchors.

Powder-Actuated Fasteners

Powder-actuated tools use energy from a gunpowder charge to drive a stud bolt, nail, or other fastener into steel or concrete. The charge can be a metal shell cartridge or powder pellet. The energy from the powder charge drives a piston, which then drives the fastener using a hammer-like blow. Some of these fasteners are only ¼" in diameter. They are made of high-strength steel, and their load handling capability is much greater than the mild steel hanger rods normally used in sprinkler fitting.

Because of the potential danger, OSHA requires that all operators of powder-actuated tools be trained and certified. Every operator is required to have an accredited operator's card issued by a qualified, authorized instructor. This card has a serial number that is registered with the employer, tool distributor, and tool manufacturer. Certification cards must be carried whenever using the tool. Employees lacking this authorization cannot use powder-actuated tools.

> **WARNING!**
> Never use a powder-actuated tool unless trained and certified. Always wear appropriate PPE and make sure that others on the jobsite are aware of the tool in use. Never point these tools at anyone or fire them into materials such as wallboard or plywood. These tools fire with the power of a handgun and can injure or kill workers on the opposite side.

When used in concrete, the quality of the concrete must meet minimum standards for the fastener to hold effectively. If the aggregate is too large, the concrete may break or chip, and there will be no gripping power. The minimum diameter mild steel rod permitted is ⅜". NFPA 13 requires attachment of an increaser coupling directly to the driven stud before mild steel rod is used. These studs are Listed for specific maximum loads and minimum concrete standards. Check the Listings for the stud to be used before installation. Always check the shop drawings before using powder-actuated tools.

Typical rules for safely operating a powder-actuated tool are as follows:

- Do not use a powder-actuated tool unless you are certified.
- Follow all safety precautions in the manufacturer's instruction manual.
- Always wear safety goggles, hearing protection, and a hard hat when operating a powder-actuated tool.
- Use the proper size pin for the job you are performing.
- When loading the driver, put the pin in before the charge.
- Use the correct booster (powder load) according to the manufacturer's instructions.
- Never hold the end of the barrel against any part of your body or cock the tool against your hand.
- Never hold your hand behind or near the material you are fastening.
- Do not shoot close to the edge of concrete.
- Never attempt to pry the booster out of the magazine with a sharp instrument.
- Always hold the muzzle perpendicular (at a 90-degree angle) to the work.

Before using a powder-actuated stud gun, be sure the specifications of the stud and charge match the weight and thickness of the structural steel or concrete. Note that powder-driven studs cannot be used in seismic areas unless they are specifically Listed for use in these locations. *Figure 20* shows powder-driven studs.

Silica Safety

Exposure to the silica in concrete can cause lung cancer, silicosis, chronic obstructive pulmonary disease (COPD), and kidney disease. According to OSHA, approximately two million construction workers are exposed to respirable crystalline silica on the job each year. This includes those who drill, cut, crush, or grind materials such as concrete and stone. The OSHA silica rule sets the permissible exposure limit (PEL) for respirable crystalline silica to 50 micrograms per cubic meter of air, averaged over an 8-hour shift.

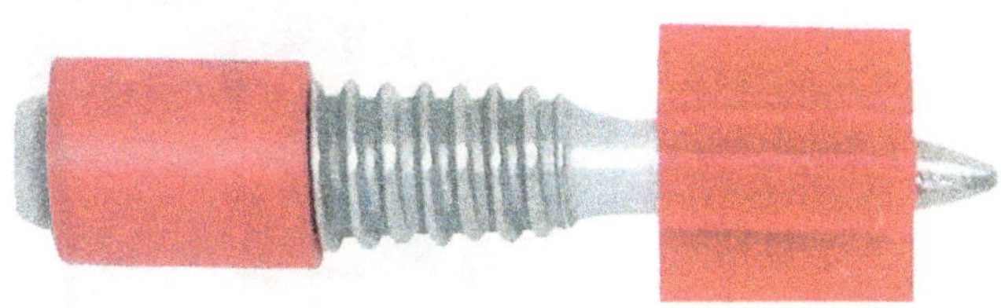

Figure 20 Powder-driven stud.

It also requires employers to use the following controls to limit worker exposure:

- Use engineering controls (e.g., tools equipped with shrouds or dust collection systems and tools equipped with integrated water delivery systems) to limit worker exposure to the PEL. *Figure 21* shows a drill with an integral dust collector.
- Provide respirators when engineering controls cannot adequately limit exposure.
- Limit worker access to high exposure areas.
- Develop a written exposure control plan.
- Offer medical exams to monitor highly exposed workers.
- Train workers on silica risks and how to limit exposures.

1.5.0 Pipe Supports and Other Attachments

Pipe supports and other attachments are used to support the weight of the system components and piping. Common pipe supports and attachments include the following:

- Ceiling flanges
- U-bolts
- Toggle bolts
- Pipe stands
- Riser clamps
- Support straps (CPVC pipe only)

1.5.1 Ceiling Flanges

A ceiling flange is a limited-support device normally used in wood construction with a maximum of 2" pipe. The screws must be a minimum of No. 18 × 1½", unless there are Listed alternate fasteners for this application.

Figure 21 Hammer drill with integral dust collector.

There are two-, three-, and four-screw flanges available. *Figure 22* shows a typical two-screw ceiling flange. *Table 8* shows screw dimensions for ceiling flanges and U-hooks.

1.5.2 U-Bolts

U-bolts are commercially available in sizes that support pipe from ½" to 36". U-bolts are used as hangers and restraints for supporting and anchoring heavy loads. The U-bolts hold pipe firmly to the structure using two nuts to secure the pipe. *Figure 23* shows a typical U-bolt.

1.5.3 Toggle Bolts

Toggle bolts are used to fasten system components in hollow tile, metal lath, or plaster. They are the most limited of all the support devices allowed in NFPA 13, and a maximum of 1½" pipe may be supported using toggle bolts. Note that pipe cannot be supported from drywall using a toggle bolt. The ceiling must be either hollow tile or metal lath and plaster.

In the sprinkler industry, toggle bolts are often used with ATR rather than the bolt supplied with the fastener (*Figure 24*). A fender washer and nut are used to provide a secure installation.

1.5.4 Pipe Stands

Pipe stands support pipe from the bottom. They are available premade but are most often field-fabricated. Pipe stands must be designed to support the weight of the water-filled pipe plus 250 pounds at each point of support. They must withstand loading from all sources, as well as allow for expansion and contraction due to temperature changes and water hammer.

Pipe stand flanges must be malleable iron or welded steel, with a minimum of four anchors used to attach the stand to the floor. Pipe stand diameters and associated anchor sizes are specified in NFPA 13. *Figure 25* shows a field-fabricated pipe stand.

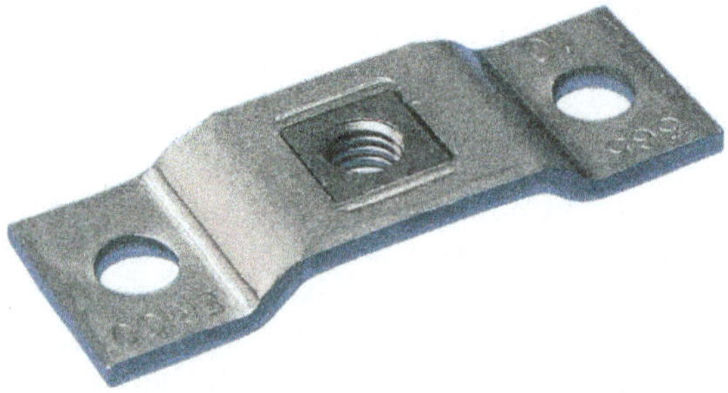

Figure 22 Typical ceiling flange.

Table 8 Screw Dimensions for Ceiling Flanges and U-Hooks (Data from NFPA 13, *Table 17.2.4.2.1*)

Pipe Size	Fitting Type
Two-Screw Ceiling Flanges	
Up to and including 2"	Wood screw No. 18 × 1½" or Lag screw 5⁄16" × 1½"
Three-Screw Ceiling Flanges	
Up to and including 2"	Wood screw No. 18 × 1½"
2½", 3"	Lag screw ⅜" × 2"
4", 5", 6"	Lag screw ½" × 2"
8"	Lag screw ⅝" × 2"
Four-Screw Ceiling Flanges	
Up to and including 2"	Wood screw No. 18 × 1½"
2½", 3"	Lag screw ⅜" × 1½"
4", 5", 6"	Lag screw ½" × 2"
8"	Lag screw ⅝" × 2"
U-Hooks	
Up to and including 2"	Drive screw No. 16 × 2"
2½", 3"	Lag screw ⅜" × 2½"
4", 5", 6"	Lag screw ½" × 3"
8"	Lag screw ⅝" × 3"

1.5.5 Riser Clamps

A riser clamp is used to support a vertical pipe riser. It is clamped tightly to a riser above a floor to support the pipe and distribute its weight to the building structure. Riser clamps are also used below the floor to prevent upward thrust when using flexible couplings in a system. Riser clamps are available to support pipe up to 20" in diameter. *Figure 26* shows a riser clamp.

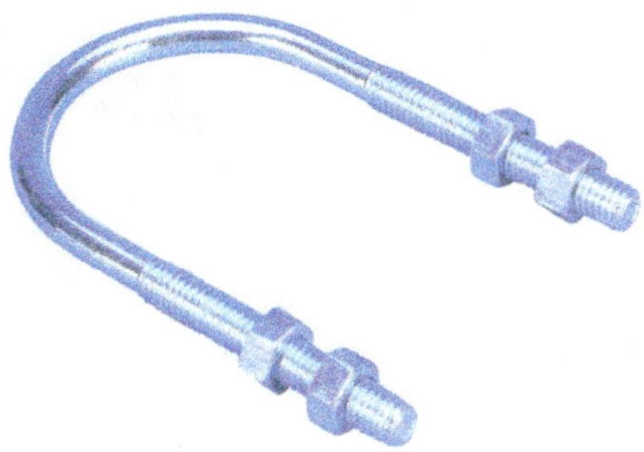

Figure 23 U-bolt.

1.5.6 Support Straps for CPVC

If a CPVC piping system runs along a wall, simple straps can be used. These lightweight straps fasten directly to the wall to hold the pipe. *Figure 27* shows typical support straps for CPVC. Some straps are also Listed for use as hangers.

> **NOTE**
> The use of CPVC pipe requires special installation considerations. Check the pipe manufacturer's guidelines for information regarding proper size and location of hangers, supports, and restraints.

Figure 24 Toggle bolt on threaded rod.

Figure 25 Field-fabricated pipe stand.

Figure 26 Riser clamp.

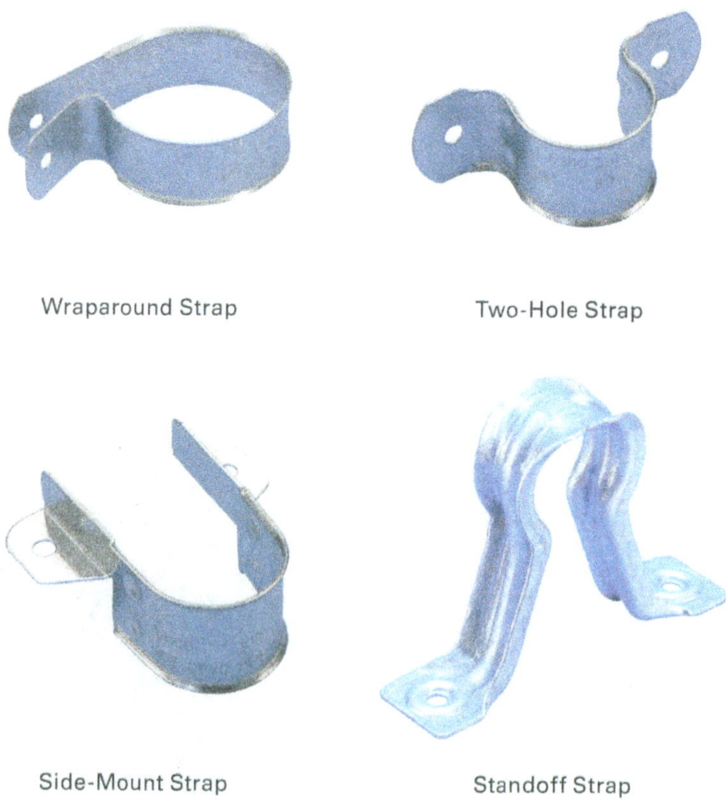

Figure 27 CPVC support straps.

1.0.0 Section Review

1. It is acceptable to suspend a drop ceiling from sprinkler piping as long as the pipe is securely attached to the building structure.
   a. True
   b. False

2. The type of pipe that requires the *closest* hanger spacing is ____.
   a. steel pipe (except threaded lightwall)
   b. threaded lightwall steel pipe
   c. copper tube
   d. CPVC

3. The *minimum* hanger rod size for a 2" pipe is ____.
   a. ⅜"
   b. ½"
   c. ⅝"
   d. ¾"

4. Which of the following connects directly to the top or bottom of a steel I-beam?
   a. Pipe clamp
   b. Riser clamp
   c. Beam clamp
   d. Purlin clamp

5. A pipe attachment that can be used to support or anchor a heavy load is a ____.
   a. ceiling flange
   b. U-bolt
   c. toggle bolt
   d. wraparound strap

SECTION TWO

2.0.0 INSTALLING SEISMIC BRACING

Objective

Install seismic bracing.

a. Identify NFPA 13 requirements for seismic bracing.
b. Identify the components in sway bracing systems.

Performance Task

4. Make up and install a seismic brace.

Trade Terms

Seismicity: The tendency of a location to be subjected to seismic forces.

Sway brace: An assembly attached to sprinkler system piping that is designed to resist earthquake (seismic) loads in two directions.

Seismic damage to sprinkler systems occurs due to the forces created when one of the Earth's tectonic plates makes a sudden shift. A mass not rigidly connected to the ground, such as overhead sprinkler piping, tends to remain stationary, while a mass rigidly connected to the ground, like a flanged spigot, tends to move with the Earth's surface. This is what causes buildings to sway, crack, and break apart when the first floor moves with the surface of the ground, while the other levels remain stationary.

The tendency of a location to be subjected to seismic forces is known as its seismicity, and it varies widely around the country. Areas on the west coast have a much higher seismicity than the central plains. *Figure 28* shows a seismicity map of the United States. The darker colors represent areas of greater seismic activity. Seismic bracing is installed to protect sprinkler systems from damage caused by these forces.

Sprinkler piping presents a unique problem because it is connected both to the ground and to flexible parts of the building. Previously, a piece of pipe used with a Listed sway brace fitting at both ends was the most common type of seismic protection used in the sprinkler industry. Currently, the two basic types of sway braces are those that act in both tension and compression,

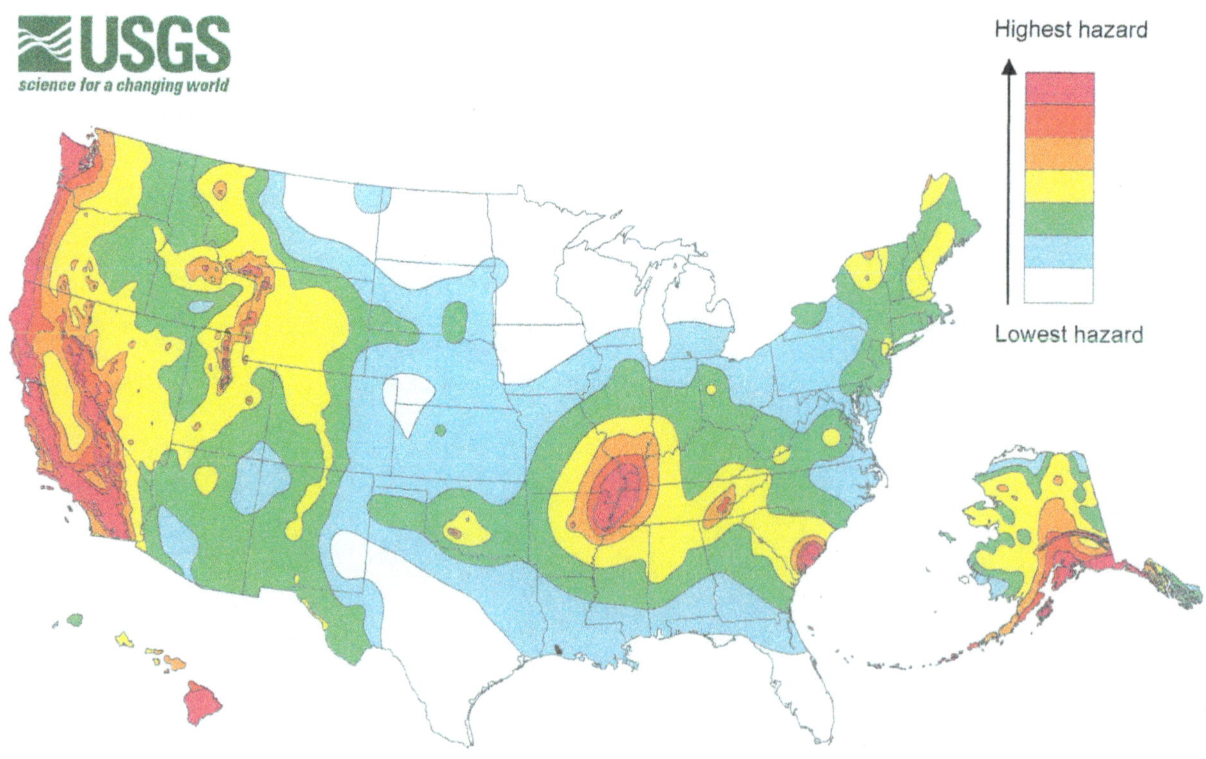

Figure 28 Seismicity map of the United States.

and those that act in tension only. Tension/compression sway braces are rigid braces and need only be attached to one side of the braced pipe. Tension-only braces (wire) braces are not rigid and must be attached to both sides of the braced pipe. Check the manufacturer's installation instructions for details.

WARNING!

Seismic requirements vary depending on the location, specific building requirements, and type of occupancy. The Engineer of Record (EoR) is responsible for specifying seismic requirements, which can be found in the contract documents and shop drawings. Strictly adhere to the manufacturer's instructions and approved shop drawings when installing seismic bracing. Failure to do so jeopardizes the safety and integrity of the piping system. If issues arise during installation that prevent adherence to the shop drawings, stop work and submit the system for redesign before proceeding.

2.1.0 NFPA Requirements for Seismic Bracing

NFPA 13, *Chapter 18* provides the minimum installation requirements for seismic protection, some of which are described in the following sections. Always refer to the standard in its entirety for complete details.

2.1.1 Floor and Wall Penetrations

A primary requirement for piping is to provide room for it to shift as it passes through floors and partitions, particularly in areas with frequent seismic activity. This begins with the flanged spigot where the entry through the floor should be sleeved and caulked with mastic for waterproofing.

Per NFPA 13, clearance is calculated such that for pipe sizes 1" to $3\frac{1}{2}$", the hole or sleeve must be 2" larger than the pipe diameter. For pipe sizes 4" and larger, the hole or sleeve must be 4" larger than the pipe diameter.

In specific locations, pipe sizes $2\frac{1}{2}$" and larger require the use of Listed flexible couplings, per NFPA 13. This allows individual sections of pipe to move with the sections of the building where they are attached. Rigid couplings do not allow for movement.

Frangible construction (e.g., sheet rock) does not require clearance as it will easily crumble in the event of seismic activity without harming the pipe.

2.1.2 Sway Braces

Sway braces using pipe, angle iron, rods, and flats are rigid tension/compression braces, while sway braces using aircraft cable are tension-only braces. Neither type offers compression resistance to upward vertical movement, but they resist the horizontal loads equally. NFPA 13 requires that supplemental supports be used to resist the resulting upward movement caused by large horizontal forces coupled with small brace angles (off the vertical). Supplemental supports can be hanger rods located within 24" of braces and strengthened against failure by pipe sheathing or other means. *Figure 29* shows sway bracing. Lateral bracing provides support perpendicular to the pipe, while longitudinal bracing provides support parallel to the pipe.

2.1.3 Seismic Separation Assemblies

A seismic separation assembly is used to cross a seismic joint in a building. Seismic joints are structurally designed to permit three-dimensional movement and can be constructed using an arrangement of flexible couplings or an approved

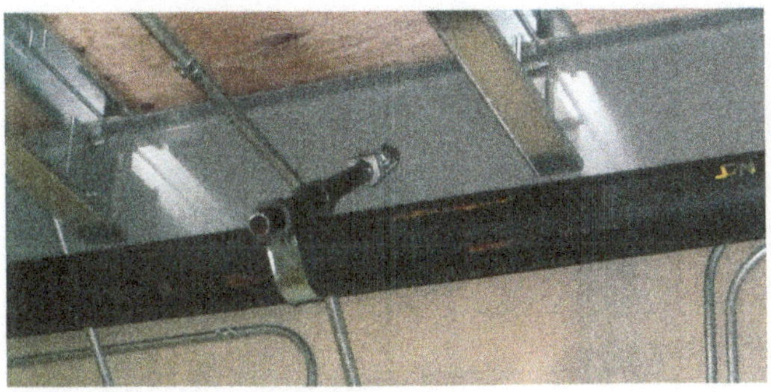

Lateral Bracing

Longitudinal Bracing

Figure 29 Sway bracing.

seismic separation assembly, as described in NFPA 13. *Figure 30* shows an arrangement of flexible grooved couplings that allow for seismic movement in all directions.

> **NOTE**
> A seismic separation assembly is not required to cross an expansion joint (which only allows concrete to expand and contract without cracking).

2.2.0 Installing Sway Bracing Systems

Without bracing, piping supported by conventional ATR/hanger assemblies tends to act like a pendulum when a seismic shock occurs. This means that the ATR deflects as it is torn between the force of the building suddenly moving and the pipe remaining fixed. Tremendous forces are then placed on the ATR, requiring it to bend, stretch, and take most of the inertial forces produced by the piping trying to remain in place.

Lateral sway braces provide restraint strength in the direction that the ATR hangers have very little strength. In the longitudinal direction, a ring hanger has no strength because of the sliding motion that makes for easy installation. Sway bracing is needed especially for mains, both longitudinally and laterally. A Listed clamp similar to a riser clamp is used to attach sway bracing to the pipe. *Figure 31* shows sway bracing.

A four-way brace can consist of two separate two-way braces used in combination. This provides a more efficient four-way bracing configuration utilizing Listed applications.

Brace spacing is based on the zone of influence analysis (per NFPA 13) or by a dynamic seismic analysis certified by a registered professional engineer. Brace spacing is related to brace size and fasteners. Fasteners are usually the limiting factor when spacing braces because the load that the fastener can hold is usually less than the capacity of the pipe or pipe clamp. In some cases, braces are spaced at less than the maximum because of the fastener or structure. Refer to the shop drawings for brace spacing and attachment information.

Braces must be connected directly to the feed on cross mains and typically not to the branch lines. The size and type of fasteners are shown on the approved shop drawings. The fasteners do not correspond to hangers. They are based on the angle of the brace, the size of the brace, its orientation, and the type of fastener being used. The brace pipe is attached to the structure and the main piping using special attachments, such as the swivel attachment shown in *Figure 32* or the bar joist attachment in *Figure 33*. Attachments to concrete may require heavy-duty wedge anchors meeting the standards of ACI-318, *Building Code Requirements for Structural Concrete*. Note that NFPA 13 prohibits the use of powder-driven fasteners for bracing/hanger attachments in areas prone to seismic activity.

Branch line restraints typically have a lower resistance to seismic loads and may use either threaded rod (*Figure 34*) or wire brace lines. The typical branch line restraint is 40' or less, with a configuration of No. 12, 440-pound wire or rod at 45 degrees or more per NFPA 13.

> **NOTE**
> If approved, wire brace lines can also be used as longitudinal or lateral sway braces; consult with the manufacturer for application information.

Figure 35 shows branch line restraint fittings for various substrates.

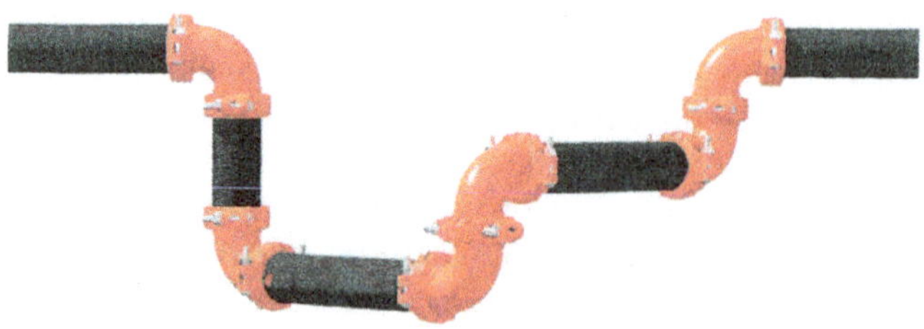

Figure 30 Flexible grooved couplings used to create a seismic separation assembly.

Figure 31 Sway bracing.

Figure 33 Bar joist sway brace attachment.

Figure 32 Sway brace swivel attachment.

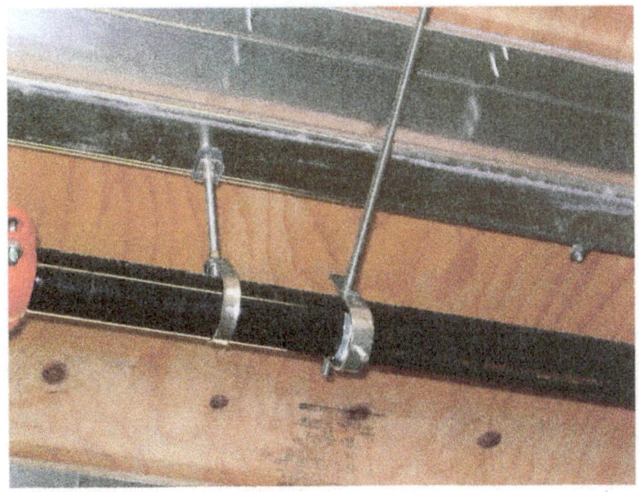

Figure 34 Branch line restraint.

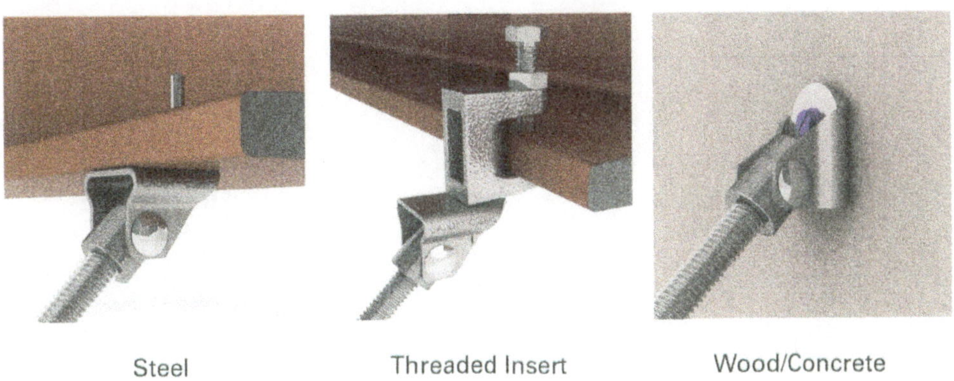

Figure 35 Branch line restraint fittings.

2.0.0 Section Review

1. Per NFPA 13, the hole size for a 2" pipe penetration through a floor is _____.
   a. 2"
   b. 3"
   c. 4"
   d. 5"

2. The type of seismic bracing *most* likely to include wire brace lines is a _____.
   a. lateral brace
   b. seismic separation assembly
   c. sway brace
   d. branch line restraint

SECTION THREE

3.0.0 IDENTIFYING FIRESTOPPING ASSEMBLIES

Objective

Identify firestopping assemblies.

a. Identify various firestopping products and their ratings.
b. Install firestopping in floor and wall penetrations.

Trade Terms

Endothermic: A property of a material that causes it to release moisture when heated.

Firestopping: A means of cutting off the oxygen supply so that fire and smoke cannot readily move from one location to another.

Intumescent: A property of a material that causes it to expand when heated, acting as an insulator in the event of a fire.

Effective **firestopping** is a critical element in fire safety. The term *firestopping* refers to cutting off the oxygen supply so that fire and smoke cannot readily move from one location to another. Sprinkler fitters install firestopping per the shop drawings where sprinkler piping penetrates walls and floors.

You will hear the term *firestop* used in two different ways:

- In frame construction, a firestop is a piece of wood or fire-resistant material inserted into an opening such as the space between studs. This firestop acts as a barrier to block airflow that would allow the space to act as a chimney, carrying fire rapidly to upper floors. It does not put out the fire, but it slows the fire's progress.
- In commercial construction and some residential applications, firestopping materials are used to close floor and wall penetrations such as those created to run conduit, piping, and ductwork. If these openings are not sealed, fire will travel through the openings in its search for oxygen. The firestopping methods used for this purpose must be Listed assemblies, and are classified with a one-hour, two-hour, three-hour, or four-hour rating, depending on the code and facility requirements. The hour rating refers to the amount of time it would take for a fire to burn through the wall.

Firestopping materials are typically applied around all types of piping, electrical conduit, ductwork, electrical and communication cables in floor slabs, walls, and other fire-rated building partitions and assemblies (*Figure 36*).

Firestopping materials are classified as **intumescent** or **endothermic**. Both are formulated to help control the spread of fire before, during, and after exposure to open flames. When subjected to the extreme heat of a fire, intumescent materials expand (typically up to three times their original size) to form a strong insulating material that seals the opening for three to four hours. Should the insulation on the cables or pipes passing through the penetration become consumed by the fire, the expansion of the

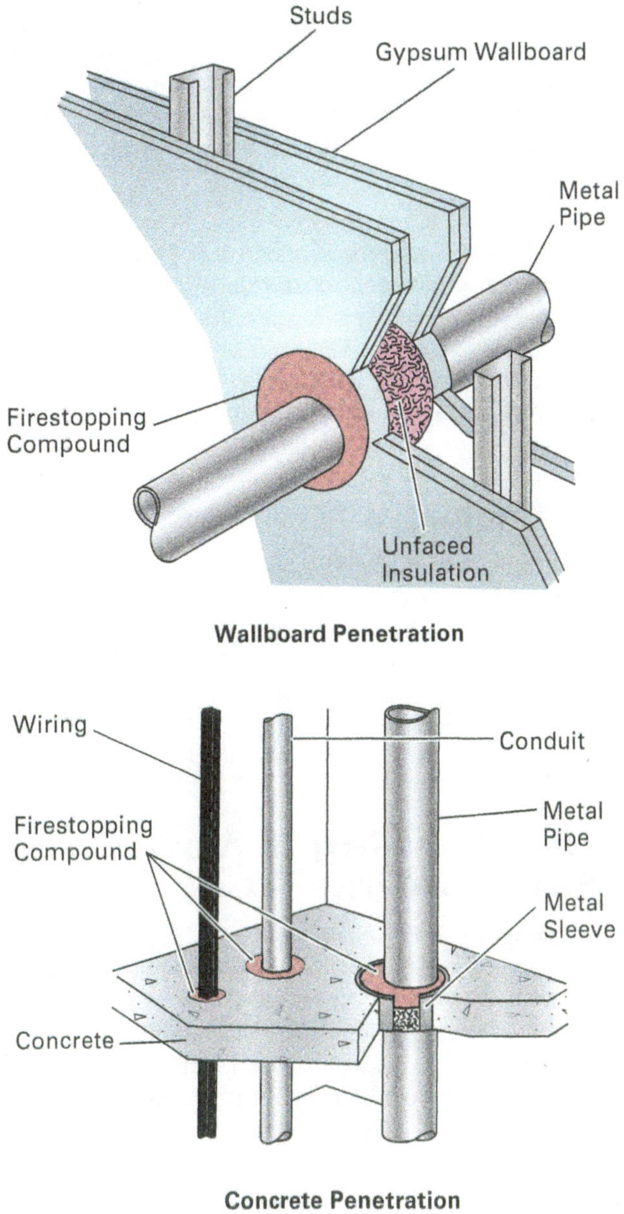

Figure 36 Fireproof openings.

firestopping material also acts to fill the void in the floor or wall in order to help stop the spread of smoke and other toxic products of combustion. Endothermic materials respond to fire by releasing chemically bound water, which absorbs heat.

Some firestopping materials are ablative, which means that they form a hard char when exposed to flame. The char inhibits the passage of flame and smoke. Firestopping materials are formulated in such a way that when activated, they are free of toxic or corrosive gases, reducing the risks to building occupants and sensitive equipment.

Firestopping materials are made in a variety of forms, including sealants, moldable putty, mortars, foams, sprays, tapes, sheets, wraps, pillows, bricks, collars, and sleeves. They come in both one-part and two-part formulations and must be installed in accordance with applicable building codes and the manufacturer's instructions. Depending on the product, firestopping materials can be applied by hand or using spray equipment, conventional caulking guns, pneumatic pumping equipment, or a putty knife.

CAUTION

Some firestopping products may produce toxic fumes or stir up dust during the installation process. Always read and follow the instructions for safe use on the product's Safety Data Sheet (SDS) and wear appropriate PPE, including eye, skin, and respiratory protection.

3.1.0 Firestopping Materials and Ratings

The most common firestopping material used in sprinkler work is firestop sealant. Firestop sealant is a ready-to-use material with the consistency of soft caulk (*Figure 37*). It is available in tubes, cartridges, and pails, and applied using a standard caulk gun or trowel.

Firestopping Is Not Fireproofing

Firestopping and fireproofing are not the same. Firestopping is typically a flexible reddish caulk or sealant that prevents the spread of fire from room to room through openings in walls and floors. Fireproofing is a rough, gray cementitious material sprayed onto steel I-beams and structural support columns. It acts as a thermal barrier that insulates structures against the high temperatures produced in a fire, preserving their shape and strength.

Firestop sealant is often applied over mineral wool or other fire barrier packing material (*Figure 38*). Backer boards or flexible backer rods may also be used. These materials are used to fill gaps and limit the amount of compound that would otherwise be required.

The combination of firestop sealant and packing material is Listed as a firestop system or assembly in the UL online Building Materials Directory. For example, a $1\frac{1}{2}$" thickness of firestop sealant may produce a two-hour rated seal. If the sealant is backed with fire-resistant mineral wool, a $\frac{1}{2}$" thickness of sealant may achieve the same two-hour fire rating. Always follow the manufacturer's instructions for the products in use to obtain the required fire rating.

After installing firestopping, apply warning labels indicating that the material must not be disturbed, or if disturbed, must be replaced by a material with the same rating (*Figure 39*).

CAUTION

For CPVC pipe, consult the pipe manufacturer for approved firestopping products. Not all products are compatible with CPVC.

Figure 37 Firestop sealant.

Figure 38 Packing material.

> **NOTE**
>
> In some instances, the shop drawings may specify the brand and type of firestopping material. Always follow the shop drawings when selecting and installing firestopping.

Sprinkler fitters may also work with cast-in-place firestops, such as the one shown in *Figure 40*. This device is installed in concrete, and the height is adjusted using pull tabs to remove the excess material. It acts as both a concrete form to create the pipe opening and as a firestop to seal the penetration.

Any firestopping materials used must meet the criteria of standard ASTM E814, *Standard Test Method for Fire Tests of Through-Penetration Fire Stops*, as tested under positive pressure. Underwriters Laboratories has a similar standard entitled UL 1479, *Standard for Fire Tests of Penetration Firestops*. Firestopping materials must have an hourly rating that is equal to or greater than the hourly rating of the floor or wall being penetrated. One of four ratings, measured in time, may be applied to firestopping materials and systems. These ratings are currently described as follows:

- *F rating* — A firestopping system meets the requirements of an F rating if it remains in the opening during the fire test for the rating period without permitting the passage of flames through the opening or the occurrence of flaming on any element of the unexposed side of the assembly.
- *FT rating* — A firestopping system meets the requirements of an FT rating if it remains in the opening during the fire test within the limitations as specified for an F rating. In addition, the transmission of heat through the firestopping system during the rated period shall not have been such as to raise the temperature of any thermocouple on the unexposed surface of the firestopping system by more than 347.8°F above its initial temperature.
- *FH rating* — A firestopping system meets the requirements of an FH rating if it remains in the opening during the fire test within the limitations for an F rating. In addition, during a hose stream test, the firestopping system shall not develop any opening that would permit a projection of water from the stream to the unexposed side.
- *FTH rating* — A firestopping system shall be considered as meeting the requirements of an FTH rating if it remains in the opening during the fire and hose stream tests within the limitations as described for FT and FH ratings.

> **CAUTION**
>
> It is essential that firestopping be installed exactly as shown on the shop drawings in order to preserve the designated fire rating of the assembly or system.

Figure 39 Firestop labels.

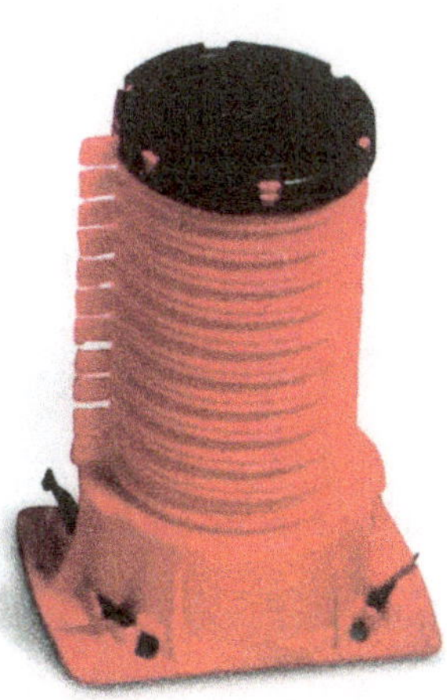

Figure 40 Cast-in-place firestop device.

3.2.0 Sealing Floor and Wall Penetrations

A key fire protection concept in multistory construction is to confine a fire to the floor of origin. If sprinkler piping penetrates fire-rated floors, these openings must be sealed to prevent the passage of heat or flames and restore the fire rating of the original floor.

The floor of a multistory building is considered a fire-rated assembly. An example of a Listed two-hour fire-rated assembly is shown *Figure 41*. Note that the firestopping requirements vary by manufacturer, pipe size and type, and wall/floor construction. Always refer to the approved shop drawings.

Figure 42 shows an example of firestopping for a concrete block wall. Note that mineral wool or other fire barrier packing material is used to seal the opening and provide a surface on which to apply the firestop sealant.

Firestopping requirements are included in all building codes. In addition, fire marshals are enforcing these requirements more strictly than in the past. Many sprinkler contractors use specialty companies to install firestopping.

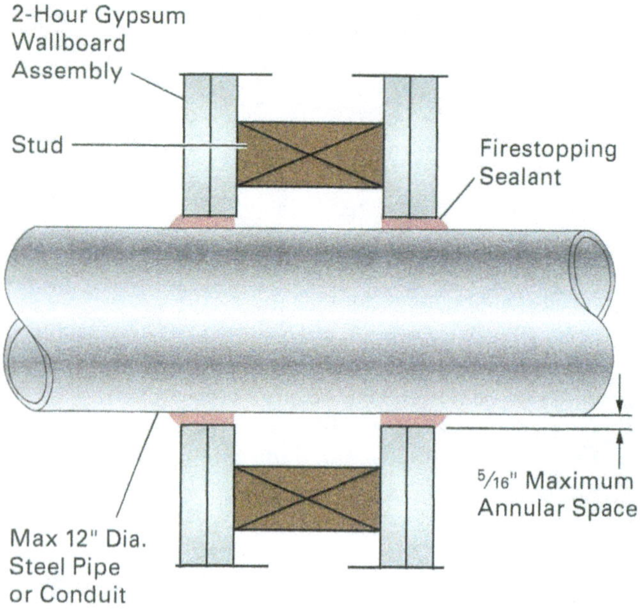

Figure 41 Two-hour fire-rated assembly.

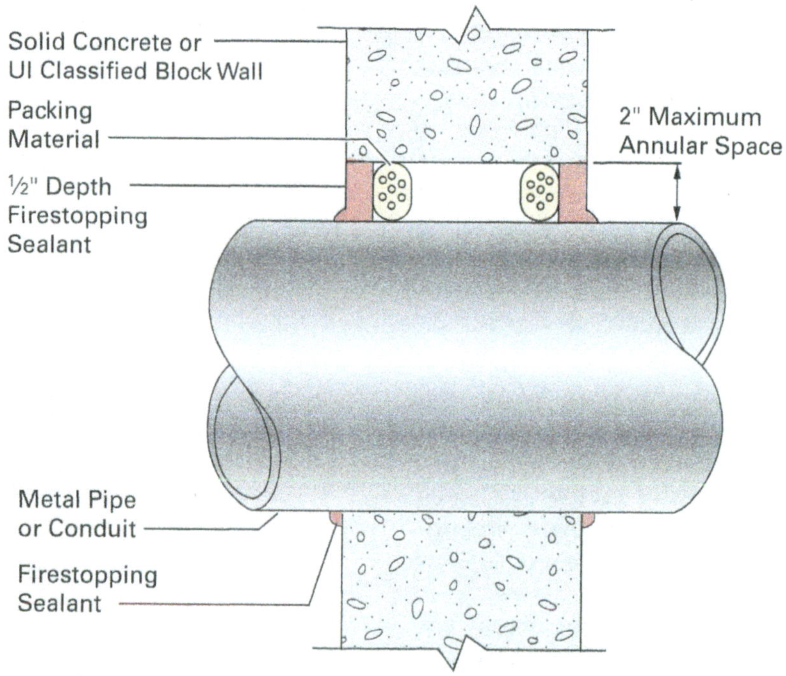

Figure 42 Penetration firestop for a 12" maximum diameter pipe/conduit through a concrete wall.

3.0.0 Section Review

1. Firestopping that meets the requirements of a temperature test as well as a hose test has an _____.
   a. F rating
   b. FT rating
   c. FH rating
   d. FTH rating
2. Many sprinkler contractors use specialty companies to install firestopping.
   a. True
   b. False

Review Questions

1. A hanger rated at 600 pounds *must* be able to support _____.
   a. 500 pounds
   b. 600 pounds
   c. 2,750 pounds
   d. 3,250 pounds

2. A trapeze hanger is designed to _____.
   a. support primarily 6" and above pipe
   b. be fabricated only in the shop
   c. distribute the load of water-filled pipe over at least two joists or purlins
   d. distribute the load of water-filled pipe over only two attachment points

3. The *minimum* hanger rod size for a 6" pipe is _____.
   a. 3⁄8"
   b. 1⁄2"
   c. 5⁄8"
   d. 3⁄4"

4. To use a powder-actuated tool, you *must* have training and an operator's card issued by _____.
   a. the site superintendent
   b. a qualified, authorized instructor
   c. your supervisor
   d. any qualified operator

5. The *minimum* diameter mild steel rod permitted with powder-driven fasteners is _____.
   a. 3⁄8"
   b. 1⁄2"
   c. 5⁄8"
   d. 3⁄4"

6. Powder-driven studs *cannot* be used in _____.
   a. areas prone to flooding
   b. buildings over five stories
   c. tornado prone areas
   d. seismic areas (unless Listed for such use)

7. Which is the *most* limited of all the support types?
   a. U-bolts
   b. U-hooks
   c. Beam clamps
   d. Toggle bolts

8. Sprinkler piping is likely to be damaged by seismic activity because it is _____.
   a. lighter than other types of pipe
   b. attached to hangers rather than directly to the structure
   c. attached to both the ground and to flexible parts of the building
   d. filled with water

9. Where seismic protection is required, the nominal clearance around a 4" pipe extending through a wall, floor, platform, or foundation *must* be _____.
   a. 1"
   b. 2"
   c. 3"
   d. 4"

10. A suitable backing material for firestopping is _____.
    a. kraft paper
    b. mineral wool
    c. rubber sealant
    d. polystyrene foam

Trade Terms Introduced in This Module

All-thread rod (ATR): Rod that is fully threaded end to end and often cut to size and used to suspend hangers and fasteners.

Anchor: A device used to attach a hanger to a structure.

Endothermic: A property of a material that causes it to release moisture when heated.

Firestopping: A means of cutting off the oxygen supply so that fire and smoke cannot readily move from one location to another.

Hanger: A device from which a pipe is suspended that is used to carry the weight of the pipe from above.

Intumescent: A property of a material that causes it to expand when heated, acting as an insulator in the event of a fire.

Load rating: The usable weight capacity of a hanger or support.

Pipe riser: A vertical pipe extending from an opening in a floor or ceiling.

Restraint: A device used to prevent pipe from shifting in any direction other than downward.

Section modulus: A measurement of the load capacity of a trapeze hanger or other support.

Seismic bracing: Bracing installed on a piping system to protect it from damage caused by seismic activity, such as an earthquake.

Seismicity: The tendency of a location to be subjected to seismic forces.

Support: A structure or saddle on which a pipe rests that is designed to support the weight of the pipe from below.

Sway brace: An assembly attached to sprinkler system piping that is designed to resist earthquake (seismic) loads in two directions.

Takeout: The distance from the center of a pipe to the bottom of the threaded rod in a hanger assembly.

Water hammer: The pressure or shock wave within piping caused by a sudden change in water flow, such as when starting a pump or quickly closing a valve.

Additional Resources

This module presents thorough resources for task training. The following reference material is suggested for further study.

ACI-318, *Building Code Requirements for Structural Concrete*, Latest Edition. Farmington Hills, MI: American Concrete Institute.

ASTM E814, *Standard Test Method for Fire Tests of Through-Penetration Fire Stops*, Latest Edition. West Conshohocken, PA: ASTM International, **www.astm.org**.

NFPA 13, *Standard for the Installation of Sprinkler Systems*, Latest Edition. Quincy, MA: National Fire Protection Association.

NFPA 13D, *Standard for the Installation of Sprinkler Systems in One- and Two-Family Dwellings and Manufactured Homes*, Latest Edition. Quincy, MA: National Fire Protection Association.

NFPA 13R, *Standard for the Installation of Sprinkler Systems in Low-Rise Residential Occupancies*, Latest Edition. Quincy, MA: National Fire Protection Association.

Figure Credits

Courtesy of nVent Electrical and Fastening Solutions, Figures 1–3, 8, 12, 15–17, 22–23, 26–27, 31–33, 35

Courtesy of Hilti, Inc., Figures 4, 19, 38

Photo courtesy of Milwaukee Tool, Figure 6

Courtesy of ASC Engineered Solutions, Figure 7

Courtesy of ITW Buildex, Figure 9

Strickland Fire Protection, Inc., Figure 10

Courtesy of ISAT, Figure 18

Image property of Stanley Black & Decker. Used with permission., Figures 20–21

Taylor Schumacher, Figure 24

Reliable Automatic Sprinkler Co., Inc., Figure 25

USGS | US Department of the Interior, Figure 28

Chuck Bamford, Figures 29, 34

Courtesy of Victaulic Company, Figure 30

Courtesy of 3M (3M.com/firestop), Figures 37, 39–40

© 2018 NFPA, reproduced with permission. This material is not affiliated with nor has it been reviewed or approved by the NFPA., Table 1

Reproduced with permission of NFPA from NFPA 13, *Standard for the Installation of Sprinkler Systems*, 2019 edition. © 2018, National Fire Protection Association. For a full copy of NFPA 13, please go to www.nfpa.org., Tables 2–8

Section Review Answer Key

Section 1.0.0

Answer	Section Reference	Objective
1. b	1.1.0	1a
2. d	1.2.0; *Table 1*	1b
3. a	1.3.0; *Table 2*	1c
4. c	1.4.2	1d
5. b	1.5.2	1e

Section 2.0.0

Answer	Section Reference	Objective
1. c	2.1.1	2a
2. d	2.2.0	2b

Section 3.0.0

Answer	Section Reference	Objective
1. d	3.1.0	3a
2. a	3.2.0	3b

Section Review Calculations

Section 2.0.0

Question 1

2" + 2" = **4"**

Answers to Think About It

Section 1.3.0

1. Distance from the bottom of the top deck to the bottom of the pipe hanger:

   20'-6" – 16'-3" = **4'-3"**

2. Distance between the bottom of the top deck and the top of the hanger:

   4'-3" – 10" = **3'-5"**

3. How long the hanger rod must be to extend into the hanger $2\frac{1}{2}$":

   1" into anchor + $2\frac{1}{2}$" into hanger = **$3\frac{1}{2}$"**

   3'-5" + $3\frac{1}{2}$" = **3'-$8\frac{1}{2}$"**

User Update

NCCER makes every effort to keep its textbooks up-to-date and free of technical errors. We appreciate your help in this process. If you find an error, a typographical mistake, or an inaccuracy in NCCER's curricula, please submit a User Update form by visiting **https://www.nccer.org/olf**. You can also scan the QR code using the camera on your phone or mobile device to access the form.

General Purpose Valves

Overview

This module explains the valves used to start, stop, direct, and regulate water flow in sprinkler systems. It covers the operation, applications, and basic maintenance of various valves, including indicating control valves, trim valves, check valves, hose valves, pressure control valves, and air venting valves.

Module 18202

18202

General Purpose Valves

Objectives

Successful completion of this module prepares you to do the following:

1. Identify the function and operation of the indicating control valves used in sprinkler systems.
   a. Identify gate valves and describe how they operate.
   b. Explain the purpose of post-indicating valves.
   c. Identify butterfly valves and describe how they operate.
2. Identify the function and operation of the trim valves used in sprinkler systems.
   a. Identify ball lever valves and describe how they operate.
   b. Identify globe valves and describe how they operate.
3. Identify the function and operation of the supply system valves used in sprinkler systems.
   a. Identify check valves and describe how they operate.
   b. Identify hose valves and describe how they operate.
4. Identify the function and operation of the pressure control and air venting valves used in sprinkler systems.
   a. Identify pressure-reducing valves and describe how they operate.
   b. Identify pressure-regulating devices and describe how they operate.
   c. Describe the function of air venting valves.

Performance Tasks

Under supervision, you should be able to do the following:

1. Install a gate valve.
2. Install a supervisory (tamper) switch.
3. Install a butterfly valve.

Trade Terms

Control valve
Distort
Erode
Globe valve
Hose header
Machined
Pressure-regulating device
Pressurized system
Residual pressure
Sealed valves
Sprinkler lead-in
Sprinkler riser
Static pressure
Throttling
Water hammer

Industry Recognized Credentials

If you are training through an NCCER-accredited sponsor, you may be eligible for credentials from NCCER's Registry. The ID number for this module is 18202. Note that this module may have been used in other NCCER curricula and may apply to other level completions. Contact NCCER's Registry at 1.888.622.3720 or go to **www.nccer.org** for more information.

You can also show off your industry-recognized credentials online with NCCER's digital badges. Transform your knowledge, skills, and achievements into badges that you can share across social media platforms, send to your network, and add to your resume. For more information, visit **www.nccer.org**.

> **NOTE**
>
> Unless otherwise specified, references in parentheses following figure and table numbers refer to NFPA 13.

Contents

Figures

SECTION ONE

1.0.0 INDICATING CONTROL VALVES

Objective

Identify the function and operation of the indicating control valves used in sprinkler systems.

a. Identify gate valves and describe how they operate.
b. Explain the purpose of post-indicating valves.
c. Identify butterfly valves and describe how they operate.

Performance Tasks

1. Install a gate valve.
2. Install a supervisory (tamper) switch.
3. Install a butterfly valve.

Trade Terms

Control valve: A valve controlling flow to water-based fire protection systems.

Distort: To bend, warp, or damage a system component.

Erode: To deteriorate the surfaces in a valve due to water flow.

Sealed valves: Valves that are wired or strapped in the open position.

Sprinkler lead-in: The pipe supplying water to the sprinkler system.

Sprinkler riser: The vertical supply pipe in a sprinkler system.

Throttling: Using a valve to regulate flow through a system.

Water hammer: A damaging force that occurs when the flow of water is started or stopped, or when the direction of flow has changed.

Valves are devices that control the flow of water through a piping system. While the designs of valves vary, all valves have a passageway for water flow and a component that opens and closes the passageway. A valve can be used to provide on-off service only, or it can be used as a regulating or throttling device. It can also relieve excess pressure in a piping system or prevent flow reversal through a line.

One of the most important types of valves is the system control valve, which controls the flow to water-based fire protection systems. The most common reason for the failure of a sprinkler system is a closed control valve. For this reason, NFPA requires that control valves be indicating valves. This means that the valve's position must be clearly indicated as open or closed using a visual inspection. Common indicators include the following:

- A window that reads "OPEN" or "SHUT"
- A valve stem that is out (open) or in (closed)
- A valve handle that is aligned with the piping (open) or at a 90-degree angle (closed)

Indicating valves should be installed at any point in the piping where closing a valve would disable the sprinkler system. Common areas in a sprinkler system where indicating valves are located are valve pits, hose headers, and zone control valves. A valve must also be located at the sprinkler riser (the vertical supply pipe in a sprinkler system) and the sprinkler lead-in (the pipe supplying water to the sprinkler system).

Listed indicating valves are required by NFPA 13 for all applications that control the water supply to a sprinkler system or part of the system. The basic types of indicating valves used in sprinkler systems include the following:

- Gate valves
- Post-indicating valves (PIVs)
- Butterfly valves

NOTE

NFPA 13 permits a Listed backflow preventer to be used as a system control valve, provided both control valves are Listed for fire protection systems and an additional control valve is not required.

1.1.0 Gate Valves

Gate valves are one of the most common types of valves and can be installed in the vertical or horizontal position. Gate valves are ideal for use in systems that do not require frequent valve operation. A gate valve should not be used as a flow control device because it will erode the gate and prevent proper sealing. Gate valves must have a minimum rated working pressure of 175 psi for all sizes 12" or smaller. Valves are sized based on the nominal diameter of the piping connected to the valve and must be clearly marked to indicate their function in the system. *Figure 1* shows a solid-wedge gate valve with flanged connections.

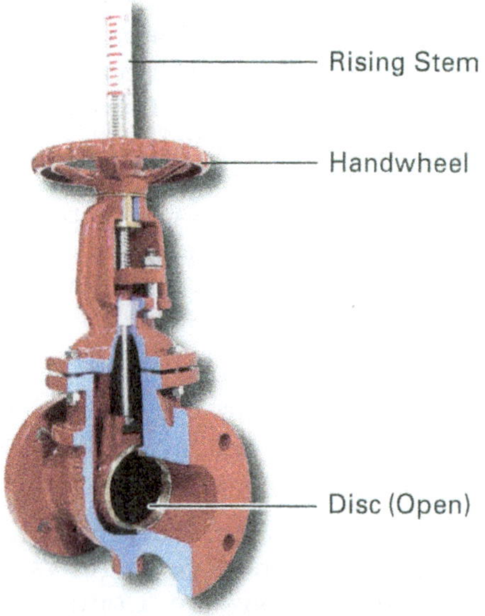

Figure 1 Solid-wedge gate valve components.

Two common types of gate valves are the iron-body, bronze-mounted (IBBM) valve, also referred to as a *double-disc valve*, and the resilient wedge gate valve. IBBM valves have a metal-to-metal seal that tends to wear out quickly with repeated use. Resilient wedge gate valves contain cast-iron gates covered with an elastic material such as rubber that prevents metal-to-metal contact and provides a tight seal. This reduces wear and increases valve life. For this reason, resilient wedge gate valves are becoming more common in sprinkler systems.

1.1.1 Outside Screw and Yoke (OS&Y) Gate Valves

The outside screw and yoke (OS&Y) gate valve is the most common type of indicating gate valve. This valve is operated only in the fully open or fully closed position and is used where minimal pressure drop is important. The OS&Y gate valve is designed so the stem threads are located outside the valve body. The wheel is held in place on the valve by a yoke and drives the stem in and out of the valve body. This prevents the stem from coming in contact with water that could corrode the stem and ruin the valve. As the handwheel turns, the stem moves up through the wheel. *Figure* 2 shows a closed OS&Y gate valve with grooved connections.

An OS&Y gate valve must be installed with enough clearance to prevent a person's hand from being jammed between the handwheel and any obstruction. The handwheel must also have the word "OPEN" clearly visible on the wheel with an arrow indicating the proper valve opening direction. *Figure 3* shows an open OS&Y gate valve with one flanged connection and one grooved connection.

A supervisory switch can be installed on an OS&Y valve to monitor the valve position and warn of any tampering (*Figure* 4). These switches communicate with proprietary supervisory services or central stations. Supervisory switches must be installed so they do not interfere with the normal operation of the valve and actuate with minimal movement of the valve's control. When installing supervisory switches, be aware of switch orientation. If installed in the wrong orientation, the valve may leak and water might enter the switch box, resulting in false signals, corrosion, and switch failure. Switches are designed to fit on specific valve sizes. Make sure the switch you are installing is the right one for the valve. Follow the approved shop drawings and the manufacturer's instructions for the switch in use. Detailed information regarding valve supervision can be found in NFPA 13 and NFPA 72®, *National Fire Alarm and Signaling Code*®.

After installation, indicating valves with public access are often locked in their normal operating position using a chain and a lock with a breakaway shackle for emergency access. Instead of a lock and chain, wire and strap seals may also be used. Consult the Authority Having Jurisdiction (AHJ) for valve sealing requirements.

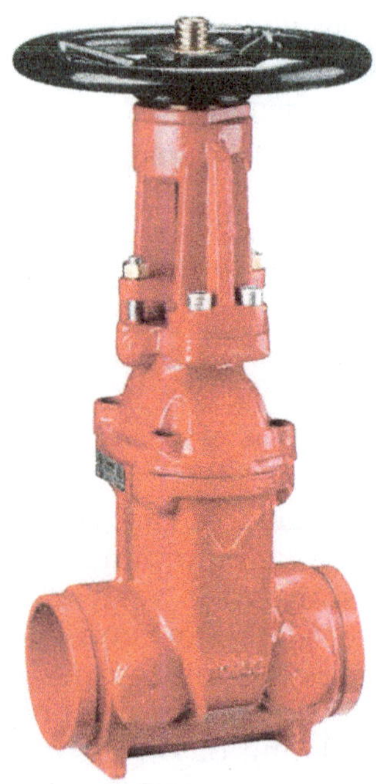

Figure 2 OS&Y valve in the closed position.

> **CAUTION**
>
> While the majority of valves are normally open, some are normally closed. If valves are locked in the wrong position, the system may leak or fail to operate in an emergency. Always check the approved shop drawings for valve positions and orientation before installing or locking any sprinkler system valves.

1.1.2 Installation and Signage Requirements

Aboveground indicating valves commonly use threaded or grooved connections, while underground installations use mechanical joints. The procedure for installing a valve is the same as that for installing a fitting.

One critical item to check when installing valves is the direction of flow. Some valves have arrows indicating the direction of flow stamped on the side, while others label the inlet and outlet ports. Always install the valve in the direction indicated. Another installation consideration is the position of the valve stem. Valves installed with the stem in the upright position tend to work best. The stem can also be installed in the horizontal position, but it should never point down. If a valve is installed in the downward position, the valve top or *bonnet* acts like a trap for sediment, which may damage the stem. Also, if water is trapped in the bonnet in cold weather, the valve may freeze and crack.

Per NFPA 13, all indicating control valves must include a weatherproof, permanently marked metal or rigid plastic identification sign to indicate the valve function and building area served by the valve. Missing signage may result in an inspection failure. Identification signs must be attached using corrosion-resistant chain, wire, or other approved means.

Figure 5 shows typical identification signs. In addition to signs at all indicating control valves, a hydraulic design information sign, or *data plate*, is required on all systems (both hydraulically calculated and pipe schedule). This sign must be placed at each zone control valve, alarm valve, dry pipe valve, preaction valve, and deluge valve supplying the system. A general information sign is also required at each system control riser, antifreeze loop, and auxiliary system control valve.

> **NOTE**
>
> When a space is not provided on the front of the sign, the building area served by a control valve is written in permanent marker on the back of the sign (e.g., FIRST FLOOR – WEST SIDE).

Figure 3 OS&Y valve in the open position.

Figure 4 Gate valve with supervisory switch.

Indicating Control Valve Sign

Hydraulic Design Information Sign

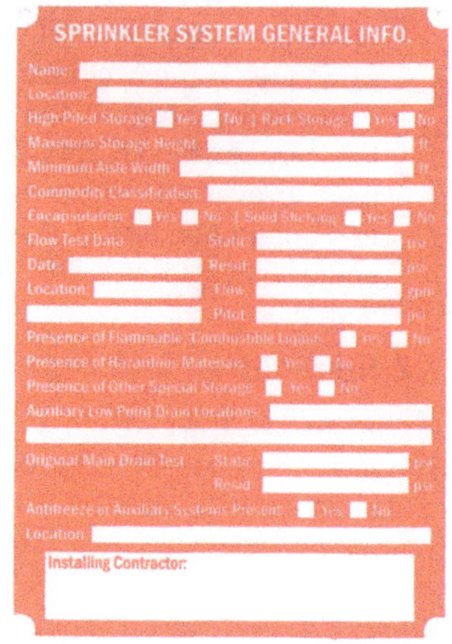

General Information Sign

Figure 5 Indicating control valve signs.

1.1.3 Gate Valve Maintenance

Routine maintenance extends the life of a valve and ensures proper operation of the fire protection system. Permission must be obtained from the proper authorities before shutting down any system or portion of a system for maintenance or repairs, and all persons affected by the shutdown must be notified. Always follow your company's impairment procedure when shutting down a system. Repairs that require system shutdown must be completed as quickly as possible.

Always open and close gate valves slowly to minimize water hammer. Closing the valve slowly can help flush out debris that has accumulated in the disc seating area. If a valve is hard to close or will not close completely, check for debris. To clear the valve, back off the handwheel and close the valve slowly. Then open the valve and repeat if necessary. If the valve continues to stick, consult the manufacturer's recommended service instructions. The following visual checks should be performed once a month for locked valves and once a week for sealed valves:

- Valves are in the normal open or closed position.
- Valves are sealed, locked, or supervised.
- Valves are accessible.
- Valves are free from external leaks.
- Valves are provided with applicable identification.

The threaded portion of the gate valve stem should be lubricated once a year per the manufacturer's instructions. After the stem is lubricated, test its ease of operation by fully closing and opening the valve. Refer to NFPA 25, *Standard for the Inspection, Testing, and Maintenance of Water-Based Fire Protection Systems*, for more information on care and maintenance of gate valves. In some cases, the AHJ may observe field testing and/or adjustments.

Test the operation of all supervisory switches before they are placed into service and at least semiannually, or as required by the AHJ. See NFPA 25 for additional requirements.

> **CAUTION**
>
> Before testing any supervised valve, notify the monitoring agency ahead of time to prevent the Fire Department from being sent to the location.

1.2.0 Post-Indicating Valves (PIVs)

Post-indicating valves (PIVs) provide access to the water supply control valve to quickly turn the water supply on or off. PIVs include windows (also known as *target plates*) that indicate whether the valve is open or shut. Post indicators are typically installed on the operating nuts of nonindicating gate valves to create an approved indicating valve. The valves being operated are usually control valves for water supplies in riser assemblies. PIVs can be fitted with electric supervisory (tamper) switches to monitor the position of the switch. *Figure 6* shows wall and yard PIVs.

Yard PIVs are used to control sprinkler systems and wet standpipe systems where the main water supply valves are located underground. They should be protected by bollards when installed in traffic areas. Yard PIVs have an operating handle stored on the side of the valve. The PIV is locked in position by installing a padlock through the handle loop or using a wire seal.

To operate a yard PIV, remove the seal or padlock and lift the handle from its stored position on the valve. The handle is used like a wrench to turn the operating nut of the PIV (similar to a hydrant wrench).

Wall post indicators are installed through a wall and used to operate a valve on the other side. Extension rods are sometimes used for greater wall thicknesses. These rods provide an adjusting range between the post head base and the operating nut and should include a pipe sleeve around the rod. Note that extension rods and pipe sleeves require cutting in the field to ensure accuracy, and metal buildings may require mounting support for the indicator.

1.2.1 PIV Maintenance

PIVs have a permanently lubricated operating stem and do not require lubrication. Inspect sealed posts weekly and locked posts monthly.

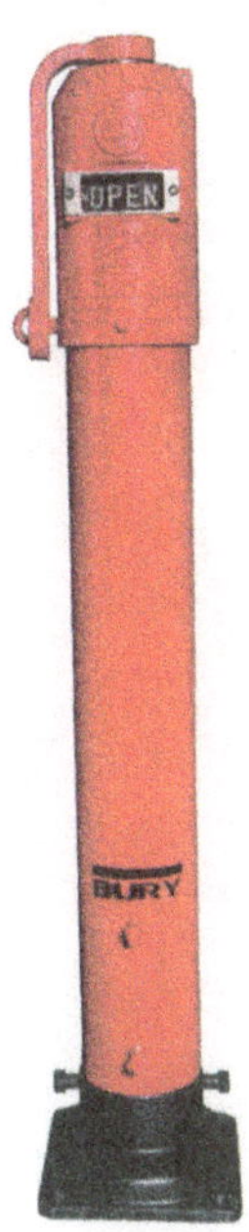

Yard Post

Wall Post

Figure 6 Post indicators.

Control valves that are electrically supervised by tamper switches must be inspected quarterly. These inspections should include the following:

- Check for mechanical damage to the handwheel, housing, or windows.
- Check the target plates to ensure that the valve is open.
- Check the posts to ensure they are properly sealed or locked fully open. If locks or seals have been broken, they must be replaced.
- Check for signs of tampering.
- Make sure the valve is not obstructed by shrubbery or other vegetation.

To help prevent a PIV from locking up, operate the valve through its full range (typically open-closed-open) at least once a year. This is known as *exercising the valve*.

1.3.0 Butterfly Valves

Butterfly valves are less expensive, require less space, and are easier to install than OS&Y gate valves. A butterfly valve has a round disc that fits tightly in its mating seat and rotates 90 degrees in one direction to open and allow water to pass through the valve. This creates greater friction loss than an OS&Y gate valve. To operate a butterfly valve, turn the handwheel.

Butterfly valves are marked with an arrow on the side to indicate the direction of flow. They must be installed in the correct orientation or the seat will not seal properly, resulting in leaks. Gear operators may be installed on butterfly valves to prevent the water hammer that can result from opening or closing the valve too fast.

PIV Valve Rotations

When installing a PIV, it is a good practice to count the number of full rotations of the handle it takes to fully open the valve and note it on the valve sign. The number of turns indicates the size of the underground piping. For example, suppose a typical 8" valve has about 26 turns. When exercising the PIV during system maintenance, count the rotations again. If you only get 10 rotations on an 8" valve, you know the valve is not fully open and there is a problem requiring further investigation.

Butterfly valves are often used as shutoff valves in sprinkler risers. They are also useful as zone or sectional control valves when a visual indication of the valve position is required. Most manufacturers offer models with built-in supervisory tamper switches. *Figure 7* shows a typical butterfly valve. The position of the handle or a colored flag (also known as a *tattletale*) indicates whether the valve is open or closed. If the handle or flag is aligned with the pipeline, the valve is open. If it is at a 90-degree angle, the valve is closed. All butterfly valves used as indicating valves must be the slow-close type that take longer than five seconds to close.

> **NOTE**
> Butterfly valves can be used anywhere that an OS&Y valve would be used, except on the suction side of a fire pump in most instances. A fire pump line normally uses an OS&Y valve.

Most sprinkler system butterfly valves are designed to be normally open, which means they allow water flow under normal operating conditions. However, some valves are normally closed, which means they prevent water flow under normal operating conditions. If the wrong type of valve is installed, it will interfere with system operation and prevent the supervisory switch from working properly.

1.3.1 Butterfly Valve Maintenance

The most common problem with butterfly valves is debris lodged in the sealing area. This problem is indicated when the valve is difficult to close. To clear the valve, open a downstream valve (such as a test or drain valve) to flow water through the butterfly valve. Then slowly open and close the valve to flush the area. Repeat if necessary. Never force a valve closed by using a wrench on the handle. Applying this type of pressure will distort the valve components and damage the sealing surfaces. If the valve will not close, shut the system down following company procedures and remove the valve. Inoperative butterfly valves are typically replaced.

> **NOTE**
> Valves that are Listed for indoor use only tend to degrade quickly in humid or salt-air environments. In these cases, make sure the valve is Listed for both indoor and outdoor use.

Slow-Close Butterfly Valves

All butterfly valves used in sprinkler systems must be the slow-close type, such as the butterfly valve shown here. Controlling valve closure helps to eliminate water hammer.

Figure Credit: Milwaukee Valve Company, LLC

Figure 7 Butterfly valve with supervisory switch.

Normally Open and Normally Closed Butterfly Valves

In most cases, manufacturers color code and/or label valves to indicate whether they are normally open or normally closed, as shown here. Always check the approved shop drawings to ensure the correct valve is installed.

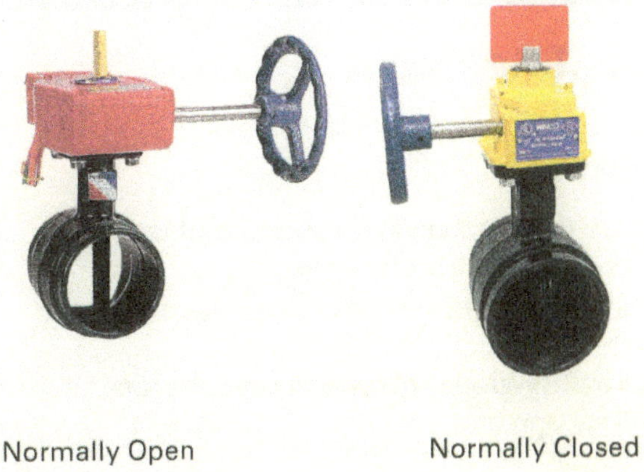

Figure Credit: NIBCO INC.

QR Codes

Some manufacturers mark devices with QR codes. Scanning the code using a smartphone or tablet provides a link to the manufacturer's online installation and maintenance instructions.

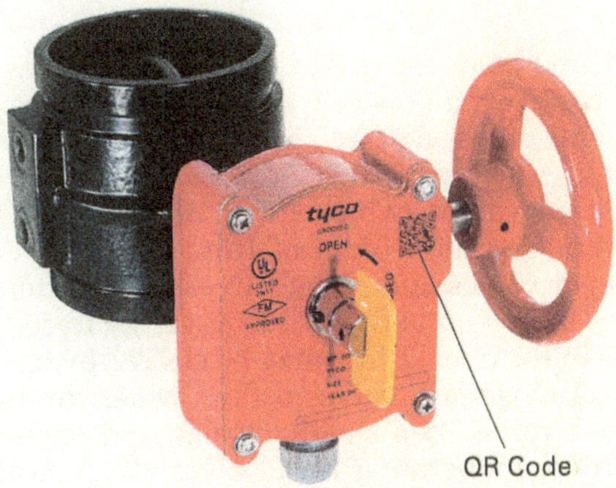

Figure Credit: Used with permission of Johnson Controls

1.0.0 Section Review

1. A gate valve is designed for frequent operation and is often used as a flow control device.
   a. True
   b. False
2. Yard post indicators are *most* likely to be used with underground _____.
   a. ball valves
   b. globe valves
   c. butterfly valves
   d. gate valves
3. If a butterfly valve is installed in the wrong direction, it _____.
   a. will not operate
   b. may not seal properly
   c. will increase water hammer when the valve is operated
   d. is unlikely to affect system operation

2.0.0 TRIM VALVES

Objective

Identify the function and operation of the trim valves used in sprinkler systems.

a. Identify ball lever valves and describe how they operate.
b. Identify globe valves and describe how they operate.

Trade Term

Globe valve: A type of valve consisting of a movable plug or disc and a stationary seat.

Trim valves are used in sprinkler systems to control the flow of water, provide drainage, and allow for the installation of pressure gauges. All valves must be Listed except trim and drain valves, which must be approved by the AHJ. Common types of trim valves include ball lever valves and globe valves. Like butterfly valves, most trim valve problems can be traced to debris in the line. Debris may be cleared by opening a downstream valve to flow water through the trim valve, then slowly closing and opening the valve. Inoperative trim valves are typically replaced.

All test and drain valves (including every low-point drain) must include an identification sign stating the function of the valve and what it controls in the system. A label for a main drain valve is shown in *Figure 8*. The main drain is the primary drain connection located on the system riser.

Figure 8 Main drain valve label.

2.1.0 Ball Lever Valves

A ball lever valve has a ball with a hole through its center that acts as the control element (*Figure 9*). This ball is located in the center of the valve body and attached to the valve stem. It can either be a full-bore ball, with an opening the same size as the inside pipe diameter, or a standard-bore ball, with an opening smaller than the inside pipe diameter. The handle is attached to the stem and moves only one-quarter of a turn to align the hole with the pipe opening and start the flow through the valve. Turning the handle in the opposite

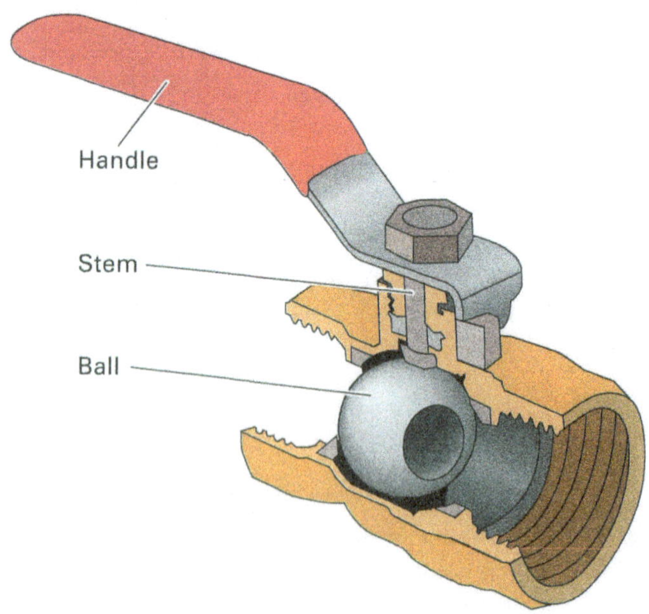

Figure 9 Ball lever valve components.

direction blocks the pipe opening, stopping the flow through the valve.

Ball lever valves must be fully open or fully closed. The position of the handle on a ball lever valve indicates whether the valve is open or closed. The valve is open if the handle is aligned with the system piping and closed if the handle is at a right angle to the piping. Ball lever valves in fire protection systems are typically used for trim and drain purposes or as backflow preventers. They are not used as control valves as they close too quickly, increasing the possibility of water hammer. *Figure 10* shows ball lever valves.

2.2.0 Globe Valves

A globe valve is used to regulate flow. It has a movable plug or disc that can be partially opened with very little friction, allowing for frequent operation with minimal seat damage (*Figure 11*). Globe valves are often installed in ½" to 2" lines but are available in sizes up to 12". These valves permit close flow control because the size of the seat-to-globe opening is directly proportional to the number of turns of the handwheel. The fewer the number of turns, the smaller the opening. Globe valves are ideal for throttling and often used as drain valves and drum drips.

Globe valves are available in straight, angle, and three-way types, depending on the flow of water through the valve. *Figure 12* shows a straight globe valve.

CAUTION

Globe valves must be installed so that the pressure enters below the plug or disc. Globe valves have an arrow that indicates the direction of flow.

Globe valves may be either Listed or approved. Listed globe valves are more expensive than unlisted valves. They may also be more difficult to find and must be ordered well in advance of installation. All valves must be handled carefully to avoid damage before or during installation.

2.2.1 Angle Valves

Angle valves are globe valves with an outlet at a 90-degree angle from the inlet. There is less resistance to flow in an angle valve than in a standard globe valve, and angle valves offer a change of direction in a pipeline without using an elbow. Angle valves can reduce the number of joints in a line and save installation time. They are used as main drain valves, as trim valves, and in hose cabinets. *Figure 13* shows an angle globe valve.

2.2.2 Side-Outlet Globe Valves

Side-outlet globe valves are used to attach a pressure gauge to the trim line. These valves are also referred to as *three-way valves*, *three-port valves*, and *gauge-test valves*. When used as a gauge outlet, the gauge can be isolated for repair or replacement without having to shut down the system. Side-outlet globe valves are commonly used in trim for alarm, deluge, and dry pipe valves. *Figure 14* shows a side-outlet globe valve. *Figure 15* shows a side-outlet globe valve with an attached pressure gauge.

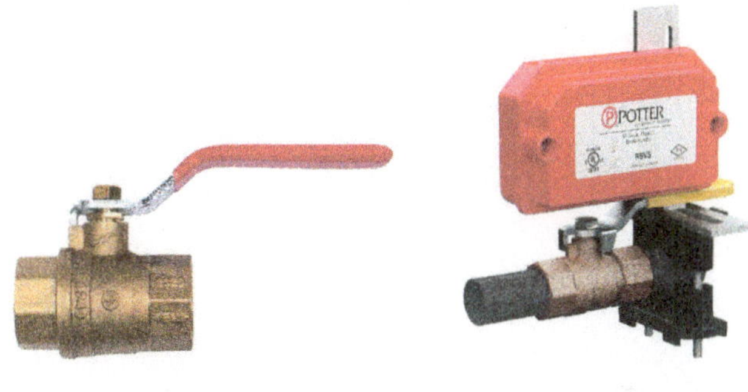

Ball Lever Valve

Ball Lever Valve with Supervisory Switch

Figure 10 Ball lever valves.

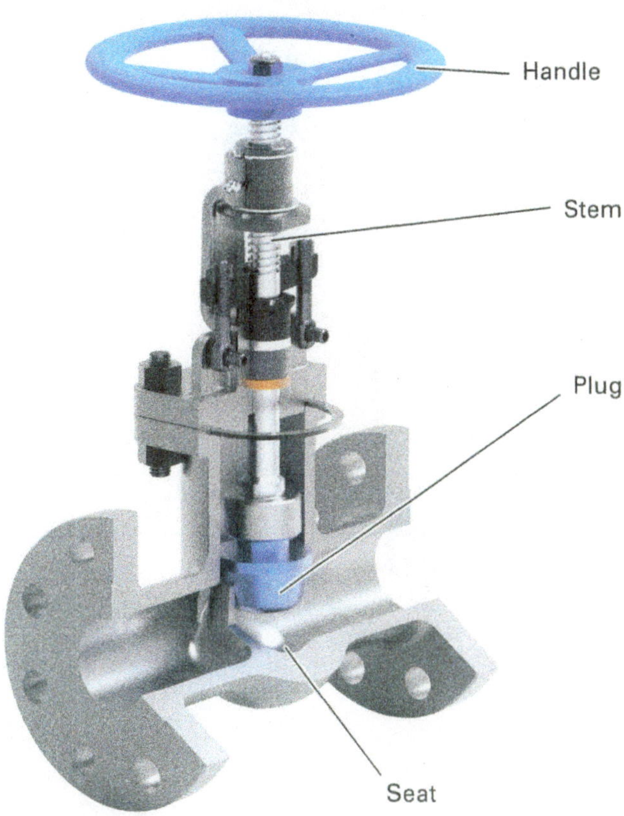

Figure 11 Globe valve components.

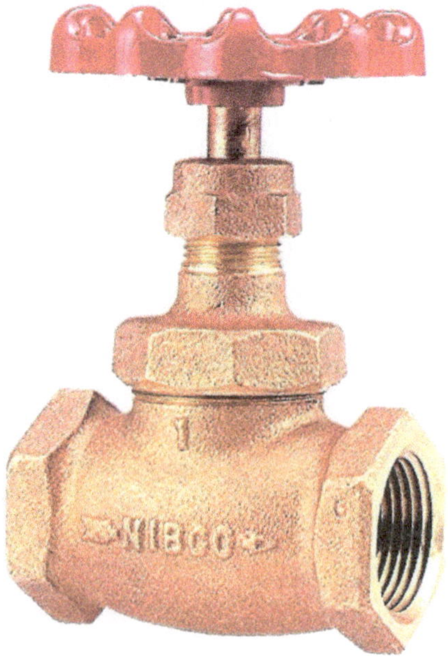

Figure 12 Straight globe valve.

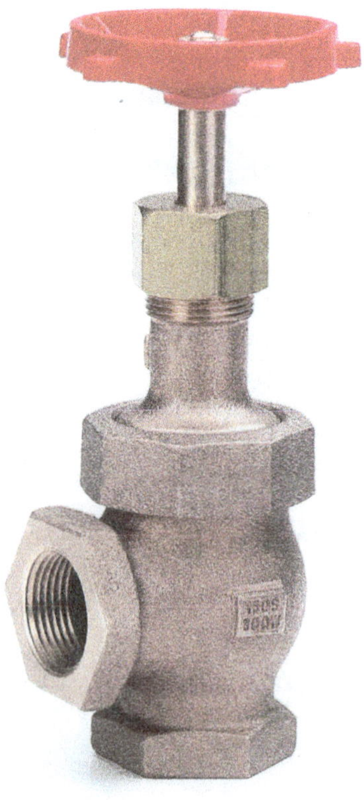

Figure 13 Angle globe valve.

Figure 14 Side-outlet globe valve.

Figure 15 Side-outlet globe valve with attached pressure gauge.

2.2.3 Drain Valves

Many of the globe valves in sprinkler systems are used for drains. Guidelines for installing drain lines include the following:

- *Connection to sewers* — NFPA 13 prohibits the direct connection of sprinkler system drains to sewers. Connection to sewers could allow bacteria to migrate from the sprinkler system into the city water supply. An air gap is required between a sprinkler system drain and the sewer to prevent the movement of bacteria. Refer to the approved shop drawings for installation requirements.
- *Buried drain lines* — All piping for drain lines buried underground must be approved, corrosion-resistant pipe. This can be copper, galvanized steel, ductile iron pipe, or CPVC pipe. Because some soils are very corrosive, the pipe may require a covering of polyethylene for protection. An air gap is required between the discharge and any area that could contaminate the water supply. The discharge pipe cannot terminate into the soil.
- *Termination of drains and freeze protection* — Drains cannot be terminated in blind spaces, such as underneath the floor in pier-and-beam construction. Blind spaces are defined as any area where water cannot evaporate, which may promote the growth of bacteria and fungus, as well as damaging the building foundation. Most drainpipes are terminated over a paved area to prevent soil erosion. These installations require a turned down 45-degree elbow on the end of the pipe. This configuration prevents dust, dirt, and other debris from accumulating in the outlet and allows the discharge water to hit the pavement at an angle. The 45-degree angle of impact reduces the force of the water and prevents damage to the pavement. All drains piped to the outside of a building provide a path for heat loss. If the distance between the outside of a building and a water-filled pipe is too short, the pipe could freeze. Allow at least 4' between the valve and the building wall to prevent freezing.

CAUTION

In freezing climates, terminate drains away from walkways to prevent ice buildup and a slipping hazard.

2.0.0 Section Review

1. Ball valves can be used in a partially open position.
   a. True
   b. False

2. Pressure gauges are typically attached using a(n) ______.
   a. straight globe valve
   b. side-outlet globe valve
   c. angle globe valve
   d. wye-type globe valve

SECTION THREE

3.0.0 SUPPLY SYSTEM VALVES

Objective

Identify the function and operation of the supply system valves used in sprinkler systems.

a. Identify check valves and describe how they operate.
b. Identify hose valves and describe how they operate.

Trade Terms

Hose header: A pipe that acts as a central connection point for one or more hoses.

Machined: A metal surface that has been held to precise tolerances during the manufacturing process.

Pressurized system: A piping system that is filled with pressure and has a pressure source.

Sprinkler supply system valves control the supply of water in a sprinkler system. Two types of supply system valves include check valves and hose valves. Check valves prevent backflow of system water into the potable water supply. Hose valves control the water to fire hose outlets.

3.1.0 Check Valves

Check valves are intended to permit flow in one direction only. The purpose of a check valve in a sprinkler system is to prevent reverse flow. An arrow stamped on the side of the check valve indicates the direction of flow. It is very important to install check valves in the proper direction to prevent system failure. The force of flow and the action of gravity cause the valve to open and close. To be effective, the check valve must offer quick and automatic reaction to flow changes. NFPA 13 requires that check valves be installed in accordance with the manufacturer's Listing. The following check valves are used in sprinkler systems:

- Swing check valves
- Riser check valves
- Double-check valves
- Alarm check valves
- Wafer check valves

3.1.1 Swing Check Valves

The most common type of check valve used in sprinkler systems is the swing check valve. This valve has a disc or clapper that swings or pivots to open and close the valve (*Figure 16*). Swing check valves can be mounted horizontally or vertically with the direction of flow upward. Some swing check valves have a removable cap or a port to allow for internal inspection without removing the valve from the system. These valves are available in plain and spring-loaded clapper configurations. *Figure 17* shows a typical swing check valve.

The disc in a swing check valve is hinged at the top and seals against a machined seat in the opening. The disc swings freely from the fully closed to the fully open position. Flow through the valve causes the disc to remain open, with the opening size determined by the volume of water moving through the valve. Gravity or reverse flow will cause the disc to close, preventing backflow through the valve. The seat of a swing check valve is at a 45-degree angle inside the valve. The angle applies one-half the clapper weight force for seating. This allows the sealing weight to be applied in either the horizontal or vertical position.

CAUTION

Check valves must be installed in either the horizontal (hinge pin on top) or vertical up position (hinge pin on left). Per NFPA 13, never install these valves in a vertical down position (hinge pin on right) as this will prevent gravity from closing the valve.

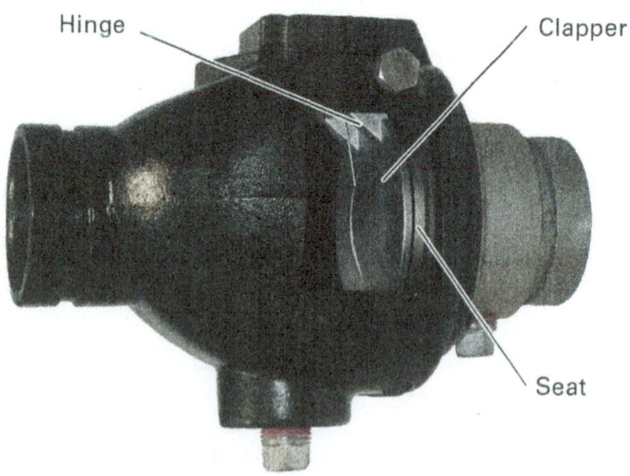

Figure 16 Swing check valve components.

The basic internal parts of a swing check valve are the clapper or disc, the seat, and the hinge. The internal components are usually made of bronze, brass, or stainless steel to prevent corrosion. When installed in the horizontal position, the clapper hangs lightly against the seat under no-flow conditions. In the vertical position, the full weight of the clapper rests against the seat under no-flow conditions. Without upward flow at sufficient pressure, the valve will remain closed.

Some swing check valves are equipped with springs to help hold the clapper shut. These valves are referred to as *spring-loaded clapper valves*.

A soft-faced clapper is normally made of rubber and provides a tighter seal than a metal clapper. Soft-faced clappers are required when air is checked instead of water. They should also be used for Fire Department Connection (FDC) valves.

3.1.2 Riser Check Valves

Check valves can be combined with pressure gauges and trim valves in a riser check valve assembly. When combined with a waterflow detector to provide an electric alarm, this assembly provides a compact and less expensive alternative to an alarm check valve. *Figure 18* shows a riser check valve assembly.

3.1.3 Double-Check Valves

Check valves are also used to prevent backflow, which is the reverse flow of contaminated water into the drinking water supply. Backflow cannot happen unless there is a direct link between the potable (drinking) water supply and another source. This is called a *cross-connection* and can be caused by conditions such as a pump failure or water main break. Backflow preventer assemblies protect against cross-connections by providing a gap or barrier to keep backflow from entering the water supply.

There are many types of backflow preventer assemblies. One type is a double-check valve backflow preventer assembly (*Figure 19*). This assembly has two spring-loaded check valves mounted side by side or in series. During normal operation, both check valves are open; however, backflow causes the valves to seal tightly. The use of two check valves in series provides additional protection. If one valve becomes fouled open by debris, protection is still provided by the second valve. Backflow preventer assemblies are discussed in more detail in NCCER Module 18106, *Underground Pipe*.

3.1.4 Alarm Check Valves

Fire protection systems draw water from the public water system to supply the fire protection system. NFPA 13 requires a check valve in the water supply of every sprinkler system (*Figure 20*). These check valves are typically $2^{1}/_{2}$" and larger. A Listed backflow prevention device is considered a check valve and an additional check valve is not required.

Figure 17 Swing check valve.

Figure 18 Riser check valve assembly.

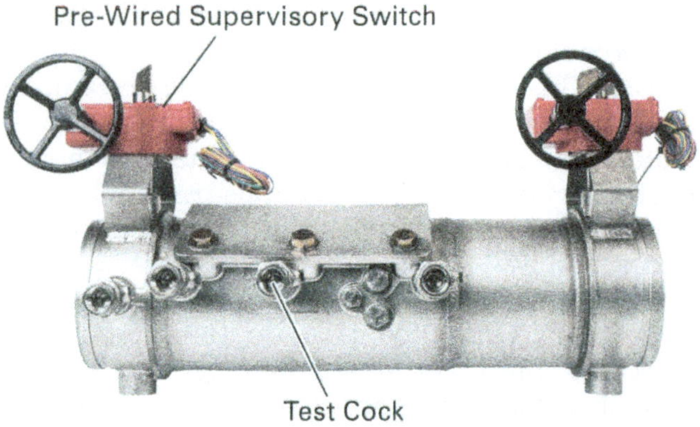

Figure 19 Double-check valve backflow preventer assembly.

> **Did You Know?**
>
> Brass is an alloy (mixture) of copper and zinc. The first brass was used around 700 BCE (Before the Common Era) by the people of the Middle East. During the 18th and 19th centuries, hundreds of copper alloys were developed. Most of them were brasses. Today, brass is the most widely used copper alloy.

In some installations, an alarm check serves as the check valve. An additional check valve is required if there is a pit located in the yard with a freestanding FDC. The function of these check valves is to block reverse flow and prevent water from a fire pumper from entering the potable water supply. These check valves also allow the pumper to increase the pressure of the water in the sprinkler system above that of the normal water supply. *Figure 21* show an alarm check valve in a riser assembly.

Check valves are made of cast iron or malleable iron with internal parts made of stainless steel or bronze. Valves that are $2^{1}/_{2}$" and larger are available in flanged, grooved, and mechanical joint configurations. These types of valves are commonly installed next to the main control valve near the property line. The check valve and main control valve must remain accessible for service. In mild climates, check valves are often installed aboveground.

When subjected to freezing temperatures, the piping between the FDC and the check valve must be equipped with a Listed automatic drain valve, sometimes referred to as an *automatic ball drip valve*. An automatic drain valve is a valve with a ball in its body that partially closes off the discharge side of the body when water pressure is applied to it. When the pressure is lost, the ball falls into a recess that allows the trapped water to drain.

This portion of pipe is trapped between the check valve and the FDC, and unless a drain valve is provided, the water will remain trapped, presenting corrosion and freeze concerns. The valve must be installed according to the manufacturer's instructions and be piped to a drain, which may be through a wall. *Figure 22* shows an automatic drain valve.

3.1.5 Wafer Check Valves

Check valves that are sandwiched between two flanges are known as *wafer check valves*. Swing-type clappers must have enough clearance beyond the valve body for the clapper to open. This may prevent these valves from being used close to other valves where clearance cannot be obtained. Pipe inserts known as *spool pieces* can

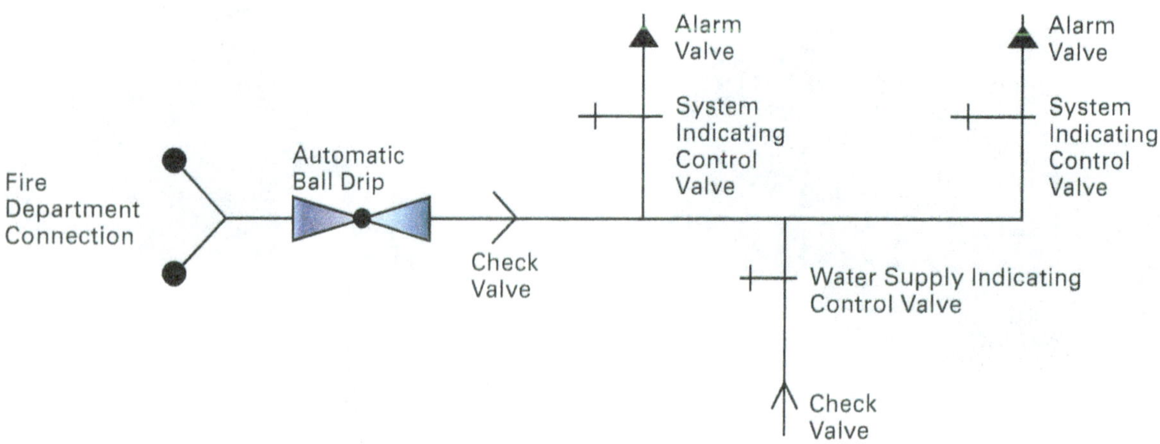

Figure 20 Water supply valves on a shop drawing.

Figure 21 Alarm check valve in a riser assembly.

be used to provide the required clearance. O-ring seals are used on each end of the spool piece to provide a tight seal.

A split-clapper wafer check valve depends on spring loading for closure. These valves are lightweight and easy to install, inspect, and repair. They are also less expensive than flanged valves and offer rapid closure under no-flow conditions. The spring-loaded disc speeds the closing of the valve and minimizes reverse flow, slam, and water hammer. These valves use gaskets and O-rings to seal each end and also require clearance beyond the valve body for opening. Some wafer check valves are available with extended bodies to avoid clearance issues. *Figure 23* shows a spring-loaded clapper in a wafer check valve.

3.1.6 Check Valve Maintenance

Check valves should be checked for proper operation at least annually. In addition, NFPA 25 requires an internal inspection of check valves every five years. If the inspection indicates that the valve requires service, refer to the manufacturer's instructions.

Due to their location in the system, swing check valves tend to collect debris between the clapper and the seat. Those installed horizontally are especially prone to this problem. If the valve becomes fouled, open the inspection port, then clean the seat and clapper. Check the seat and clapper for damage after cleaning. If the seat or clapper has been damaged, the valve will not operate properly and should be repaired or replaced.

NOTE

Rust, scale, and other foreign material can foul sprinkler orifices and other system devices and prevent them from operating properly. A Listed fire service strainer may be installed upstream of devices to protect them from damage caused by large particles.

Figure 22 Automatic drain valve.

nnifire.com

Figure 23 Spring-loaded clapper in a wafer check valve.

3.2.0 Hose Valves

A hose valve is used to control the water to an individual fire hose outlet. The fire hose outlet is where the hose connects to the water source or supply. Hose valves can be part of a fire hydrant, a code-required hose station, a simple outlet on a standpipe system, or a hose header used for testing fire pumps. *Figure 24* shows a test header.

Gate-style hose valves are normally used on fire pump test headers and fire hydrants. Hose gate valves with metal gates and seats are designed for use in normally unpressurized dry-pipe systems. Metal-to-metal seals are not drip-tight and should not be used with any pressurized system.

Hose valves are available in a variety of patterns, depending on the installation. On standpipe systems, straight- or angle-pattern globe valves are typically used. Straight-pattern gate-style hose valves are used when full-flow capability of the line is required. These valves are available with either a threaded or a grooved inlet. They use a soft disc closure and can be closed drip-tight.

Angle-pattern hose valves are standard devices on standpipes and in most hose stations. These valves are constructed like a standard angle valve with a soft disc closure and are used to throttle flow. *Figure 25* shows an angle-pattern hose valve.

Hose valves are available in two common sizes, $1\frac{1}{2}$" and $2\frac{1}{2}$" (3" hose valves are also available but are not common). These sizes refer to the nominal outlet size on the valve.

3.2.1 Hose Valve Threads

The threads on a hose valve must be identified as female or male, and the threads-per-inch must be specified. These threads are matched to the local hose pattern. Most municipalities have adopted the American National Fire Hose Connection Screw Threads (NH) system as specified in NFPA 1963, *Standard for Fire Hose Connections*. NH is also known as the *National Hose Thread (NHT)* or *National Standard Thread (NST)*. Be aware that some municipalities use the NH pattern on $2\frac{1}{2}$" valves but require a National Pipe Thread (NPT) on $1\frac{1}{2}$" valves. In areas where the fire protection system has been in place for more than a century, proprietary hose threads may be in use. This may require special outlet threads or standard valves with hose adapters to make the connection. Consult local fire codes to determine requirements. *Figure 26* shows hose adapters.

Figure 24 Test header.

Figure 25 Angle-pattern hose valve.

3.2.2 Hose Valve Maintenance

NFPA 25 requires that hose valves be inspected every three months to check for obstructions, leaks, and damage (including the threads). Hose valves must also be exercised annually by opening and closing each valve. Any valves that leak or are difficult to operate must be lubricated, repaired, or replaced per the manufacturer's instructions.

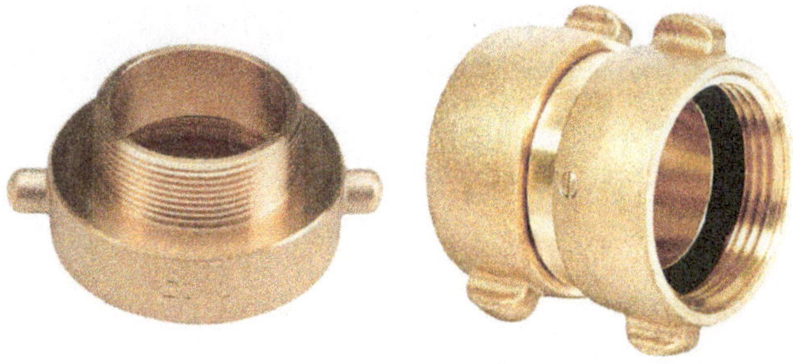

Figure 26 Hose adapters.

3.0.0 Section Review

1. Check valves that are sandwiched between two flanges are called _____.
   a. double-check valves
   b. wafer check valves
   c. riser check valves
   d. swing check valves
2. Most municipalities have adopted the American National Fire Hose Connection Screw Threads (NH) system for hose threads.
   a. True
   b. False

SECTION FOUR

4.0.0 PRESSURE CONTROL AND AIR VENTING VALVES

Objective

Identify the function and operation of the pressure control and air venting valves used in sprinkler systems.

a. Identify pressure-reducing valves and describe how they operate.
b. Identify pressure-regulating devices and describe how they operate.
c. Describe the function of air venting valves.

Trade Terms

Pressure-regulating device: A device designed for the purpose of reducing, regulating, controlling, or restricting water pressure.

Residual pressure: Pressure in a piping system under flow conditions.

Static pressure: Pressure in a piping system under no-flow conditions.

Pressure control valves are used to control the pressure in a sprinkler system. Two common types of pressure control valves are pressure-reducing valves and pressure-regulating devices. In addition to controlling the supply of water, wet pipe systems also require a method of removing air from the system. This is provided by an air venting valve.

4.1.0 Pressure-Reducing Valves

Pressure-reducing valves are also called *pressure-regulating valves*. In most cases, they are simply referred to as *PRVs*. These valves are normally used to reduce pressure in piping systems to an allowable range. Pressure-reducing valves are used at the outlet of any sprinkler system, standpipe system, or combined system where the supply pressure exceeds a safe operating pressure for the equipment or people in the area. Normally, NFPA 14, *Standard for the Installation of Standpipe and Hose Systems*, limits standpipe outlet pressure to a maximum of 175 psi for Class I standpipe systems. These limits are important in systems for high-rise buildings where pumps are used to raise pressures to the proper level on upper floors but may over-pressurize the system on lower floors. PRVs are required by NFPA 13 to be installed in a system where all components are not Listed for the maximum pressure generated by the water supply.

> **CAUTION**
>
> PRVs are designed to reduce pressure at a specific elevation and must be installed at this elevation to ensure proper operation.

There are two types of PRVs used in the sprinkler industry. A **pressure-regulating device** controls the water pressure under both flow and no-flow conditions, while a pressure-restricting device reduces the downstream pressure under flow conditions only. While they may be seen in existing installations, pressure-restricting devices are no longer used in new installations. The term **residual pressure** refers to outlet pressure under flow conditions. **Static pressure** refers to pressure under no-flow conditions. Devices that control pressure under both flow and no-flow conditions are used with hose nozzles that have a shutoff feature.

4.1.1 Pressure-Regulating Valves (PRVs)

Nonadjustable, rising-stem regulating valves are used to reduce pressure under flow and no-flow conditions. When the valve is opened, the handwheel rises with the stem and provides a visual indication of the valve position. *Figure 27* shows a typical factory-set PRV.

Standpipe System Classes

There are three classes of standpipe systems:

- *Class I* – A Class I standpipe system has a 2½" hose connection for use by the Fire Department.
- *Class II* – A Class II standpipe system has a 1½" hose connection for use by the Fire Department or other trained personnel. The hose connection is sometimes located in a cabinet with 100' of hose.
- *Class III* – A Class III standpipe system has a 1½" hose connection to supply water for use by trained personnel and a 2½" hose connection to supply a larger volume of water for use by the Fire Department.

Pressure-reducing valves are available in $2\frac{1}{2}$" and $1\frac{1}{2}$" configurations and are factory set to conform to a specific performance curve. Valve setting selection charts showing performance curves are used to select the proper valve setting for the specific installation. Refer to the manufacturer's instructions for details.

Some sprinkler system installations require field-adjustable PRVs. These valves are available in several different configurations. Some types have the pressure adjustment nut located inside the valve assembly. *Figure 28* shows a field-adjustable PRV with a protective shroud that is removed to adjust the valve using the provided adjustment rod. Other valves have accurate adjustments engraved on the upper stem for setting the pressure. Refer to the manufacturer's instructions for adjustment information.

4.1.2 Drains and Test Risers

NFPA 14 requires a 3" drain riser to be installed for every standpipe that uses pressure-reducing devices. Each riser must have a 3" × $2\frac{1}{2}$" tee with an internal swivel fitting installed on every other floor. The fitting must have standard NH threads and a plug. Refer to NFPA 1963 for NH thread sizes.

> **NOTE**
> Contact the local AHJ to determine whether local Fire Departments use threads other than those designated in NFPA 1963.

4.1.3 PRV Maintenance

PRVs must be maintained in accordance with the manufacturer's instructions. Debris is the most common cause of PRV malfunction or failure. PRVs regulate outlet pressure according to the separation between the disc and the valve seat. If debris lodges between the disc and the seat, the valve will not reduce pressure properly and will apply full-source pressure under no-flow conditions. It is important to flush all debris from the sprinkler and standpipe system before installing PRVs in the system.

In areas where these valves are critical to the system, they should be checked on the outlet side. To perform this check, ensure that the valve is in the fully open position in no-flow conditions to determine the status of its regulating ability. If the system is not equipped with a pressure gauge, the PRV will have to be closed completely so the gauge can be installed. Use caution when closing the PRV to prevent damage to the seat or disc if there is debris in the valve.

Figure 27 Typical factory-set PRV.

Figure 28 Field-adjustable PRV.

NFPA 25 inspection requirements include the following:

- All valves must be inspected quarterly to verify that the valves are in the open position, not leaking, maintaining downstream design pressures, and have operating handwheels installed.
- A partial flow test adequate to move the valve from its seat must be conducted annually.
- A full flow test must be conducted every five years and compared to previous test results.
- Any adjustments must be made in accordance with the manufacturer's instructions.

4.2.0 Pressure-Regulating Devices

Sprinkler systems are closed-loop systems subject to pressure increases due to temperature changes. In systems where air is present, the air in the system compensates for higher pressures by compressing. This prevents a significant increase in the system pressure. In wet pipe systems, there is very little air in the system to compress. Water does not compress, and as it expands when the temperature rises, it can easily exceed the system operating pressure and break fittings, produce leaks, and damage the sprinklers. To prevent this pressure buildup, NFPA 13 requires the use of a pressure-regulating device to relieve excess pressure caused by surges or temperature changes. Common types of pressure-regulating devices include relief valves and expansion tanks.

4.2.1 Relief Valves in Wet Pipe Sprinkler Systems

NFPA 13 requires that wet pipe systems be equipped with a Listed relief valve not less than ½" in size and operating at 175 psi or 10 psi in excess of the maximum system pressure, whichever is greater. NFPA 13 also requires that a relief valve be installed downstream of any pressure-reducing valves in the system. *Figure 29* shows typical relief valves.

Pressure relief valves are operated by pressure acting directly on the bottom of a disc that is held on its seat by a spring. To avoid debris collecting in the seat, these valves work best when installed in a vertical position. When the pressure becomes high enough to overcome the force of the spring on the disc, the disc rises and allows water to escape. After enough water has escaped to allow the pressure to drop, the spring pushes the disc back onto its seat and stops the flow of water through the valve.

Pressure relief can also be provided by certain types of test and drain valves. *Figure 30* shows a wet system riser manifold with pressure relief.

The relief port on these valves should always be piped to an area where the water can drain without causing damage. Proper drainage is especially important for antifreeze systems.

Requirements for relief valves are covered in NFPA 13. Note that if an auxiliary air reservoir is installed to absorb pressure increases, a relief valve is not required.

> **NOTE**
> Like indicating, test, and drain valves, all venting valves must include an identification tag stating the function of the valve and what it controls in the system.

4.2.2 Relief Valves in Antifreeze Systems

NFPA 13 requires that antifreeze systems installed without a backflow prevention device have a check valve with a $\frac{1}{32}$" hole in the clapper, except where a Listed expansion tank or other approved means of relief is provided. The hole allows pressure relief in the portion of the system on the outside of the building. All air is bled out as this type of system is charged. This results in a complete system fill at normal temperatures. As the weather gets warmer, the antifreeze solution expands. Without the hole in the clapper, the pressure can increase significantly. Some areas do not allow this hole in the clapper due to concerns that the antifreeze solution will back up in the system and contaminate the public water supply. In these cases, a relief valve is installed to prevent damaging pressure buildup. Consult NFPA 13 for additional information.

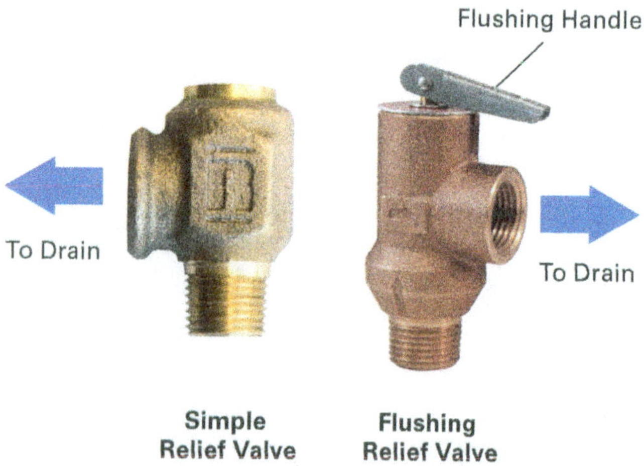

Figure 29 Relief valves.

When antifreeze systems are installed with reduced-pressure backflow preventers, a relief valve is replaced with a Listed expansion tank located on the system side of the backflow preventer. A Listed ½" relief valve is permitted in lieu of an expansion tank where the antifreeze system volume does not exceed 40 gallons.

4.2.3 Expansion Tanks

An expansion tank, also known as an *expansion chamber*, provides additional space for water expansion due to temperature changes. Expansion tanks can be used in both wet pipe sprinkler systems and antifreeze systems. When the water or antifreeze solution expands due to a temperature increase, the excess volume flows into the expansion chamber, maintaining the system pressure at the desired level. These tanks are available in various sizes depending on the volume of the piping system. *Figure 31* shows typical expansion tanks.

4.2.4 Relief Valves Used with Pressure-Reducing Valves

NFPA 13 requires the use of a relief valve whenever a pressure-reducing valve is used in a system. Pressure-reducing valves can malfunction if debris becomes lodged in the regulating mechanism; therefore, a relief valve of not less than ½" is required on the discharge side of the pressure-reducing valve to relieve pressure in case of a malfunction. The relief valve selected must operate at no greater than the maximum operating pressure of the system. Some suppliers recommend selecting a relief valve that provides a pressure of 10 to 15 psi above the static pressure at the PRV outlet.

4.2.5 Relief Valve Maintenance

Relief valves require periodic maintenance to ensure proper operation. These valves are prone to periodic leakage. When a relief valve opens, water is released and even though the quantity of water leakage is small, it can still be a source of water damage. This is a greater problem when the seat has collected debris. Relief valves should be tested regularly to make sure they operate at the required pressure without leaks. If a leaking valve has debris on the seat, purge the valve to clear the seat. If a simple purge does not clear the seat, the valve is typically removed and replaced. Refer to NFPA 25. Inspection requirements include the following:

- All valves must be inspected quarterly to verify that the valves are in the open position, not leaking, maintaining downstream design pressures, and have operating handwheels installed.

Figure 30 Riser manifold including test and drain valve with pressure relief.

Figure 31 Expansion tanks.

- A partial flow test adequate to move the valve from its seat must be conducted annually.
- A full flow test must be conducted every five years and compared to previous test results.
- Any adjustments must be made in accordance with the manufacturer's instructions.

4.3.0 Air Venting Valves

Air trapped in a metal wet pipe sprinkler system will result in corrosion. Air pockets may form when a system is initially filled or when a system is opened for repair and maintenance. Air pockets tend to collect at high points in the system and can also restrict the flow of water when the system is activated, resulting in water hammer. To reduce these issues, NFPA 13 requires that all wet pipe sprinkler systems using metal piping include an air vent near the high point of the system. This vent can be a minimum ½" manual valve, an automatic air vent, a remote inspector's test valve, or other approved means. The specific type and location of the air venting valve will be shown on the approved shop drawings.

Figure 32 shows typical air venting valves. The manual air vent is installed vertically and opened when filling the system. It contains a ball check that allows air to vent from the system while it is filling with water and prevents water from escaping when it reaches the valve. This reduces labor costs because a second person is not required to monitor the valve while filling the system. After the system is filled, the valve is closed.

The combination valve allows for rapid purging of air as the system is filled and continuous venting of excess air after the system is in operation. The purge valve is opened during system filling and closed during normal operation. The isolation valve remains open during system operation to provide continuous venting and is only closed for hydrostatic testing. Follow the manufacturer's instructions for testing and maintenance.

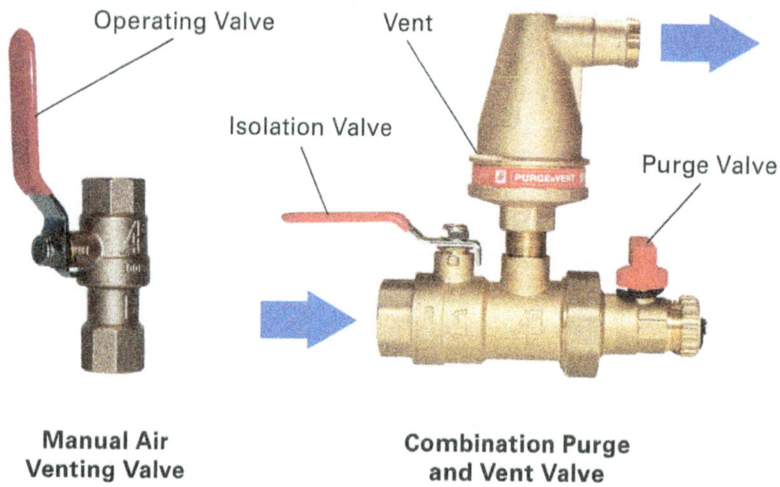

Figure 32 Air venting valves.

4.0.0 Section Review

1. NFPA 14 limits outlet pressure on a Class I standpipe valve to a *maximum* of _____.

   a. 65 psi
   b. 100 psi
   c. 175 psi
   d. 200 psi

2. NFPA 13 requires the use of a relief valve whenever a pressure-reducing valve is used in a system.

   a. True
   b. False

3. Air trapped in a metal pipe sprinkler system can be removed using a(n) _____.

   a. pressure-reducing valve
   b. factory-set PRV
   c. field-adjustable PRV
   d. air venting valve

Review Questions

1. A visual indicator of whether a valve is open or closed is required by NFPA 13 on all _____.
   a. globe valves
   b. control valves
   c. ball valves
   d. trim valves

2. Valves work *best* if they are installed with their stem _____.
   a. removed
   b. horizontal
   c. upside down
   d. upright

3. Which of the following is *true* with regard to butterfly valves?
   a. Butterfly valves take up more space than OS&Y gate valves.
   b. Butterfly valves are more expensive than OS&Y gate valves.
   c. Butterfly valves can be installed in any flow direction.
   d. Butterfly valves are frequently used as shutoff valves in sprinkler risers.

4. All valves must be Listed, *except* _____.
   a. butterfly valves
   b. OS&Y valves
   c. trim and drain valves
   d. relief valves

5. Globe valves are typically used as drain valves and _____.
   a. check valves
   b. drum drips
   c. ball drips
   d. main control valves

6. A valve used to provide an outlet for the installation of a pressure gauge is the _____.
   a. OS&Y valve
   b. side-outlet globe valve
   c. butterfly valve
   d. angle valve

7. In order to prevent the passage of harmful bacteria, which of the following is required between a sprinkler system drain and the sewer?
   a. Flex pipe
   b. Air gap
   c. Welded connection
   d. Filter

8. Which of the following is *true* regarding drain terminations?
   a. Drains should be terminated in blind spaces wherever possible.
   b. Most drainpipes are terminated over a paved area to prevent soil erosion.
   c. Drainpipe terminations require a 90-degree elbow on the pipe end.
   d. Sprinkler system drains should connect directly to the city sewer system.

9. Valves that regulate the direction of flow are called _____.
   a. globe valves
   b. OS&Y valves
   c. gate valves
   d. check valves

10. When subjected to freezing temperatures, the piping between the FDC and a check valve *must* be equipped with a Listed _____.
    a. straight 2" nipple
    b. side-outlet globe valve
    c. automatic drain valve
    d. flexible connection

11. A puck-shaped check valve that can be inserted between two pipe ends is a _____.
    a. wafer check valve
    b. riser check valve
    c. swing check valve
    d. double-check valve

12. Check valves should be checked for proper operation *at least* every _____.
    a. 3 months
    b. 6 months
    c. 9 months
    d. 12 months

13. Which of the following hose valves are used when full-flow capability is required of the line?
    a. Straight-pattern
    b. Angle-pattern
    c. Cross-pattern
    d. 90-degree

14. As specified by NFPA 1963, *Standard for Fire Hose Connections*, most municipalities use the _____.
    a. National Standard Thread (NST) system
    b. American National Fire Hose Connection Screw Threads (NH) system
    c. National Fire Hose Thread (NFHT) system
    d. National Fire Thread (NFT) system

15. If the *maximum* operating pressure is 180 psi, a relief valve *must* operate at _____.
    a. 175 psi
    b. 190 psi
    c. 200 psi
    d. 215 psi

Trade Terms Introduced in This Module

Control valve: A valve controlling flow to water-based fire protection systems.

Distort: To bend, warp, or damage a system component.

Erode: To deteriorate the surfaces in a valve due to water flow.

Globe valve: A type of valve consisting of a movable plug or disc and a stationary seat.

Hose header: A pipe that acts as a central connection point for one or more hoses.

Machined: A metal surface that has been held to precise tolerances during the manufacturing process.

Pressure-regulating device: A device designed for the purpose of reducing, regulating, controlling, or restricting water pressure.

Pressurized system: A piping system that is filled with pressure and has a pressure source.

Residual pressure: Pressure in a piping system under flow conditions.

Sealed valves: Valves that are wired or strapped in the open position.

Sprinkler lead-in: The pipe supplying water to the sprinkler system.

Sprinkler riser: The vertical supply pipe in a sprinkler system.

Static pressure: Pressure in a piping system under no-flow conditions.

Throttling: Using a valve to regulate flow through a system.

Water hammer: A damaging force that occurs when the flow of water is started or stopped, or when the direction of flow has changed.

Additional Resources

This module presents thorough resources for task training. The following reference material is suggested for further study.

NFPA 13, *Standard for the Installation of Sprinkler Systems*, Latest Edition. Quincy, MA: National Fire Protection Association.

NFPA 14, *Standard for the Installation of Standpipe and Hose Systems*, Latest Edition. Quincy, MA: National Fire Protection Association.

NFPA 25, *Standard for the Inspection, Testing, and Maintenance of Water-Based Fire Protection Systems*, Latest Edition. Quincy, MA: National Fire Protection Association.

NFPA 72®, *National Fire Alarm and Signaling Code®*, Latest Edition. Quincy, MA: National Fire Protection Association.

NFPA 1963, *Standard for Fire Hose Connections*, Latest Edition. Quincy, MA: National Fire Protection Association.

Figure Credits

Mac McGaffagan, Module Opener
Copyright Henry Pratt Company, LLC. Used with permission, Figure 1
Courtesy of Victaulic Company, Figures 2, 16–17
Reliable Automatic Sprinkler Co., Inc., Figures 3, 10 (Ball Lever Valve), 18, 25, 29–30
Used with permission of Johnson Controls, Figure 7
Courtesy of Potter Electric Signal, Figure 10 (Ball Lever Valve with Supervisory Switch)
COOPER® VALVES, Figure 11
NIBCO INC., Figures 12, 14
Courtesy of Milwaukee Valve, Figure 13
Courtesy of www.wattswater.com, Figure 19
Courtesy of General Air Products, Figure 21
Courtesy of Dixon Valve & Coupling Company., Figures 22, 26–28
Courtesy of NNI Fire Equipment, Figure 23
AMTROL Inc., Figure 31
AGF Manufacturing, Figure 32

Section Review Answer Key

Section 1.0.0

Answer	Section Reference	Objective
1. b	1.1.0	1a
2. d	1.2.0	1b
3. b	1.3.0	1c

Section 2.0.0

Answer	Section Reference	Objective
1. b	2.1.0	2a
2. b	2.2.2	2b

Section 3.0.0

Answer	Section Reference	Objective
1. b	3.1.5	3a
2. a	3.2.1	3b

Section 4.0.0

Answer	Section Reference	Objective
1. c	4.1.0	4a
2. a	4.2.1	4b
3. d	4.3.0	4c

User Update

NCCER makes every effort to keep its textbooks up-to-date and free of technical errors. We appreciate your help in this process. If you find an error, a typographical mistake, or an inaccuracy in NCCER's curricula, please submit a User Update form by visiting **https://www.nccer.org/olf**. You can also scan the QR code using the camera on your phone or mobile device to access the form.

Math for Sprinkler Fitters

Overview

This module reviews basic mathematical principles and explains how to apply them to various sprinkler fitting calculations, including floor areas, pitch, offsets, sprinkler spacing, pressure, and volume.

Module 18203

18203
MATH FOR SPRINKLER FITTERS

Objectives

Successful completion of this module prepares you to do the following:

1. Use basic math operations to solve problems.
   a. Make calculations using addition, subtraction, multiplication, and division.
   b. Use fractions, decimal equivalents, and percentages to express different values.
   c. Identify the exponents used to express area and volume.
   d. Identify the types of angles found in a circle.
2. Demonstrate practical calculations related to sprinkler systems.
   a. Calculate floor areas.
   b. Perform volume calculations.
   c. Make right-triangle calculations.
   d. Calculate offsets.
   e. Determine sprinkler spacing by area.
   f. Make pressure calculations.

Performance Tasks

This is a knowledge-based module. There are no Performance Tasks.

Trade Terms

Above the finished floor (A.F.F.)
Decimal equivalents
Exponents
Factor
Hypotenuse
Mathematical operator
Pi
Pressure
Scale
Static pressure

Industry Recognized Credentials

If you are training through an NCCER-accredited sponsor, you may be eligible for credentials from NCCER's Registry. The ID number for this module is 18203. Note that this module may have been used in other NCCER curricula and may apply to other level completions. Contact NCCER's Registry at 1.888.622.3720 or go to **www.nccer.org** for more information.

You can also show off your industry-recognized credentials online with NCCER's digital badges. Transform your knowledge, skills, and achievements into badges that you can share across social media platforms, send to your network, and add to your resume. For more information, visit **www.nccer.org.**

> **NOTE**
> Unless otherwise specified, references in parentheses following figure and table numbers refer to NFPA 13.

Contents

Figures and Tables

Section One

1.0.0 Mathematical Operators

Objective

Use basic math operations to solve problems.

a. Make calculations using addition, subtraction, multiplication, and division.
b. Use fractions, decimal equivalents, and percentages to express different values.
c. Identify the exponents used to express area and volume.
d. Identify the types of angles found in a circle.

Trade Terms

Decimal equivalents: Numbers that represent fractions, with every decimal place indicating a multiple of 10.

Exponents: Numbers or symbols, such as the 2 in 10 ft^2, that indicate the number of times an expression is multiplied by itself.

Hypotenuse: The longest side of a triangle that contains a right angle.

Mathematical operator: A symbolic instruction to perform a specific type of calculation. The four basic mathematical operators are addition (+), subtraction (−), multiplication (×), and division (÷).

Pi: A mathematical constant defined as the ratio of a circle's circumference to its diameter. Pi is equal to approximately 3.1416 or $^{22}/_{7}$ and is represented by the symbol π.

Scale: A ratio representing the relationship between drawing dimensions and actual dimensions. Sprinkler shop drawings are often drawn to a scale of $^{1}/_{8}$" = 1'.

Mathematics is used in sprinkler fitting on a daily basis. From determining how many boxes of sprinklers you might need on a job to making measurements between runs, it is essential that you have a solid understanding of basic calculations and how they are used.

A mathematical operator is simply a symbolic instruction to perform a specific type of calculation. Some of the mathematical operators used in sprinkler fitting include the following:

- Addition, subtraction, multiplication, and division (+, −, ×, and ÷)
- Fractions and decimal equivalents (e.g., $^{1}/_{2}$ and 0.5)
- Exponents (e.g., 10^2 or 2^3)
- Square roots and cube roots (e.g., $\sqrt{4}$ or $\sqrt[3]{8}$)

1.1.0 Addition, Subtraction, Multiplication, and Division

The four basic mathematical operators are addition (+), subtraction (−), multiplication (×), and division (÷). Most of the math you will do as a sprinkler fitter will involve these basic functions. Follow along as we work through some of the math you might encounter on a typical job.

Example 1: You have a system with an operating pressure of 160 pounds per square inch (psi) and you know that the hydrostatic test pressure must be 50 psi over the operating pressure. What is the hydrostatic test pressure for this system?

$$\text{Hydrostatic test pressure} = \text{operating pressure} + 50 \text{ psi}$$
$$\text{Hydrostatic test pressure} = 160 \text{ psi} + 50 \text{ psi} = 210 \text{ psi}$$

The hydrostatic test pressure for this system is 210 psi.

Example 2: You have a 21' length of pipe and have been instructed to use it to make as many 1' drum drips as possible (*Figure 1*). How many 1' drum drips can you make? How many valves will be required to complete the job?

Step 1 Divide the length of each drum drip by the total length of the pipe:

$$\text{Number of 1' drum drips} = 21' \div 1'/\text{drum drip} = 21 \text{ drum drips}$$

Step 2 Since each drum drip requires two valves, multiply the number of drum drips by two:

$$\text{Valves required} = 21 \text{ drum drips} \times 2 = 42 \text{ valves}$$

You can make 21 drum drips that are each 1' long and you will need 42 valves to complete the job.

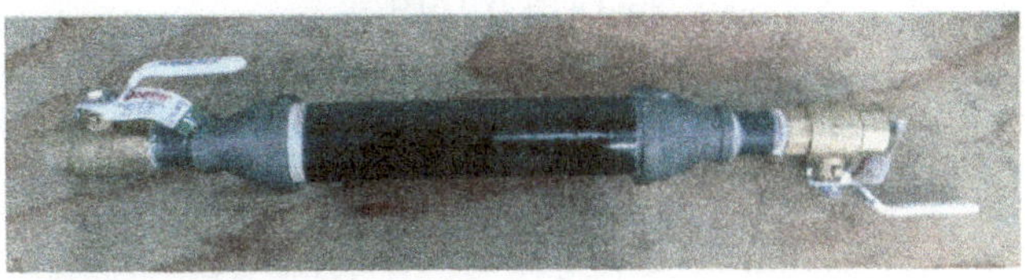

Figure 1 1' drum drip.

Example 3: You are packing up for a job that will require the use of 400 pendent sprinklers and 160 upright sprinklers. Given that there are 80 sprinklers per box, how many boxes of each type are required for this job?

Step 1 Divide the number of pendent sprinklers required by the number in each box to determine the number of boxes required:

Pendent boxes required = 400 pendent sprinklers ÷ 80 sprinklers/box = 5 boxes

Step 2 Divide the number of upright sprinklers required by the number in each box to determine the number of boxes required:

Upright boxes required = 160 upright sprinklers ÷ 80 sprinklers/box = 2 boxes

This job will require five boxes of pendents and two boxes of uprights, for a total of seven boxes to be loaded:

$$5 + 2 = 7$$

Example 4: You are moving a bundle of ten 21' lengths of 2" Schedule 40 pipe. Given that the weight of 2" Schedule 40 pipe is 3.65 pounds/foot, how much does this bundle weigh?

Step 1 Multiply the weight per foot by 21' to get the weight of each length of pipe:

Weight of each pipe = 21' × 3.65 pounds/foot = 76.65 pounds

Step 2 Multiply the weight per pipe by the number of pipes in the bundle:

Weight of bundle = 76.65 pounds/pipe × 10 pipes = 766.5 pounds

Example 5: You are working on a remodel job in an office building and have to relocate 300 sprinklers. You have been told by your supervisor to allow 40 minutes per sprinkler, which includes everything from piping adjustments to the installation of hangers. How many hours will it take to complete this job?

Step 1 Calculate the time required in minutes:

Time = 300 sprinklers × 40 minutes/sprinkler = 12,000 minutes

Step 2 Convert minutes to hours:

Time = 12,000 minutes ÷ 60 minutes/hour = 200 hours

Example 6: Building on the previous example that required 200 hours of installation time, suppose there are five fitters on this job and they each work 40 hours per week. How many weeks will it take to complete this job?

Step 1 Multiply the number of workers by the hours worked per week:

Time = 5 fitters × 40 hours/week = 200 hours

Step 2 Divide the total number of hours required by the hours worked per week:

Time = 200 total hours required ÷ 200 hours/week = 1 week

With five fitters on the job, it can be completed in one week.

Example 7: Before leaving a job where 300 sprinklers were installed, your supervisor reminds you to include the required number of spares in a wall-mounted spare sprinklers box. You must leave at least two of each sprinkler type, and a minimum of 12 total for a job involving 300 sprinklers or more. You have three types of sprinklers used in this installation: uprights, pendents, and sidewalls, but the vast majority are pendents. What do you leave in the box?

Step 1 Even though there are far fewer upright and sidewall sprinklers, you still need to leave a minimum of two of each type:

2 upright sprinklers + 2 sidewall sprinklers = 4 sprinklers out of the 12 in the box

Step 2 Subtract from the total to find the spaces remaining for the pendent sprinklers in the box:

$$12 - 4 = 8$$

This box should contain at least 2 upright sprinklers, 2 sidewall sprinklers, and 8 pendent sprinklers.

1.2.0 Fractions, Decimal Equivalents, and Percentages

In addition to whole numbers, sprinkler fitters must be able to make calculations using fractions and their decimal equivalents. Decimal equivalents are often expressed as percentages.

1.2.1 Fractions

A fraction is obtained by dividing a whole number into a specific number of pieces. For example, an inch is $\frac{1}{12}$ of a foot.

An understanding of fractions is essential when reading shop drawings. Sprinkler shop drawings are commonly produced in $\frac{1}{8}$ scale. This means that $\frac{1}{8}$" on the drawing equals 1' in actual dimensions. Scales vary by drawing so it is important to find the scale before reading any drawing. For example, detail drawings may be scaled to $\frac{3}{8}$" = 1'. Due to their complexity, fire pump rooms may be drawn to an even larger scale, such as $\frac{3}{4}$" = 1'.

There are several ways that fractions are written in the sprinkler industry. For example, a common linear designation would be 10'-8$\frac{1}{2}$". However, it may simply be written as 10-8$\frac{1}{2}$, with feet and inches understood. In other instances, you may see it written as 10-$8^1{}_2$ with the fraction slash mark understood as well.

Example: You are installing pendent sprinklers and must determine the overall ceiling thickness in order to position the sprinklers correctly. See *Figure 2*. This ceiling consists of a $\frac{1}{2}$" metal furring channel and two layers of $\frac{5}{8}$" gypsum wallboard (GWB).

Many Symbols, Same Results

Multiplication and division are two operators that can be expressed in a number of different ways. No matter which symbol is shown, the results will be the same.

Multiplication can be indicated by the symbols "·" and "×," but also with parentheses or brackets:

$$5 \cdot 7 = 35$$
$$5 \times 7 = 35$$
$$5(7) = 35$$
$$(5)(7) = 35$$
$$[5][7] = 35$$

When letters or symbols are used in an equation, they are often simply shown next to one another to indicate multiplication (e.g., Circumference = $2\pi R$).

Division can be indicated by the symbols "÷" and "/," but other lines may also be used:

$$10 \div 2 = 5$$
$$10/2 = 5$$
$$\frac{10}{2} = 5$$

1-Hour Rated Ceiling (Exit Stair):

- Stair Structure Per Drawings
- 1/2" Metal Furring Channel
- (2) Layer 5/8" Type "X" GWB - Taped and Staggered Seams

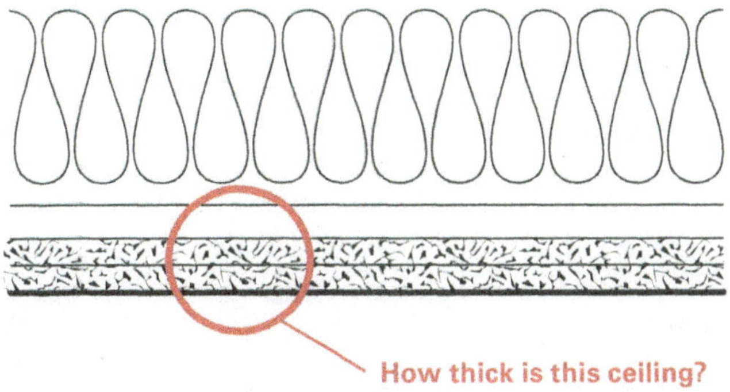

Figure 2 Determining ceiling thickness.

The total ceiling thickness can be calculated as follows:

Step 1 Add the equivalent fractions:

$$\frac{5}{8}" + \frac{5}{8}" = \frac{10}{8}" \text{ or } 1\frac{1}{4}"$$

Step 2 Add the remaining value to find the total:

$$1\frac{1}{4}" + \frac{1}{2}" = 1\frac{3}{4}"$$

1.2.2 Decimal Equivalents

Decimals are fractions to the base of 10, 100, 1,000, and so on. The decimal system is useful for both large and small numbers, and it is the system used by calculators and computers. *Figure 3* shows the abbreviated decimal system.

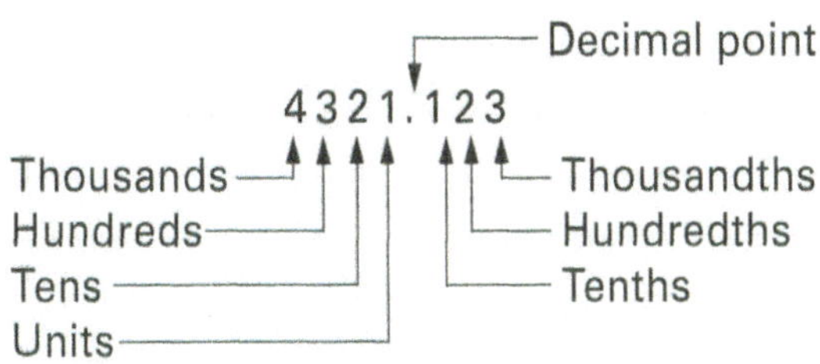

Figure 3 Abbreviated decimal system.

The most common use of the decimal system is in processing numbers dealing with money. For example, $19.27 means nineteen dollars and twenty-seven cents. Twenty-seven cents is 27/100 or twenty-seven hundredths of a dollar. The entire monetary system is based on the decimal system with 100 cents to the dollar.

For every fraction, there is a decimal equivalent. For every dimension in inches, there is a decimal equivalent in feet. Sprinkler fitters will be required to convert from decimals to inches and decimals to feet. *Table 1* shows decimal equivalents for fractions commonly used in the sprinkler industry.

The decimal equivalent is obtained by performing the division indicated in the fraction.

Example 1: To find the decimal equivalent of 7/16, perform the division operation as follows:

$$\frac{7}{16} = 16\overline{)7.0000} = \begin{array}{r} 0.4375 \\ 16\overline{)7.0000} \\ -64 \\ \hline 60 \\ -48 \\ \hline 120 \\ -112 \\ \hline 80 \\ -80 \\ \hline\hline 0 \end{array}$$

Table 1 Decimal Equivalents

Inches	Decimal Feet
1/8"	0.125" ÷ 12" = 0.0104'
1/4"	0.25" ÷ 12" = 0.0208'
1/2"	0.50" ÷ 12" = 0.0417'
3/4"	0.75" ÷ 12" = 0.0625'
1"	1" ÷ 12" = 0.0833'
1 1/2"	1.5" ÷ 12" = 0.1250'
2"	2" ÷ 12" = 0.1667'
2 1/2"	2.5" ÷ 12" = 0.2083'
3"	3" ÷ 12" = 0.2500'
3 1/2"	3.5" ÷ 12" = 0.2917'
4"	4" ÷ 12" = 0.3333'
4 1/2"	4.5" ÷ 12" = 0.3750'
5"	5" ÷ 12" = 0.4167'
5 1/2"	5.5" ÷ 12" = 0.4583'
6"	6" ÷ 12" = 0.5000'
6 1/2"	6.5" ÷ 12" = 0.5417'
7"	7" ÷ 12" = 0.5833'
7 1/2"	7.5" ÷ 12" = 0.6250'
8"	8" ÷ 12" = 0.6667'
8 1/2"	8.5" ÷ 12" = 0.7083'
9"	9" ÷ 12" = 0.7500'
9 1/2"	9.5" ÷ 12" = 0.7917'
10"	10" ÷ 12" = 0.8333'
10 1/2"	10.5" ÷ 12" = 0.8750'
11"	11" ÷ 12" = 0.9167'
11 1/2"	11.5" ÷ 12" = 0.9583'
12"	12" ÷ 12" = 1.0000'

Did You Know?

Most conversions will result in numbers with four or more decimal places to the right of the decimal point. For example, ⅛" has a decimal equivalent of 0.0104'. Accuracy to 0.01' (one-hundredth of a foot) is usually sufficient for sprinkler work. This means that a number such as 0.0104' is rounded off to two places to the right of the decimal point. If the third place is less than 5, round down. If the third place is 5 or more, round up. In this example, the third place is a 0, so 0.0104' rounds down to 0.01'. Practice rounding using the values shown in *Table 1*.

It may be necessary to convert mixed units of feet and inches into a single type of unit in order to use them in calculations, such as for floor areas or volumes.

Example 2: Suppose you must convert 12'-3" into decimal feet to complete an area equation.

Step 1 Convert the inches into decimal feet:

3" ÷ 12" per foot = 0.25'

Step 2 Add the equivalent values:

12' + 0.25' = 12.25'

1.2.3 Percentages

Both fractions and decimal equivalents are often expressed as percentages. For example:

½ = 0.50 = 50%
⅓ = 0.33 = 33%
⅔ = 0.67 = 67%

Percentages are easy to use in calculations by simply converting them to decimal equivalents.

Example 1: Suppose you are in charge of maintenance on a large fire pump with a diesel engine and need to call in a refill to your diesel supplier when the tank is less than two-thirds full. The fuel tank label in *Figure 4* shows that the tank on

Getting Paid

One of the math problems everyone wants to get right is their paycheck. To calculate the net pay you receive when you cash your check, you need to know how much you are paid per hour, how much tax you owe, and how many deductions you have each pay period. For example, if your pay is $15.00 per hour and you work a 40-hour week, your weekly pay is calculated as follows:

Pay rate × hours worked = gross pay
$15.00/hour × 40 hours = $600.00 per week gross (gross pay is pay before taxes and deductions)
Your annual pay is $600.00 × 52 weeks = $31,200.00

Next, calculate the taxes that are automatically deducted from your paycheck to pay the federal government. Those taxes are income taxes, Social Security, and Medicare. The percentage of income tax you pay, called a *tax bracket*, depends on a number of factors, but primarily on the amount of your income. We will assume you pay 12% of your gross salary for federal income taxes. The Social Security tax is 6.2% of gross pay for everyone and the Medicare tax is 1.45%. To calculate the total taxes withheld each pay period (called *withholding*), add the three tax rates, then multiply the total by your gross pay, as follows:

Income tax rate = 12.0% (depending on salary)
Social Security tax rate = 6.2%
Medicare tax rate = 1.45%
Total taxes withheld = 19.65% of gross pay

Next, multiply the total withholding rate (19.65%) by gross pay ($600.00) to get the amount deducted from your check for taxes.

19.65% or 0.1965 × $600.00 = $117.90
$600.00 − $117.90 = $482.10 net pay per week

There may be more tax withheld from your check if you live in a state that has state income taxes. You may also have additional deductions for insurance and retirement funding. If so, subtract those deductions to calculate your net pay.

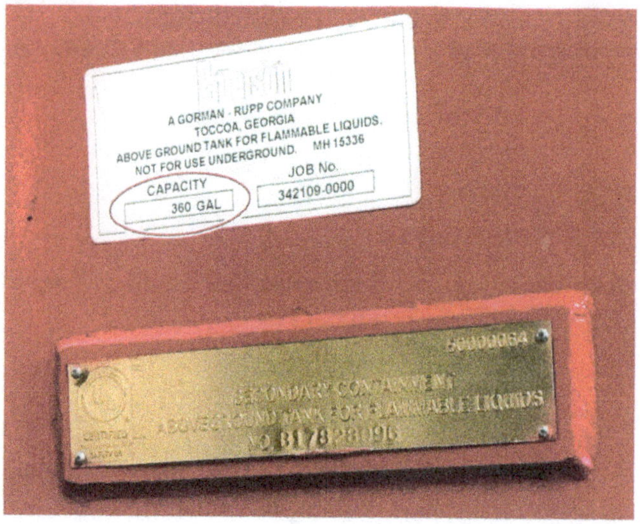

Figure 4 Diesel fire pump tank label.

this site has a capacity of 360 gallons. Rounding to the nearest 10 gallons, at what level must you call the supplier for a refill?

$$0.67 \times 360 \text{ gallons} = \text{about } 240 \text{ gallons}$$

Example 2: Per NFPA 20, *Standard for the Installation of Stationary Pumps for Fire Protection*, except where equipped with automatic shutoff devices, fuel tanks may only be filled to a maximum of 90% of the tank capacity due to the need to account for expansion and sump. Building on the previous example, if the gauge on a 360-gallon tank reads 240 gallons, and it may only be filled to 90%, how much fuel will it take to fill this tank?

$$\text{Maximum fuel} = (\text{tank volume} \times 0.90) - \text{tank gauge reading}$$
$$\text{Maximum fuel} = (360 \text{ gallons} \times 0.90) - 240 \text{ gallons}$$
$$\text{Maximum fuel} = 324 \text{ gallons} - 240 \text{ gallons} = 84 \text{ gallons}$$

It will take 84 gallons of fuel to fill this tank. Now if the price of diesel fuel is \$3.65 per gallon, how much will it cost to fill the tank?

$$\text{Fuel cost} = \text{gallons required} \times \text{price/gallon}$$
$$\text{Fuel cost} = 84 \times \$3.65/\text{gallon} = \$306.60$$

1.3.0 Exponents

Exponents are used when calculating area, volume, and other values. Common exponents include squares, cubes, and roots (fractional exponents).

1.3.1 Squares

Squaring a number means to multiply it by itself, such as $2 \times 2 = 2^2 = 4$. In the value of 2^2, the small 2 is called an *exponent*. Squaring a number is used when calculating an area with equal sides. For example, *Figure 5* shows a square with 10' sides. The area is calculated by squaring both the number 10 ($10 \times 10 = 100$), and the unit (feet × feet = square feet):

$$10' \times 10' = (10')^2 = 100 \text{ ft}^2$$

This produces 100 ft^2 as the area. The number was squared, and the unit was also squared. If the room had unequal dimensions (e.g., 8' by 10'), the numbers are multiplied, not squared, but the units are still squared ($8' \times 10' = 80 \text{ ft}^2$).

> **NOTE**
> When working with values expressed in both feet and inches (e.g., 1' × 1'-6"), one or the other must be converted so that they are in the same units (in this case, 1' × 1.5' or 12" × 18"). It does not matter which units are selected as long as they are the same type.

1.3.2 Cubes

Cubing a number means to multiply it by itself twice, such as $2 \times 2 \times 2 = 2^3 = 8$. It is indicated by an exponent of 3 and can be used to determine the volume of a square block or cube. See *Figure 6*. The volume of this cube is calculated as follows:

$$12'' \times 12'' \times 12'' = (12'')^3 = 1{,}728 \text{ in}^3$$

The volume of this cube is calculated by cubing the number 12 ($12 \times 12 \times 12 = 1{,}728$), then cubing the units (inches × inches × inches = cubic inches). If the shape had unequal dimensions (e.g., 8" by 10" by 12"), the numbers are simply multiplied, but the units are still cubed ($8'' \times 10'' \times 12'' = 960 \text{ in}^3$).

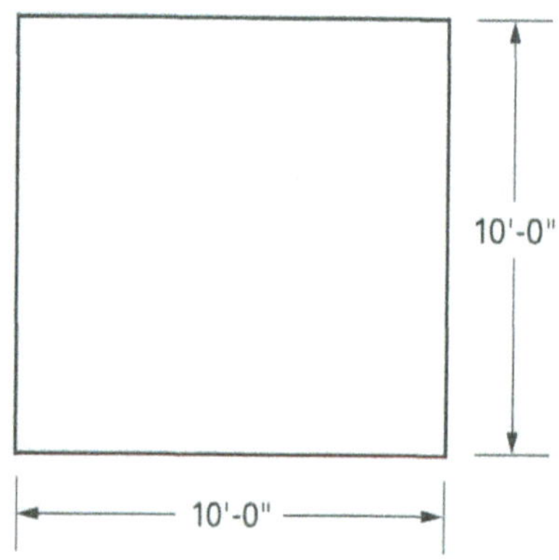

Figure 5 Square.

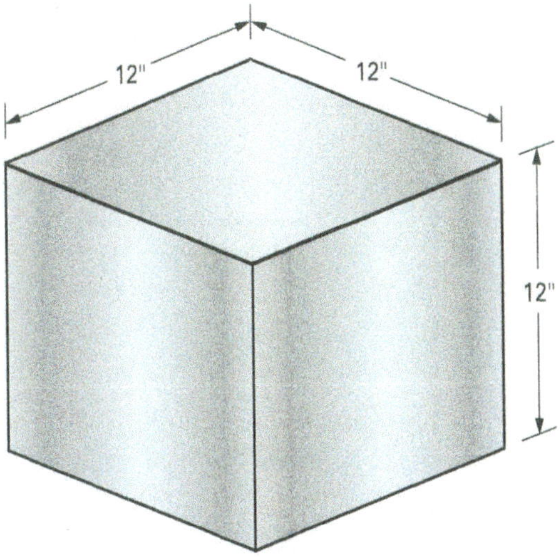

Figure 6 Cube.

1.3.3 Roots

Exponents do not have to be whole numbers. In a square root, the exponent is one-half. A square root is simply the reverse of a square. It is the number that when multiplied by itself will produce the initial number. For example, the square root of 4 is 2 ($2 \times 2 = 4$). Square roots can be indicated by the following notations:

$$\sqrt{4} = 4^{1/2} = 4^{0.5}$$

Some layout technicians and engineers prefer to express roots as fractional coefficients because that way they can be typed without the need for the square root symbol.

A cube root is a number that will produce the original number when multiplied by itself twice. For example, the cube root of 8 is 2 ($2 \times 2 \times 2 = 8$). A cube root can be expressed as follows:

$$\sqrt[3]{8} = 8^{1/3}$$

Did You Know?

You probably have a scientific calculator right in your pocket. Most iPhone calculators can be converted into scientific calculators when the orientation lock is off by simply opening the calculator app and turning the phone on its side. Android phones have a similar feature.

Using Square Roots

One use for square roots is in calculating sprinkler discharge rates. The equation for this is:

$$Q = K \times \sqrt{P}$$

Where:

Q = the quantity of water discharged
K = discharge coefficient of the sprinkler
P = pressure applied to the sprinkler

For example, suppose a storage sprinkler with a K-factor of 25.2 is accidentally knocked off during material movement. Assuming a pressure of 100 psi, how many gallons of water per minute will be discharged through this sprinkler?

$$Q = K \times \sqrt{P}$$
$$Q = 25.2 \times \sqrt{100}$$
$$Q = 25.2 \times 10 = 252 \text{ gpm}$$

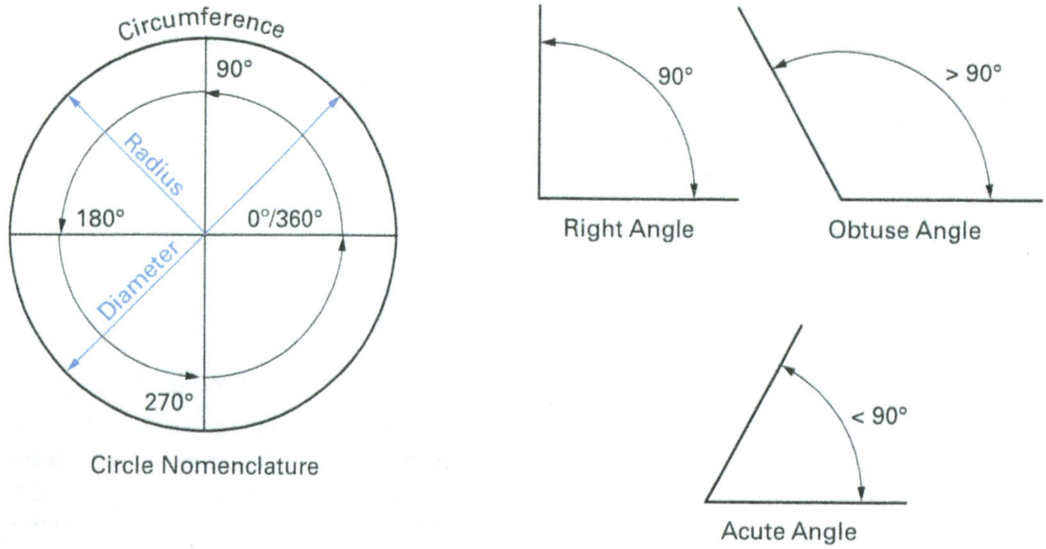

Figure 7 Circle and angles.

1.4.0 Circles and Angles

Figure 7 shows a circle. The circumference is the length of the line around the outside. The distance between the center and any point on the outside is the radius. A circle is divided into 360 increments called *degrees* (°). Any angle from 0° to 360° can be represented in a circle.

1.4.1 Circles

Two common equations involving circles are those used to calculate circumference and area.

Circumference

The circumference of a circle is found using the equation:

$$C = 2\pi R \text{ or } C = \pi D$$

Where:

$$\pi = 3.1416$$
$$R = \text{radius}$$
$$D = \text{diameter}$$

The symbol π or pi represents a constant equal to approximately 3.1416. That is the ratio of the circumference of any circle divided by its diameter.

Example: If a circle has a radius of 13', the circumference can be found as follows:

$$C = 2\pi R$$
$$C = (2)(3.1416)(13') = 81.68' \text{ (rounded to two decimal places)}$$

Area

The area of a circle is found using the equation:

$$A = \pi R^2$$

Example: If a circle has a radius of 10', the area is found as follows:

$$A = \pi R^2$$
$$A = (3.1416)(10')^2 = (3.1416)(100\ ft^2) = 314.16\ ft^2$$

Note that in the process of solving an equation, the factors with exponents are calculated first. Multiplication and division (from left to right) are performed next, and then addition and subtraction (again from left to right). This is known as the *order of operations*.

1.4.2 Angles

Figure 7 showed common angles. A right angle is 90°. An acute angle is less than 90°. An obtuse angle is greater than 90°.

There are a total of 180° in a triangle. This is divided between the three angles according to the type of triangle. *Figure 8* shows a right triangle and some of the equations related to it.

A right triangle has a 90° or right angle between the two sides. The line connecting the ends of the two sides (and opposite the right angle) is called the **hypotenuse**.

The area of a right triangle is one-half the product of the two sides. The dimensions of the two sides must be in the same units.

The slope or *pitch* of the hypotenuse is used to calculate the length of hangers or pipe as required when dealing with pitched roofs. This is the ratio of rise to run, with both dimensions in the same units.

The Pythagorean theorem, $a^2 + b^2 = c^2$, provides the formula to find any side of a right triangle if the other two sides are known.

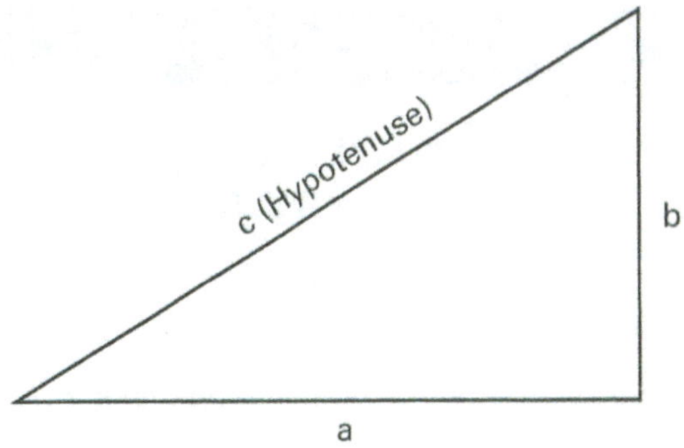

$$\text{Area} = \tfrac{1}{2}\, a \times b$$

$$\text{Slope} = \frac{\text{Rise}}{\text{Run}} = \frac{b}{a}$$

$$a^2 + b^2 = c^2$$

Figure 8 Right triangle equations.

1.0.0 Section Review

1. You have a 21' length of pipe and need to make seven 1' drum drips. How much pipe would you have left over? ______________
2. Demonstrate how to convert 1'-9" into decimal feet. ______________
3. Demonstrate how to calculate the area of a 6'-6" square. ______________
4. Demonstrate how to calculate the area of a circle with a diameter of 36". ______________

Section Two

2.0.0 Practical Calculations for Sprinkler Fitters

Objective

Demonstrate practical calculations related to sprinkler systems.

a. Calculate floor areas.
b. Perform volume calculations.
c. Make right-triangle calculations.
d. Calculate offsets.
e. Determine sprinkler spacing by area.
f. Make pressure calculations.

Trade Terms

Above the finished floor (A.F.F.): A measurement taken from the surface of the installed flooring (e.g., tile), rather than the underlayment.

Factor: An amount by which a value is multiplied to produce a specific result.

Pressure: The amount of force applied to a given unit of surface area. Pressure is often expressed in *pounds per square inch (psi)*.

Static pressure: A measurement of the pressure of still water. A column of water that is 1' high will produce a static pressure of 0.433 psi.

Basic mathematical principles can be applied to many on-the-job sprinkler fitting applications. Common applications of math in sprinkler fitting include the following:

- Floor areas
- Volumes
- Rise, run, and pitch
- Offsets
- Sprinkler spacing
- Pressures

2.1.0 Calculating Floor Areas

In sprinkler work, linear dimensions are stated in feet and inches and measured using a 25', 30', or 100' measuring tape. Laser measuring devices may also be used. *Figure 9* shows a 25' tape and a laser measuring device.

25' Tape

Laser Measuring Device

Figure 9 25' tape and laser measuring device.

Area calculations are used to determine the number of square feet covered by a sprinkler and to determine the square footage of floor areas. Floor area is calculated using the formula:

Area = length × width

Example 1: Figure 10 shows the scaled plan view of a room that is 8'-4" wide by 11'-3" long. What is the floor area?

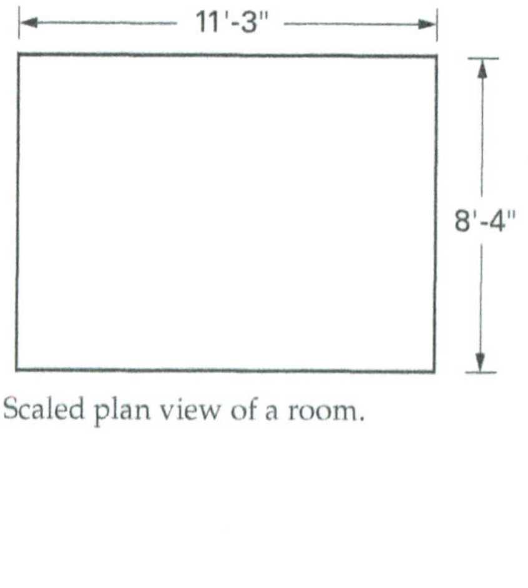

Figure 10 Scaled plan view of a room.

Step 1 As described earlier in this module, before calculating the floor area, you must first convert the dimensions to decimal feet by dividing the inches by 12:

$$8'\text{-}4'' = 8' + (4'' \div 12'' \text{ per foot} = 0.33') = 8.33'$$
$$11'\text{-}3'' = 11' + (3'' \div 12'' \text{ per foot} = 0.25') = 11.25'$$

Step 2 Calculate the area of the room:

$$\text{Area} = \text{length} \times \text{width}$$
$$\text{Area} = 11.25' \times 8.33' = 93.71 \text{ ft}^2 \text{ (rounded to two decimal places)}$$

Example 2: *Figure 11* illustrates a complex floor plan. It has been divided into two rectangular areas to simplify the calculations. What is the total area?

Step 1 Calculate Area 1:

$$\text{Area} = \text{length} \times \text{width}$$
$$\text{Area 1} = 108' \times 35' = 3{,}780 \text{ ft}^2$$

Step 2 Calculate Area 2:

$$\text{Area} = \text{length} \times \text{width}$$
$$\text{Area 2} = 73' \times 90' = 6{,}570 \text{ ft}^2$$

Step 3 Find the total area:

$$\text{Total area} = 3{,}780 \text{ ft}^2 + 6{,}570 \text{ ft}^2 = 10{,}350 \text{ ft}^2$$

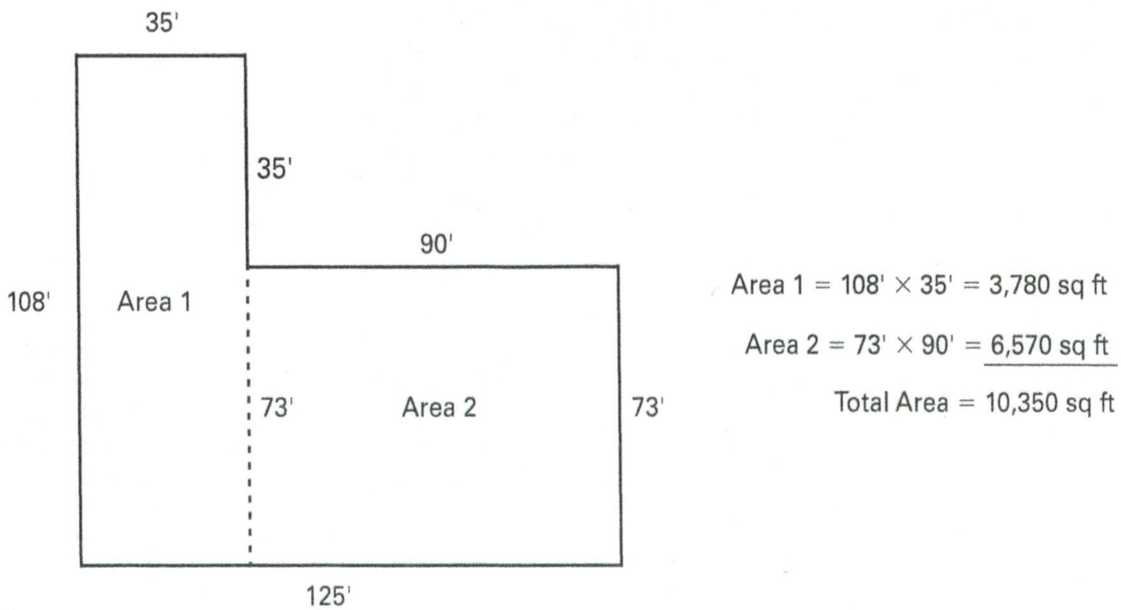

Figure 11 Subdividing an area into simple shapes.

2.2.0 Volume Calculations

Sprinkler fitters may be called upon to calculate the volumes of a variety of shapes, including rectangular solids, cylinders, and piping. Volume is generally expressed in cubic feet but may also be expressed in gallons or other fluid measures.

2.2.1 Rectangular Volume Calculations

To determine the cubic feet in a rectangular room, multiply length by width by height. *Figure 12* shows the volume of a rectangle.

Example 1: Volumes for data centers must be calculated for alternate fire extinguishing systems, such as halon-free fire protection fluids. *Figure 13* shows the dimensions of an example data center. The 8'-0" ceiling height is the elevation above the finished floor (A.F.F.).

The volume of this data center is calculated as follows:

$$\text{Volume} = \text{length} \times \text{width} \times \text{height}$$
$$\text{Volume} = 16' \times 8' \times 8' = 1{,}024 \text{ ft}^3$$

Example 2: Sprinkler fitters may be required to calculate the amount of concrete needed to construct a concrete pad, such as the one beneath the diesel engine in *Figure 14*. If the dimensions of this pad are 6' long by 3' wide by 10" tall, and each 60-pound bag of concrete yields about 0.5 ft^3, how many bags of concrete were required to complete this job?

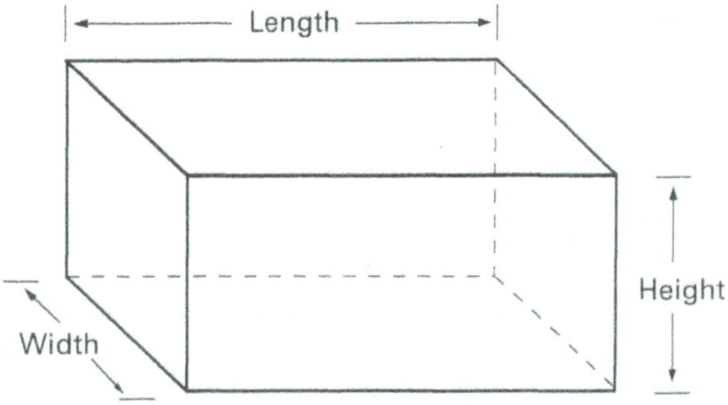

Figure 12 Volume of a rectangular solid.

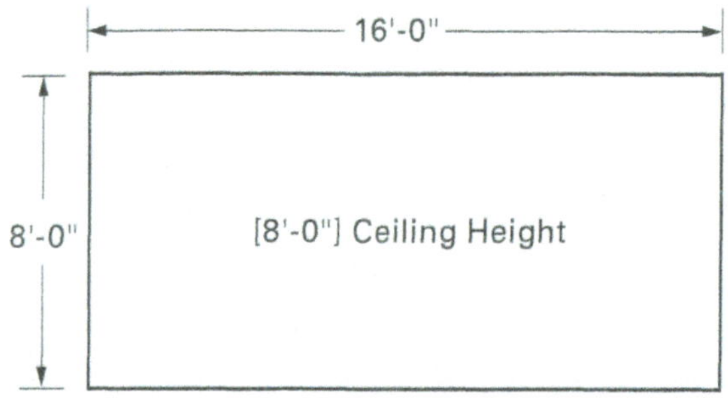

Figure 13 Example data center.

Figure 14 Large concrete pad example.

Step 1 Convert the 10" to feet:

10" ÷ 12" per foot = 0.83'

Step 2 Calculate the total volume of the rectangular area:

Volume = length × width × height
Volume = 6' × 3' × 0.83'
Volume = approximately 15 ft^3 (rounded to the nearest cubic foot)

Step 3 Calculate the number of bags required:

15 ft^3 ÷ 0.5 ft^3 per bag = 30 bags

> **NOTE**
> If you failed to convert inches to feet in the above calculation, you would end up purchasing and moving 360 bags of concrete rather than 30! Always make sure the units are equal when making calculations.

Example 3: Suppose you have been instructed to lay the concrete for a small pad, such as the one shown beneath the panel in *Figure 15*. If the dimensions of this pad are 30" long by 18" wide by 4" tall, and each 60-pound bag of concrete yields about 0.5 ft^3, how many bags of concrete would be required to complete this job?

Step 1 Since the final dimensions are in cubic feet, convert the dimensions in inches to feet:

30" ÷ 12" per foot = 2.5'
18" ÷ 12" per foot = 1.5'
4" ÷ 12" per foot = 0.33'

Step 2 Calculate the total volume of the rectangular area:

Volume = length × width × height
Volume = 2.5' × 1.5' × 0.33'
Volume = 1.2375 ft^3 (approximately 1.25 ft^3)

Step 3 Calculate the number of bags required:

1.25 ft^3 ÷ 0.5 ft^3 per bag = 2.5 bags

2.2.2 Liquid Volume Calculations

Water and diesel fuel storage tanks used in the sprinkler industry are usually specified by the volume of fluid that can be stored in them, regardless of the shape of the tank.

Figure 15 Small concrete pad example.

Rectangular Storage Tanks

One cubic foot of water is equal to 7.48 gallons. To determine the liquid capacity of a square or rectangular tank, the following formula is used:

Tank capacity in gallons = (length × width × height) × 7.48 gallons/ft^3

Example: How many gallons of water will a rectangular tank hold if it measures 12' long, 3' wide, and 8' high?

Capacity in gallons = (length × width × height) × 7.48 gallons/ft^3
Capacity in gallons = (12' × 3' × 8') × 7.48 gallons/ft^3
Capacity in gallons = 288 ft^3 × 7.48 gallons/ft^3 = 2,154.24 gallons

Cylindrical Storage Tanks

Large diesel fuel tanks are typically cylindrical. *Figure 16* shows a cylindrical tank.

The following equation can be used to find the volume of a cylinder:

$$V = \pi R^2 h$$

Where:

V = volume
π = 3.1416
R = radius
h = height

This equation will provide the tank volume in cubic feet or inches, depending on the units used in the equation. Normally, tank volumes are expressed in gallons. One gallon is equal to 231 in^3.

Example: The diesel fuel tank shown in *Figure 17* measures 66" long and has a diameter of 40". How many gallons of fuel will it hold?

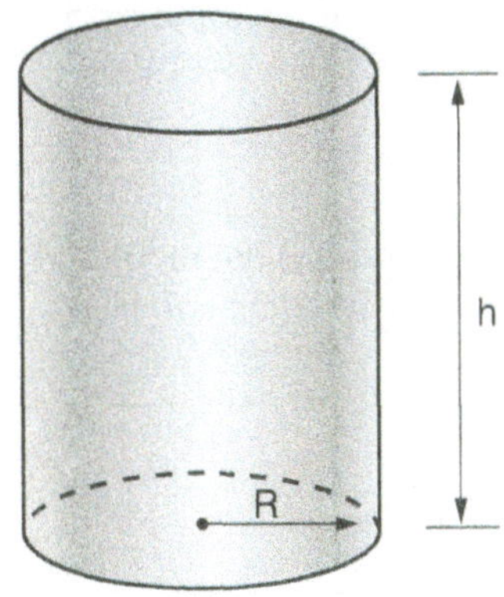

Figure 16 Cylindrical tank.

Step 1 Divide the diameter by two to find the radius:

$$40'' \div 2 = 20''$$

Step 2 Plug the known values into the equation for the volume of a cylinder:

$$V = \pi R^2 h$$
$$V = (3.1416) \times (20'')^2 \times (66'') = 82{,}938.24 \text{ in}^3$$

Step 3 Divide the result in cubic inches by 231 in^3 per gallon to find the answer in gallons:

$$V = 82{,}938.24 \text{ in}^3 \div 231 \text{ in}^3/\text{gallon} = 359.04 \text{ gallons}$$

This tank holds 359.04 gallons or roughly 360 gallons. Even though it does not hold precisely 360 gallons, it is still referred to as a *360-gallon tank*.

2.2.3 Pipe Volume Calculations

As discussed previously, piping must be pitched to avoid trapping water at low points in the system. However, even systems that are properly pitched may have trapped sections of piping, such as around ducts or other obstructions. NFPA 13 requires auxiliary drains in any location where a change in piping direction prevents drainage of system piping through the main drain valve. The auxiliary drain size depends on the volume or capacity of water in the trapped piping and the type of system. The requirements for a dry pipe system are as follows:

- Where the capacity of isolated trapped sections of pipe is more than five gallons, the auxiliary drain must consist of two 1" valves and one 2" × 12" condensate nipple or equivalent in an accessible location.
- Where the capacity of isolated trapped sections of pipe is less than five gallons, an auxiliary drain consisting of a nipple and cap or plug not less than ½" in size must be used.

NOTE

Auxiliary drains are not required for pipe drops supplying dry pendent sprinklers.

To calculate the capacity of a section of trapped piping, you must know the pipe schedule and diameter, as well as the total length of the piping around the obstruction. *Table 2* shows the capacity in gallons per foot of pipe for various sizes of Schedule 40 and Schedule 10 steel pipe.

Example: *Figure 18* shows 4" Schedule 40 steel pipe passing under a rectangular duct. What size drain would be required to prevent water from standing in the pipe surrounding the duct?

Step 1 Determine the total length of the obstructed pipe:

$$3' + 5' + 3' = 11'$$

Step 2 Refer to *Table 2* to find the capacity per foot for 4" Schedule 40 steel pipe and then multiply it by the total length of obstructed pipe to find the total capacity:

$$11' \times 0.660 \text{ gallons per foot} = 7.26 \text{ gallons}$$

Since this section of pipe is more than five gallons, the auxiliary drain must consist of two 1" valves and one 2" × 12" condensate nipple or equivalent in an accessible location.

2.3.0 Making Right-Triangle Calculations

Sprinkler fitters often have to compute the values of rise, run, and pitch. This can be done using the simple equation for pitch. In addition, sometimes fitters must calculate the precise length of an unknown side of a right triangle. This is done using the Pythagorean theorem.

Figure 17 Diesel fuel tank.

Figure 18 Schedule 40 steel pipe obstruction.

Table 2 Capacity (Gallons) of 1' Pipe (Data from NFPA 13, *Table A.8.2.3*)

Nominal Pipe Diameter	Schedule 40 Steel Pipe	Schedule 10 Steel Pipe
1"	0.045	0.049
1¼"	0.078	0.085
1½"	0.106	0.115
2"	0.174	0.190
2½"	0.248	0.283
3"	0.383	0.433
4"	0.660	0.740

2.3.1 Calculating Rise, Run, and Pitch

The following equation has already been introduced:

$$\text{Pitch} = \frac{\text{Rise}}{\text{Run}}$$

This equation can be solved for any quantity if the other two values are known. For example, a pitch of 1' in 10' means that the roof rise is 1' in a run of 10'. To calculate a range of values, substitute 0.1 for the value of pitch (pitch = 1 ÷ 10 or 0.10).

Example 1: Given a pitch of 0.10, determine the rise when the run is 15.

Step 1 Plug the known values into the equation:

$$\text{Pitch} = \frac{\text{Rise}}{\text{Run}}$$

$$0.1 = \frac{\text{Rise}}{15}$$

Step 2 Rearrange the equation to get the unknown variable on one side by itself. To do this, multiply each side by 15 as follows:

$$0.1(15) = \frac{\text{Rise (15)}}{15}$$

Step 3 Since 15 over 15 is equal to 1, this equation can be rewritten as:

$$0.1(15) = \frac{\text{Rise } (\cancel{15})}{\cancel{15}}$$

$$1.5 = \text{Rise}$$

This states that the rise is 1.5' when the run is 15', and the pitch is 0.1. By substituting various values for run, different values for the rise can be generated. *Table 3* shows the rise and run for a pitch of 0.1.

Table 4 shows how the pitch varies as the rise and run change. For a given run, such as 7, the pitch increases as the rise increases. For a given rise, such as 5, the pitch decreases as the run increases.

Table 3 Rise Versus Run for a Pitch of 0.1

Run	Rise
5	0.5
10	1.0
15	1.5
20	2.0
25	2.5
30	3.0
35	3.5

Note: All dimensions in feet.

Table 4 Rise, Run, and Pitch

		Run, Inches											
		1	**2**	**3**	**4**	**5**	**6**	**7**	**8**	**9**	**10**	**11**	**12**
Rise, Inches	**12**	—	—	—	—	—	—	60°	56°	53°	50°	46°	45°
	11	—	—	—	—	—	—	58°	54°	51°	48°	45°	42°
	10	—	—	—	—	—	59°	55°	51°	48°	45°	42°	40°
	9	—	—	—	—	—	56°	52°	48°	45°	42°	39°	37°
	8	—	—	—	—	58°	53°	49°	45°	42°	39°	36°	34°
	7	—	—	—	60°	54°	49°	45°	41°	38°	35°	32°	30°
	6	—	—	—	56°	50°	45°	41°	37°	34°	31°	29°	27°
	5	—	—	59°	51°	45°	40°	36°	32°	29°	27°	24°	23°
	4	—	—	63°	45°	39°	34°	30°	27°	24°	22°	20°	18°
	3	—	56°	45°	37°	31°	27°	23°	21°	18°	17°	15°	14°
	2	—	45°	34°	27°	22°	18°	16°	14°	13°	11°	10°	9°
	1	45°	27°	18°	14°	11°	9°	8°	7°	6°	6°	5°	5°

Did You Know?

There are many free online calculators for determining values of rise, run, and pitch.

Plotting a graph on paper using the table values is useful when the pitch of a roof is known and hanger lengths at various points along a horizontal sprinkler piping run are needed. *Figure 19* shows an installation with piping in an attic and sprinklers in the ceiling.

Dry systems must have the line piping and main piping pitched so water will drain back to the riser or an auxiliary drain.

NFPA 13 requires that branch lines in dry pipe systems be pitched at least ½" for every 10' of horizontal run, while main piping must be pitched at least ¼" for every 10' of horizontal run in nonrefrigerated areas. The longer the run of piping, the greater its elevation must be to satisfy the pitch requirement.

Example 2: What will the end elevation be for a 120' long dry system main piping if the starting elevation is 13'?

Step 1 The requirement is ¼" for every 10' of horizontal run. Divide 120' into 10' sections as follows:

120' ÷ 10' = 12 sections, each with a ¼" rise (0.25")

Step 2 Multiply the rise by the number of sections:

12 sections × 0.25" per section = 3" rise over 120'

Step 3 Add the rise to the initial elevation to find the end elevation:

13' elevation + 3" rise = 13'-3" end elevation

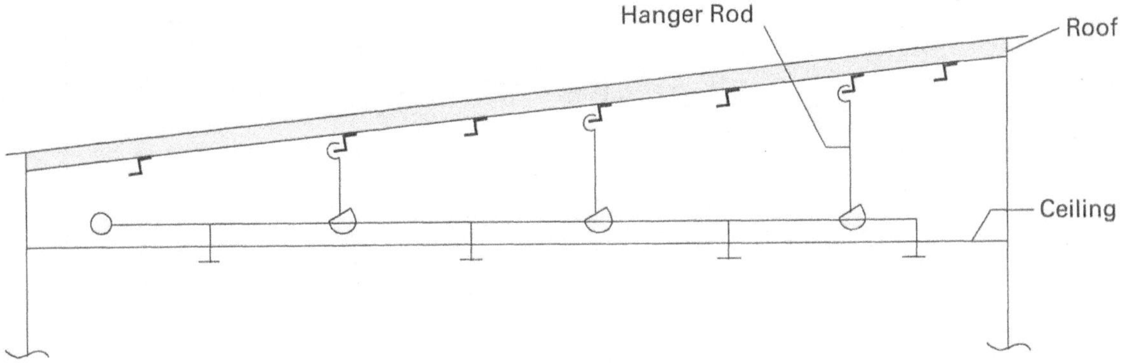

Figure 19 Sloped roof with variable hanger rods.

Example 3: What will the end elevation be for a 100' long dry system branch line if the starting elevation is 18'?

Step 1 The requirement is ½" for every 10' of horizontal run. Divide 100' into 10' sections as follows:

100' ÷ 10' = 10 sections, each with ½" rise (0.5")

Step 2 Multiply the rise by the number of sections:

10 sections × 0.5" per section = 5" rise over 100'

Step 3 Add the rise to the initial elevation to find the end elevation:

18' elevation + 5" rise = 18'-5" end elevation

2.3.2 Calculating the Third Side of a Triangle

The Pythagorean theorem ($a^2 + b^2 = c^2$) provides a way to find any side of a right triangle when the lengths of the other two sides are known (*Figure 20*).

Example 1: What is the length of side c when side a is 3' and side b is 4'?

Step 1 Plug the known values into the equation:

$$a^2 + b^2 = c^2$$
$$(3')^2 + (4')^2 = c^2$$
$$9 \text{ ft}^2 + 16 \text{ ft}^2 = c^2$$
$$25 \text{ ft}^2 = c^2$$

Step 2 To eliminate the exponent, take the square root of each side. The square root of c^2 is c and the square root of 25 is 5:

$$\sqrt{(25 \text{ ft}^2)} = \sqrt{(c^2)}$$
$$5' = c$$

This is sometimes referred to as a *3-4-5 right triangle*. It is helpful when determining whether an angle is a true 90°. Measure 3' up one side and 4' along the bottom. If the hypotenuse or distance between those points is 5' then the angle is 90°.

Example 2: *Figure 21* shows a circumstance where a fitter is required to calculate the true length of a piece of pipe under a pitched roof. Side a is 20' and side b is 2'. The values of a and b can be used to find the value of the hypotenuse (c).

Step 1 Plug the known values into the equation:

$$a^2 + b^2 = c^2$$
$$(20')^2 + (2')^2 = c^2$$
$$(400 \text{ ft}^2) + (4 \text{ ft}^2) = c^2$$
$$(404 \text{ ft}^2) = c^2$$

Step 2 To eliminate the exponent, take the square root of each side:

$$\sqrt{(404 \text{ ft}^2)} = \sqrt{(c^2)}$$
$$20.09975' = c$$

Step 3 Multiply the decimal value by 12 to convert it into inches:

$$0.09975 \times 12'' \text{ per foot} = 1.197$$
$$c = 20'\text{-}1\tfrac{3}{16}''$$

The piping running parallel to the roof will be 20'-1³⁄₁₆" in length.

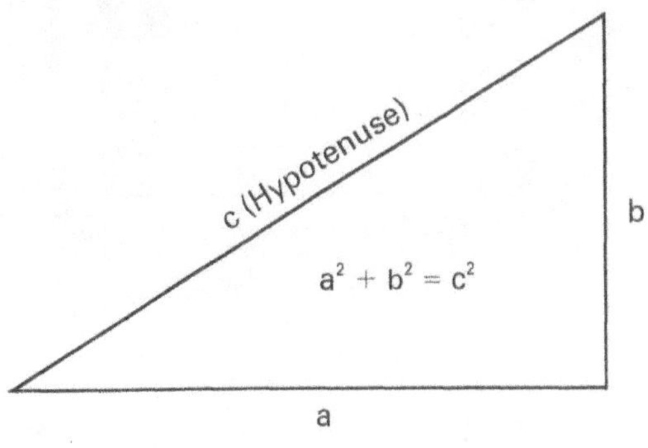

Figure 20 Pythagorean theorem.

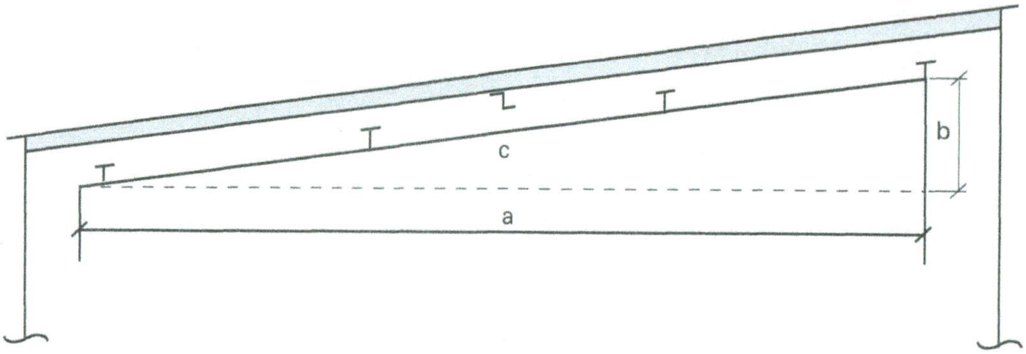

Figure 21 Sloped pipe example.

Did You Know?

Construction calculators are available to simplify such calculations as rise, run, and pitch. They can also be used to solve for equations using square roots, exponents, and more.

Figure Credit: Courtesy of Calculated Industries

2.4.0 Calculating Offsets

Sprinkler fitters commonly use 90° elbows and tees to provide right-angled pipe connections. There are occasions where a 45° elbow is needed. In grooved fittings, there are further divisions of fittings to 22½° and 11¼°. *Figure 22* illustrates the size of these angles.

A 45° triangle is an isosceles triangle, which means that the legs (rise and run) are equal in length. *Figure 23* shows an isosceles triangle. 45° elbows can be used to offset piping elevation or lateral location. They may also be used where 90° elbows are inconvenient or have too much take-out in fitting length.

Two equations can be used with 45° offsets. When the rise or run of an offset is known, multiply it by 1.414 to determine the offset. When the offset is known, multiply it by 0.707 to determine the rise or run.

Example 1: What is the length of a 45° offset if the piping elevation changes from 10' A.F.F. to 13' A.F.F.? *Figure 24* shows this offset.

$$\text{Offset} = \text{rise or run} \times 1.414$$
$$\text{Offset} = 3' \times 1.414 = 4.24' \text{ or } 4'\text{-}3''$$

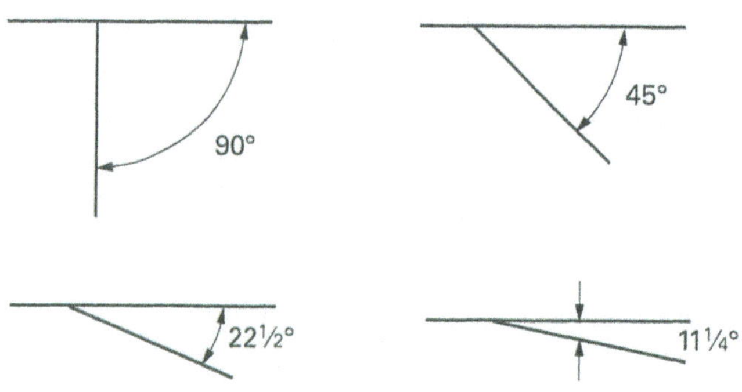

Figure 22 Common angles used in sprinkler work.

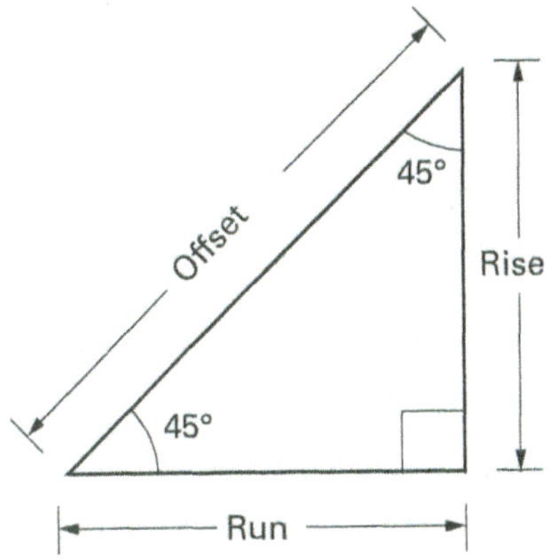

Figure 23 Isosceles triangle.

> **NOTE**
> 0.24 is multiplied by 12 to arrive at nearest whole inches.

Example 2: What would the vertical rise in elevation be if an offset on a 45° angle measured 2'? *Figure 25* shows this offset.

Offset × 0.707 = rise or run
2' × 0.707 = 1.41' or 1'-5"

Example 3: Determine the length of pipe for a 45° offset if the rise equals 5'-0". See *Figure 26.*

Offset = rise or run × 1.414
Offset = 5' × 1.414 = 7.07' or 7'-1"

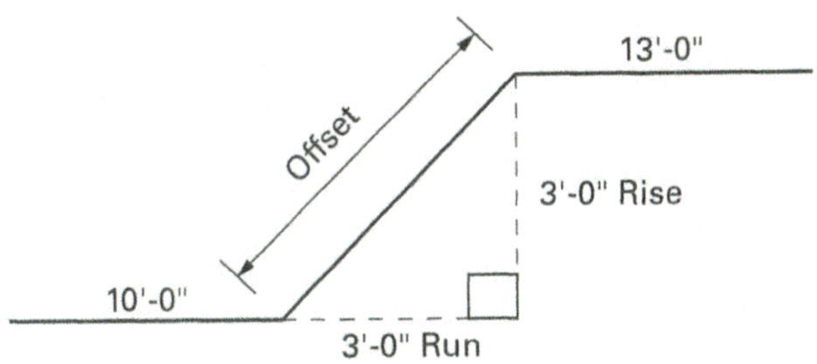

Figure 24 Length of 45° offset.

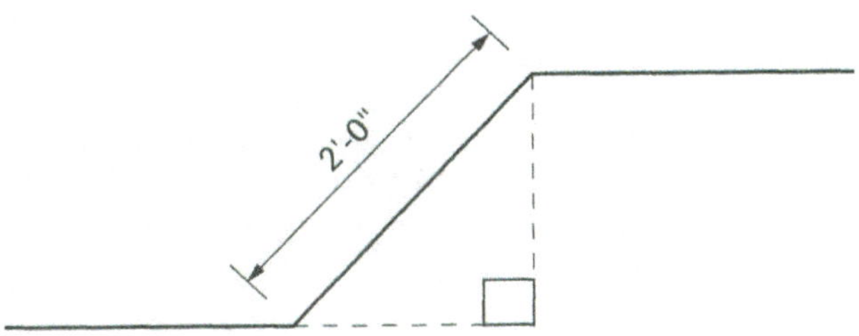

Figure 25 45° offset vertical rise in elevation.

This is the length from center to center of the 45° elbows. To determine the length of pipe between the elbows, subtract the takeout of each fitting.

(7'-1") – (3") – (3") = 6'-7" of 4" pipe

> **NOTE**
> Always read the drawing legend to make sure you understand the meaning of each symbol. The requirements for symbols used on sprinkler shop drawings can be found in NFPA 170, *Standard for Fire Safety and Emergency Symbols.*

Using 45° and 90° elbows in conjunction can aid when offsetting around piping. *Figure 27* shows a typical offset configuration.

2.5.0 Determining Sprinkler Spacing by Area

There are many variables to take into consideration when laying out sprinklers. Protection areas and maximum spacing of standard pendent and upright spray sprinklers for ordinary hazard areas are 130 ft^2 with a maximum spacing of 15'. The minimum spacing for most standard sprinklers is 6'. Note that extended coverage sprinklers or special sprinklers may differ in their spacing requirements.

The following steps must be taken prior to starting a layout:

Step 1 Identify the hazard group as identified on the shop drawings.

Step 2 Identify the maximum area for each hazard group.

Step 3 Commit to spacing between lines without exceeding maximum distances between sprinklers for a given hazard.

Step 4 Commit to distances between sprinklers on a line without exceeding the maximum area of sprinklers for a given hazard.

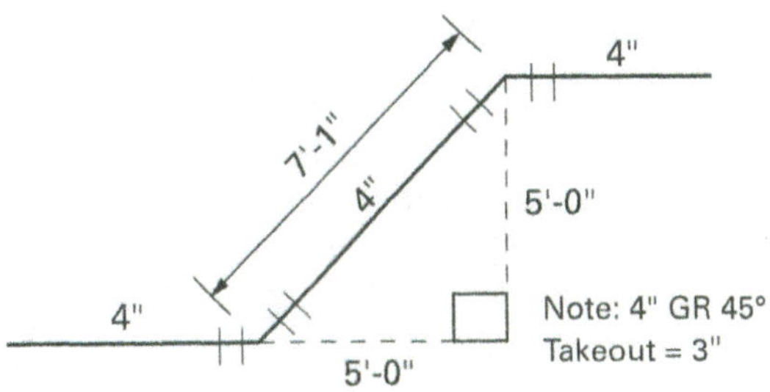

Figure 26 Length of pipe 45° offset.

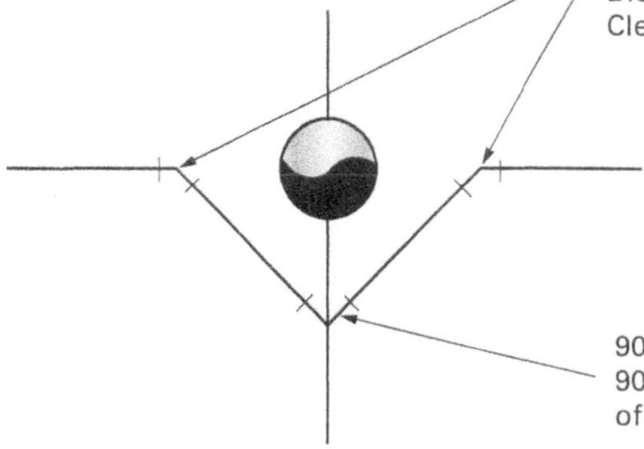

Figure 27 Typical 45° offset configuration.

NFPA 13 dictates the amount of area a sprinkler can cover based on the hazard to which it is exposed. The greater the hazard, the closer the spacing and the smaller the area the sprinkler is allowed to protect.

Sprinkler coverage area is determined by multiplying the distance between sprinklers on a line by the distance between lines. *Figure 28* shows the sprinkler coverage area for a 24' × 40' space.

The area of coverage for each sprinkler shown in *Figure 28* is 120 ft^2 (10' × 12' = 120 ft^2).

2.5.1 Coverage Next to Walls

When determining the coverage area of sprinklers next to walls or steel columns, NFPA 13 states that twice the distance off a wall or the distance between sprinklers in the same direction (such as in line), whichever is greater, must be used. *Figure 29* shows an ordinary hazard tree system.

The sprinklers with circles around them are covering exactly 130 ft^2, as follows:

Area of coverage = twice the distance off the wall × distance between in-line sprinklers
Area of coverage = (6'-6" × 2) × 10'-0"
Area of coverage = 13'-0" × 10'-0" = 130 ft^2

2.5.2 Coverage Away from Walls

The sprinklers without circles around them in *Figure 29* are covering exactly 120 ft^2, as follows:

Area of coverage = distance between sprinklers on a line × distance between sprinkler lines
Area of coverage = 12'-0" × 10'-0" = 120 ft^2

The maximum allowable area of coverage for a hazard will dictate the maximum distance from a wall or column, not the maximum allowable distance between sprinklers in a hazard. This means if the sprinkler line is located 7'-6" away from a wall in an ordinary hazard area, the greatest distance between sprinklers on that line would be 8'-8" for a maximum area of coverage per sprinkler of 130 ft^2. *Figure 30* shows the maximum area of coverage off a wall.

2.5.3 Sprinkler Layout

Figure 31 illustrates an area of ordinary hazard occupancy. Choosing to spot the lines 10' on center establishes a commitment to a maximum spacing between sprinklers on a line of 13'.

130 ft^2 ÷ 10' between lines = 13' between sprinklers on a line

Typically, in sprinkler installations, designers will plan as few lines per bay as possible. With three lines shown on *Figure 31*, the fewest number of sprinklers to cover the area is 27. *Figure 32* illustrates the same ordinary hazard area as in *Figure 31* but with fewer lines. In this case, 28 sprinklers will be required.

2.5.4 Sprinkler Layout with Obstructions

When a sprinkler is located near an obstruction, such as a column, duct, pipe, soffit, or even an emergency lighting fixture, it must be positioned at a specific distance to provide adequate coverage despite the obstruction. NFPA 13 provides detailed rules for dealing with obstructions. Two of the rules for standard pendent and upright spray sprinklers are as follows:

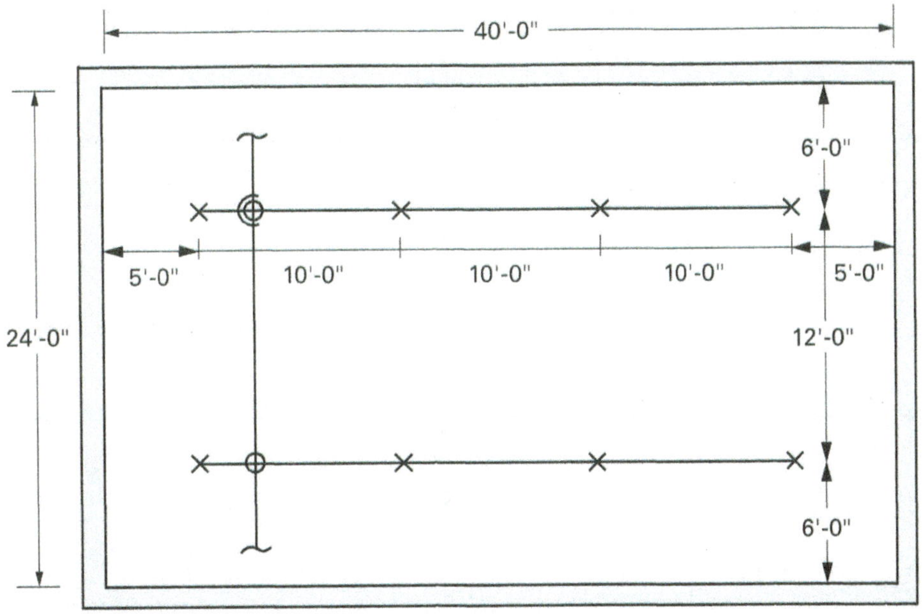

Figure 28 Sprinkler coverage area.

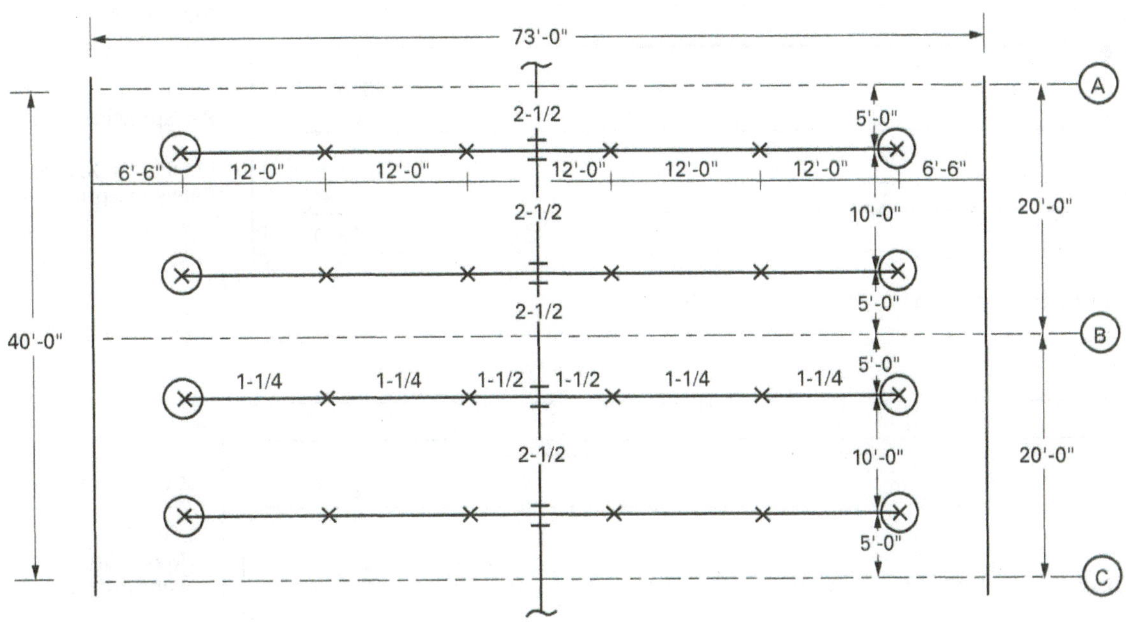

Figure 29 Ordinary hazard tree system.

- *Obstruction against a wall* — See *Figure 33*. The distance between a sprinkler and an obstruction (A) is equal to (D – 8") + B, where D is the depth of the obstruction (maximum 30") and B is the distance between the bottom of the obstruction and the surface of the deflector.
- *Obstruction not against a wall* — The distance between a sprinkler and an obstruction must be at least three times the largest dimension of the obstruction, with a maximum distance of 24".

> **NOTE**
> The above obstruction rules do not apply if the top of the obstruction is level with or above the plane of the sprinkler deflector. Other rules apply to larger obstructions and those in different locations. Refer to NFPA 13 for details.

Example 1: A mechanical soffit located against a wall is 30" wide and 12" deep with a deflector 2" from the ceiling. How far away from the edge of the soffit must a sprinkler be located to protect underneath it?

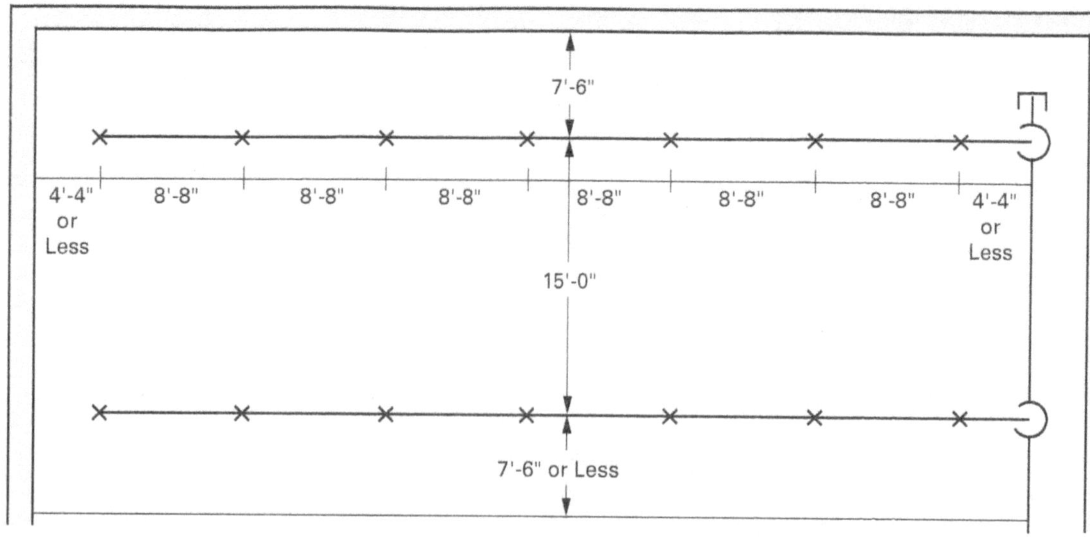

Figure 30 Maximum area of coverage off a wall.

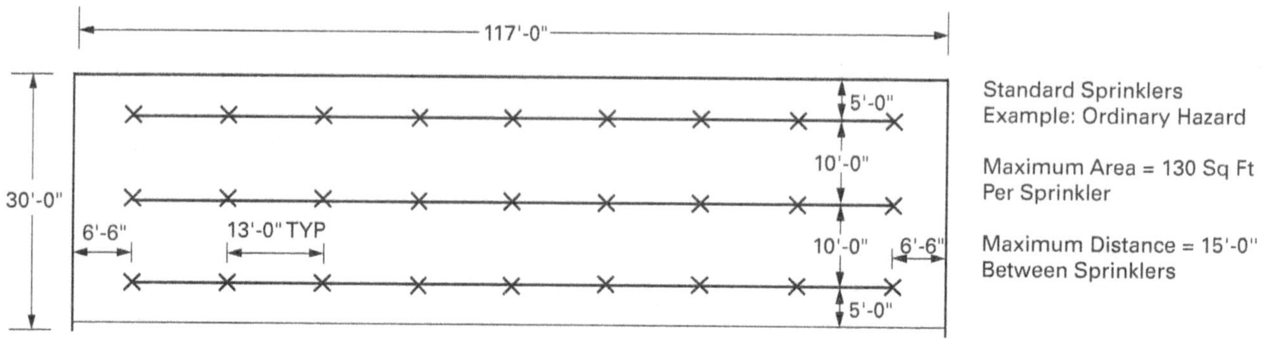

Figure 31 Ordinary hazard area.

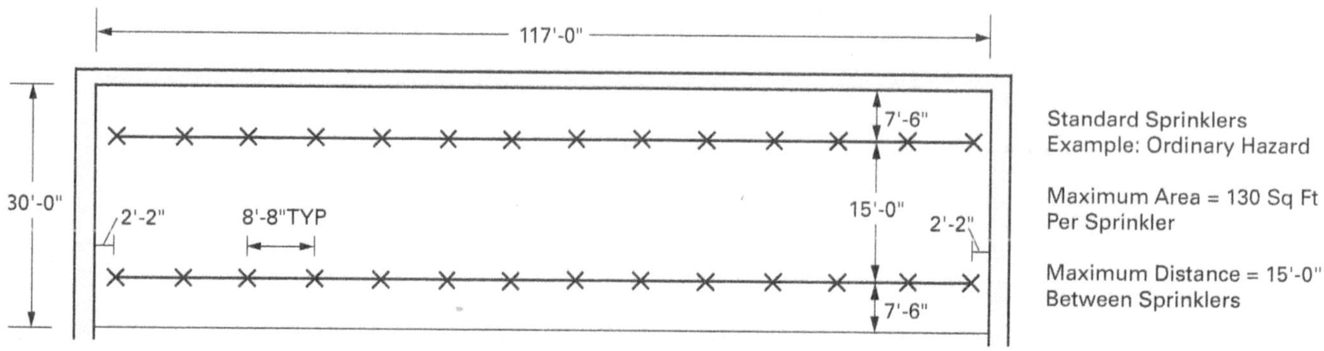

Figure 32 Ordinary hazard area with fewer lines.

Minimum distance = (D – 8") + B
Minimum distance = (30" – 8") + (12" – 2" deflector)
Minimum distance = (22") + (10") = 32"

This sprinkler must be located at least 32" from the soffit to provide adequate coverage below the obstruction.

Example 2: You are installing pendent sprinklers and notice there is a 4" by 6" column in line with the sprinklers that is not indicated on the shop drawings. What is the required distance from this obstruction?

Minimum distance = maximum dimension of obstruction × 3
Minimum distance = 6" × 3 = 18"

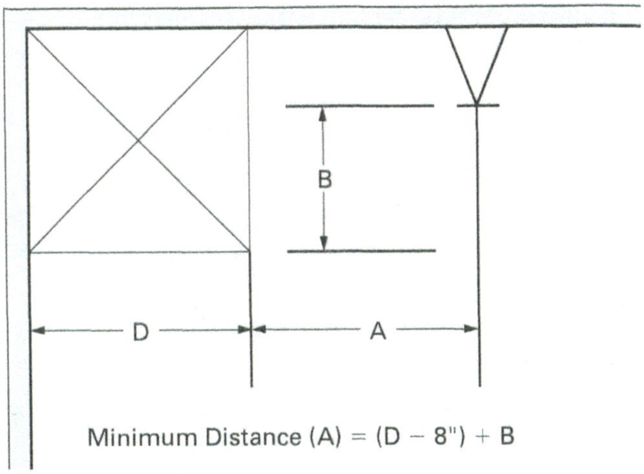

Figure 33 Obstruction against a wall.

This sprinkler must be located at least 18" from the column to provide adequate coverage beyond the obstruction.

Example 3: You are relocating pendent sprinklers in an office remodel and realize there is a 10" by 12" exit light located near the next sprinkler. What is the required distance from this obstruction?

Minimum distance = maximum dimension of obstruction × 3
Minimum distance = 12" × 3 = 36"

The calculated distance of 36" exceeds the maximum distance of 24"; therefore, this sprinkler will be located 24" from the obstruction.

2.6.0 Making Pressure Calculations

Pressure is the amount of force applied to a given unit of surface area and is often expressed in *pounds per square inch (psi)*. *Figure 34* shows a pressure gauge calibrated in psi. Most pressure gauges such as this one do not include the gravitational force of 14.7 psi produced by the atmosphere. That is, the gauge reads zero at 14.7 psi. This is sometimes referred to as *pounds per square inch gauge* or *psig*. If the gauge reads in psi, the measurement is assumed to be psig.

On occasion, you might see the term *psia*. This means *pounds per square inch absolute* and includes the gravitational force of 14.7 psi. Expressed as an equation:

psia = gauge pressure + 14.7 psi

Most sprinkler applications are only concerned with gauge readings in psi (psig).

2.6.1 Reading Pressure Gauges

The gauge shown in *Figure 34* includes numbers for every 50 psi and then has lines to indicate the values between those amounts. The shorter lines indicate units of 5 psi and the longer lines indicate units of 10 psi. With this in mind, what is the reading on this gauge? This gauge reads from left to right, so start at the first number behind the needle (150) and add 10 for each longer line between that number and the needle. In this case, there are two long lines, which takes you from 150 +10 +10 = 170. There is one short line before the needle, which means you need to add another 5 (170 + 5 = 175). This gauge reads 175 psi.

2.6.2 Calculating Head Pressures

Another way that pressure is measured in the sprinkler industry, particularly with fire pumps, is in feet of head. Head pressure is also a conversion to elevation. This system is based on the fact that a column of water 1' high will produce a pressure of 0.433 psi. This is known as the **static pressure**. Correspondingly, a 100' column of water will produce a pressure of 43.3 psi. The **factor** used to convert elevation to pressure is 0.433 psi/ft multiplied by each foot of vertical rise.

As mentioned, rated pressures for fire pumps are sometimes indicated in terms of head. This determines to what elevation the fire pump can supply water. This is something to be aware of when hydrostatically testing a system at an extreme height.

Figure 34 Standard gauge dial calibrated in pounds per square inch.

Pressure required for elevation is a fixed equation. Elevation pressure is not controlled by the number of fittings installed on a system. Elevation or head pressure is controlled by vertical rise, from either supply or demand.

Example 1: A building is 18' tall. What pressure is required to overcome the elevation or head pressure?

Head pressure = elevation × 0.433 psi/ft
Head pressure = 18' × 0.433 psi/ft = 7.8 psi

Example 2: In some water systems, water is supplied from an elevated tank or from a tank much higher than the area it supplies. *Figure 35* shows a water storage tank that is supplying water to two buildings at different elevations below it.

Building A has an elevation that is 100' lower than the water tank. The head or elevation pressure supplied to Building A would be:

100' × 0.433 psi/ft = 43.3 psi

Building B has an elevation that is 200' lower than the water tank. The head or elevation pressure supplied to Building B would be:

200' × 0.433 psi/ft = 86.6 psi

Example 3: Finding the static head at the base of a tank is sometimes necessary and can be done using the same calculation. Find the static pressure of the 29.5' tall tank in *Figure 36*:

Head pressure = elevation × 0.433 psi/ft
Head pressure = 29.5' × 0.433 psi/ft = 12.77 psi

Did You Know?

Because head pressure works out to be just under ½ psi per foot (0.433), sprinkler fitters sometimes apply a rough estimate of ½ psi per foot when determining head pressure.

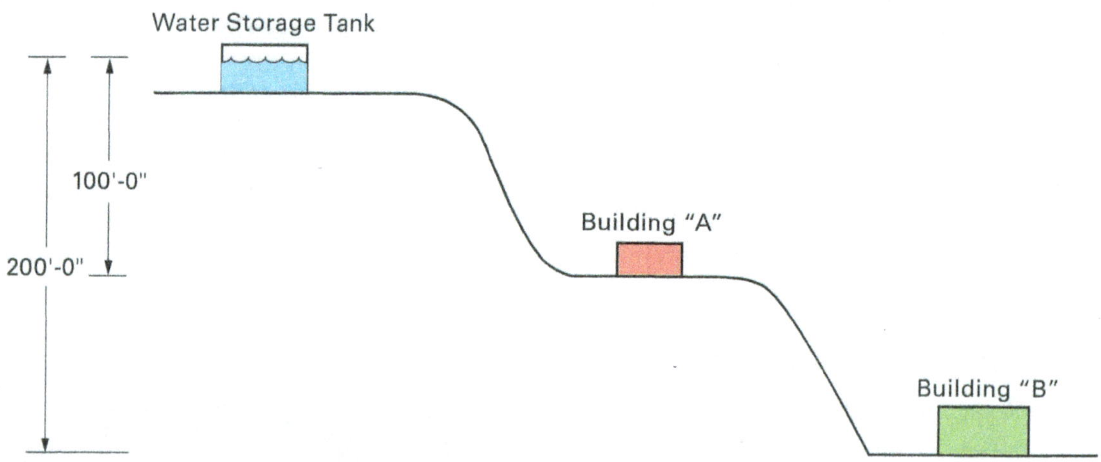

Figure 35 Water storage tank supplying water to two buildings.

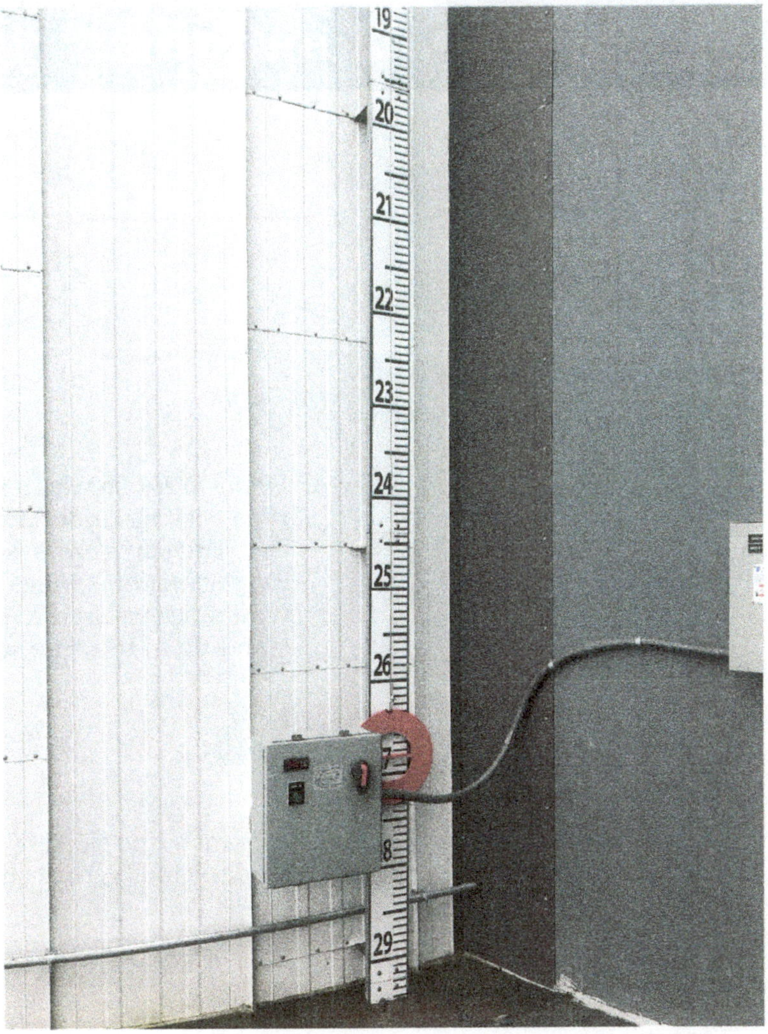

Figure 36 Water storage tank.

2.0.0 Section Review

1. What is the floor area of a room with dimensions of 10' by 15'? ______________
2. What is the volume of a room that is 25' by 20' by 10'? ______________
3. What is the pitch with a rise of 2' and a run of 10'? ______________
4. What is the length of a 45° offset if the piping elevation changes from 12' A.F.F. to 14' A.F.F.? ______________
5. The distance between a sprinkler and the wall is 6'-6" and the distance between sprinklers is 12'. Per NFPA 13, which distance is used when determining the sprinkler coverage area? ______________
6. A building is 40' tall. What pressure is required to overcome the elevation or head pressure? ______________

Review Questions

1. $4\frac{1}{2} + 2\frac{1}{4} + 7\frac{1}{2} =$ _____.
   a. 14
   b. $14\frac{1}{4}$
   c. $14\frac{1}{2}$
   d. 15

2. What is the decimal equivalent of 12'-6"?
   a. 11.25'
   b. 11.50'
   c. 12.50'
   d. 12.80'

3. $4^{0.5}$ is equal to _____.
   a. 2
   b. 4
   c. 8
   d. 16

4. What is the floor area, in square feet, of a room that is 10'-0" × 22'-0"?
   a. 220 ft^2
   b. 840 ft^2
   c. 1,760 ft^2
   d. 3,036 ft^2

5. A location that would require an auxiliary drain per NFPA 13 is _____.
   a. the high point in a dry pipe sprinkler system
   b. a straight piping section of 8' in length between two pendent sprinklers
   c. a change in piping direction that prevents drainage through the main drain valve
   d. within 6" of the main drain valve on a horizontal run

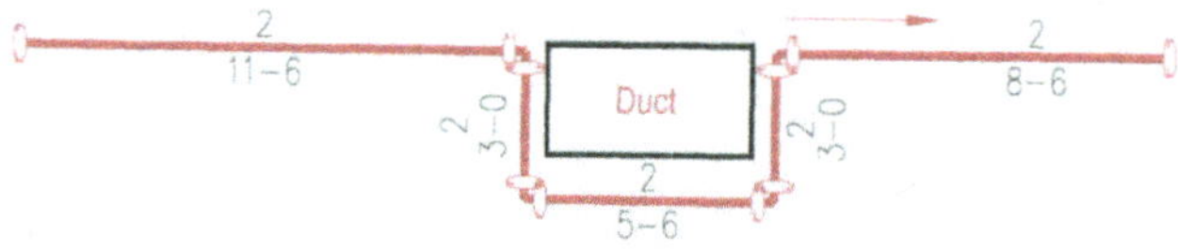

Figure Credit: AFSA Technical Department

Figure RQ01

6. *Figure RQ01* shows 2" Schedule 10 black steel pipe passing under a rectangular duct in a dry pipe sprinkler system. Given a capacity per foot of 0.190 gallons, what size drain would be required to prevent water from standing in the pipe surrounding the duct?
   a. $\frac{1}{2}$"
   b. $\frac{5}{8}$"
   c. $\frac{3}{4}$"
   d. 1"

7. A branch line slope is $\frac{1}{2}$" per 10' over a distance of 100'. What is the change in elevation?
   a. 2"
   b. 3"
   c. 4"
   d. 5"

8. If a right triangle has two sides equal to 3' and 4' respectively, how long is the hypotenuse?
   a. 5'
   b. 7'
   c. 10'
   d. 25'

9. What is the equation to find the 45° offset when the rise or run is known?

   a. Offset = rise or run × 1.414
   b. Offset = rise or run × 0.707
   c. Offset = rise – run
   d. Offset = rise × run × 0.707

10. If a 45° offset is 7', the rise or run is approximately ______.

    a. 4'-5"
    b. 4'-11$\frac{1}{2}$"
    c. 6'-5"
    d. 9'-10$\frac{3}{4}$"

11. The *maximum* sprinkler spacing for an ordinary hazard system is ______.

    a. 8'
    b. 12'
    c. 15'
    d. 18'

Figure RQ02

12. What is the reading on the pressure gauge shown in *Figure RQ02*?

    a. 1 psi
    b. 2 psi
    c. 5 psi
    d. 10 psi

13. Elevation (head) pressure is dependent on ______.

    a. pipe size
    b. vertical rise
    c. volume
    d. number of piping bends

14. The formula for calculating head pressure is ______.

    a. elevation in feet ÷ 0.433 psi/ft
    b. rise × 0.707 psi
    c. rise × 1.414 psi
    d. elevation in feet × 0.433 psi/ft

15. What is the static head at the base of a 10' tall ground level tank?

    a. 1.08 psi
    b. 2.16 psi
    c. 4.33 psi
    d. 8.66 psi

Trade Terms Introduced in This Module

Above the finished floor (A.F.F.): A measurement taken from the surface of the installed flooring (e.g., tile), rather than the underlayment.

Decimal equivalents: Numbers that represent fractions, with every decimal place indicating a multiple of 10.

Exponents: Numbers or symbols, such as the 2 in 10 ft^2, that indicate the number of times an expression is multiplied by itself.

Factor: An amount by which a value is multiplied to produce a specific result.

Hypotenuse: The longest side of a triangle that contains a right angle.

Mathematical operator: A symbolic instruction to perform a specific type of calculation. The four basic mathematical operators are addition (+), subtraction (−), multiplication (×), and division (÷).

Pi: A mathematical constant defined as the ratio of a circle's circumference to its diameter. Pi is equal to approximately 3.1416 or $^{22}/_{7}$ and is represented by the symbol π.

Pressure: The amount of force applied to a given unit of surface area. Pressure is often expressed in *pounds per square inch (psi).*

Scale: A ratio representing the relationship between drawing dimensions and actual dimensions. Sprinkler shop drawings are often drawn to a scale of $^{1}/_{8}$" = 1'.

Static pressure: A measurement of the pressure of still water. A column of water that is 1' high will produce a static pressure of 0.433 psi.

Additional Resources

This module presents thorough resources for task training. The following reference material is suggested for further study.

NFPA 13, *Standard for the Installation of Sprinkler Systems*, Latest Edition. Quincy, MA: National Fire Protection Association.

NFPA 20, *Standard for the Installation of Stationary Pumps for Fire Protection*, Latest Edition. Quincy, MA: National Fire Protection Association.

NFPA 170, *Standard for Fire Safety and Emergency Symbols*, Latest Edition. Quincy, MA: National Fire Protection Association.

Figure Credits

iStock@YanLev, Module Opener

Chuck Bamford, Figures 1–2

Image property of Stanley Black & Decker. Used with permission., Figure 9

AFSA Technical Department, Figure 18

Reproduced with permission of NFPA from NFPA 13, *Standard for the Installation of Sprinkler Systems*, 2019 edition. Copyright© 2018, National Fire Protection Association. For a full copy of NFPA 13, please go to www.nfpa.org., Table 2

Section Review Answer Key

Section 1.0.0

Answer	Section Reference	Objective
1. Leftover pipe = 21' – 7' = 14'	1.1.0	1a
2. Convert the inches into decimal feet: 9" ÷ 12" per foot = 0.75' Add the equivalent values: 1' + 0.75' = 1.75'	1.2.2	1b
3. Convert the inches into decimal feet: 6" ÷ 12" per foot = 0.50' Add the equivalent values: 6' + 0.50' = 6.50' Multiply: 6.50' × 6.50' = 42.25 square feet	1.3.1	1c
4. Divide the diameter by two to find the radius: 36" ÷ 2 = 18" Calculate the area of the circle: A = πR squared A = (3.1416) × (18 square inches) A = (3.1416) × (324 square inches) = 1,017.88 square inches (rounded to two decimal places)	1.4.1	1d

Section 2.0.0

Answer	Section Reference	Objective
1. Area = length × width Area = 10' × 15' = 150 square feet	2.1.0	2a
2. Volume = length × width × height Volume = 25' × 20' × 10' = 5,000 cubic feet	2.2.1	2b
3. Pitch = rise ÷ run Pitch = 2' ÷ 10' = 0.2	2.3.1	2c
4. Offset = rise or run × 1.414 Offset = 2' × 1.414 = 2.83' or 2'-10"	2.4.0	2d
5. NFPA 13 states that twice the distance off a wall or the distance between sprinklers in the same direction (such as in line), whichever is greater, must be used. 6'-6" × 2 = 13' 13' is larger than 12' and will therefore be used.	2.5.1	2e
6. Head pressure = elevation × 0.433 psi/ft Head pressure = 40' × 0.433 psi/ft = 17.32 psi	2.6.2	2f

User Update

NCCER makes every effort to keep its textbooks up-to-date and free of technical errors. We appreciate your help in this process. If you find an error, a typographical mistake, or an inaccuracy in NCCER's curricula, please submit a User Update form by visiting **https://www.nccer.org/olf**. You can also scan the QR code using the camera on your phone or mobile device to access the form.

Shop Drawings

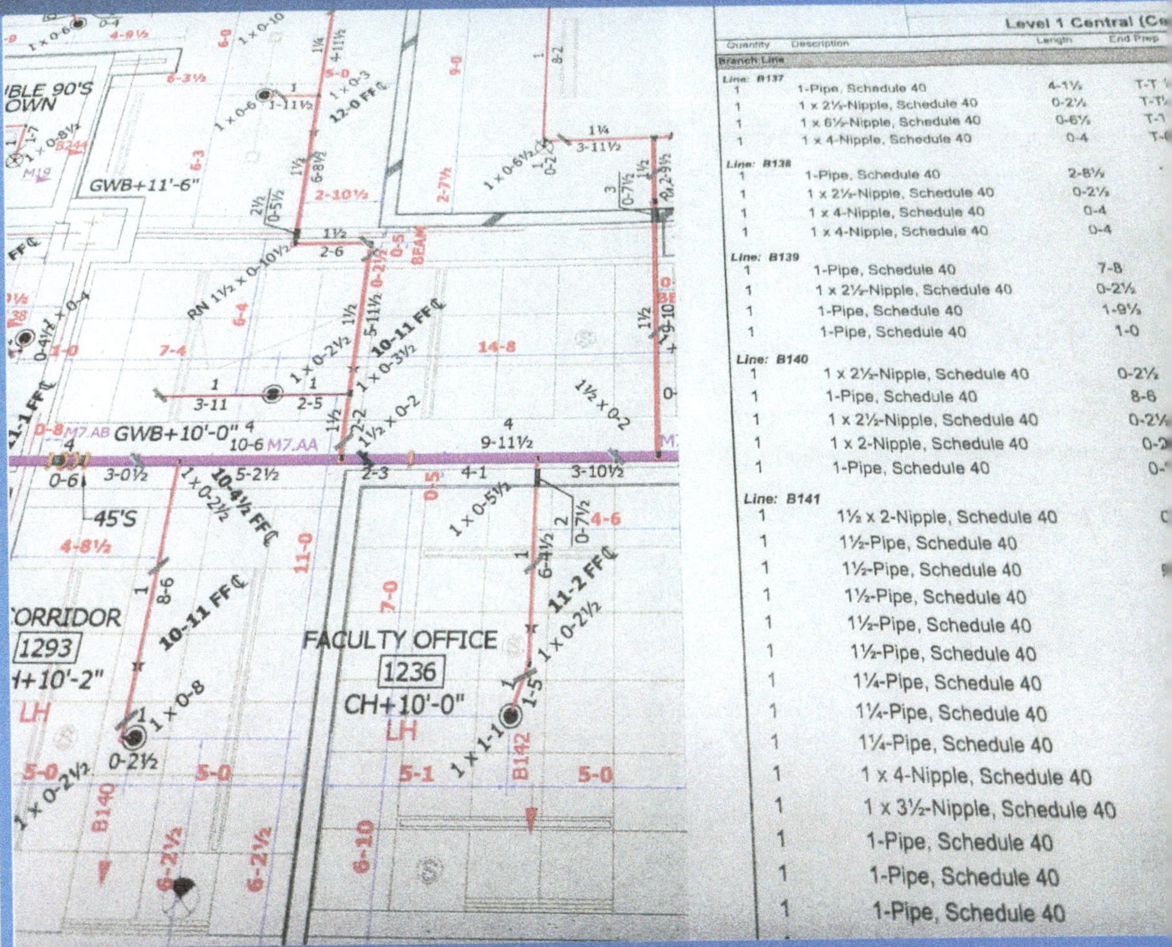

Overview

This module covers the shop drawings used in sprinkler fitting. It describes the symbols used for various sprinkler system components and explains how to use a shop drawing to identify the types and locations of piping and components. It also provides an overview of Building Information Modeling (BIM).

Module 18204

18204

Shop Drawings

Objectives

Successful completion of this module prepares you to do the following:

1. Identify the types of drawings used in sprinkler system installation.
   a. List the types of drawings contained in a typical drawing package.
   b. Identify the information that can be found in a title block, notes, and drawing legend.
   c. Describe the purpose and content of sprinkler system specifications.
   d. Point out the NFPA requirements for shop drawings that are applicable to fitters.
2. Identify common symbols used in sprinkler system shop drawings.
   a. Identify typical sprinkler symbols.
   b. Identify common valve and device symbols.
   c. Describe the symbols used to show fittings, hangers, and bracing.
   d. Identify the symbols for rises, drops, plugs, and hydraulic calculation references.
   e. Point out various components and dimensions on a sample shop drawing.
3. Identify the advantages of Building Information Modeling (BIM).
   a. Define BIM 4D through 7D.
   b. Describe how 3D scanning is used in renovation work.

Performance Task

Under supervision, you should be able to do the following:

1. Use an instructor-supplied shop drawing to identify various types of information, including the sprinkler legend, title block, north arrow, sprinkler type and temperature, and pipe size.

Trade Terms

Building Information Modeling (BIM)
Column
Elevation view
General notes
Grade beam
Keyed notes
Legend
Plan view
Reflected ceiling plan
Revision block
Site plan
Sprig
Sprinkler identification number (SIN)
Symbol
Title block

Industry Recognized Credentials

If you are training through an NCCER-accredited sponsor, you may be eligible for credentials from NCCER's Registry. The ID number for this module is 18204. Note that this module may have been used in other NCCER curricula and may apply to other level completions. Contact NCCER's Registry at 1.888.622.3720 or go to **www.nccer.org** for more information.

You can also show off your industry-recognized credentials online with NCCER's digital badges. Transform your knowledge, skills, and achievements into badges that you can share across social media platforms, send to your network, and add to your resume. For more information, visit **www.nccer.org**.

NOTE

Unless otherwise specified, references in parentheses following table numbers refer to NFPA 13.

Contents

Figures and Tables

SECTION ONE

1.0.0 DRAWING TYPES

Objective

Identify the types of drawings used in sprinkler system installation.

a. List the types of drawings contained in a typical drawing package.
b. Identify the information that can be found in a title block, notes, and drawing legend.
c. Describe the purpose and content of sprinkler system specifications.
d. Point out the NFPA requirements for shop drawings that are applicable to fitters.

Trade Terms

Column: A round structural member used to support beams and girders where loadbearing walls do not exist.

Elevation view: Usually a two-dimensional view on a vertical plane showing internal and external dimensions of buildings. This view helps sprinkler fitters determine roof contours and floor levels. This is sometimes called a *sectional view*.

General notes: A written list on a drawing that provides information such as applicable standards, types of pipe, and fittings.

Grade beam: The structural beam forming the perimeter of the building and used as the ultimate reference point for all construction dimensions.

Keyed notes: A system by which numbered notes are assigned to associated numbers on a drawing to provide detailed information without crowding the drawing itself.

Legend: A list of symbols and associated definitions used on a drawing.

Plan view: A two-dimensional layout or drawing of a building.

Reflected ceiling plan: An architectural term used to describe a plan view of a ceiling that typically shows the lights, fixtures, diffusers, and sprinklers within a given area.

Revision block: A box or table on a drawing that lists the date of every revision to the drawing.

Site plan: A bird's-eye view of a building in small scale.

Sprinkler identification number (SIN): A unique manufacturer identification number for every sprinkler type, change in K-factor or shape, deflector characteristic, pressure rating, or thermal sensitivity.

Symbol: A written or printed character or sign representing a drawing element.

Title block: A boxed area on a drawing that shows the installation location, address, contractor, owner, occupant, and other identifying information as required by NFPA 13.

Construction drawings are used to represent a structure or system before it is built. A sprinkler fitter's work revolves around the approved shop drawings. These are the documents prepared by a sprinkler system layout technician to communicate the specific details of a sprinkler system installation.

A layout technician is an individual who has achieved certification in the appropriate subfield and has the knowledge, experience, and skills necessary to lay out sprinkler systems. Layout technicians use the engineering design documents, including system design drawing(s), specifications, and nationally recognized codes and standards, to prepare shop drawings in accordance with the engineering design or as otherwise permitted by state regulations. They also perform supplemental calculations based on the engineering design.

While some layout technicians still prefer hand drawings, most construction drawings today are created by architects, engineers, and layout technicians in computer-aided drafting (CAD) software (*Figure 1*). CAD software can be used to create both two-dimensional (2D) and three-dimensional (3D) models of a building and its components. These models are used to generate and print all of the drawings in a drawing set.

There are many types of drawings that a sprinkler fitter must know how to read. These include the following:

- *Sprinkler shop drawings* — These drawings include the piping, sprinklers, fittings, valves, pumps, and other devices used in the installation. They also include the structural features of a building needed to clarify the requirements for installation.

- *Other trade drawings* — There will be times when sprinkler fitters must coordinate with other trades to review their drawings and resolve space conflicts. There is nothing written in any code book that requires sprinkler piping to be installed around other installation work; however, it is written into some specifications. One common rule is that where conflicts exist, the larger system remains where it is, and the smaller system moves. For example, with conflicting HVAC, electrical, and sprinkler installations, the sprinkler piping would move around the larger ductwork, but the smaller electrical conduit would move around the sprinkler system.
- *Architect's drawing packages* — There may be occasions on the job when it is necessary to review the entire drawing package provided by the project architect. This package includes all of the drawings prepared by the architect and engineers on the project.

1.1.0 Sprinkler Drawing Packages

Sprinkler shop drawings provide very detailed information. This information is provided for several purposes. First, these drawings permit the Authority Having Jurisdiction (AHJ) to review and approve the layout. Second, these drawings provide information to substantiate product Listings. Finally, the drawings provide the fitter with the necessary information to install the system as designed by the layout technician.

There are three elements to a sprinkler drawing package. These are the different drawing views of the jobsite. There must be a site plan that shows an overhead view of the building on the site. There must also be a plan view that displays the bulk of the information, and an elevation view, which shows the building dimensions.

1.1.1 Site Plan

The site plan shows a bird's-eye view of the building and the overall site in a small scale. *Figure 2* shows a portion of a typical site plan. To be useful, it must include the following:

- Building location relative to streets
- Water supply mains, fire hydrants, and valves
- Directional arrow (north arrow)
- Water supply information
- Title block
- Sprinkler contractor's name and address
- General contractor's name and address

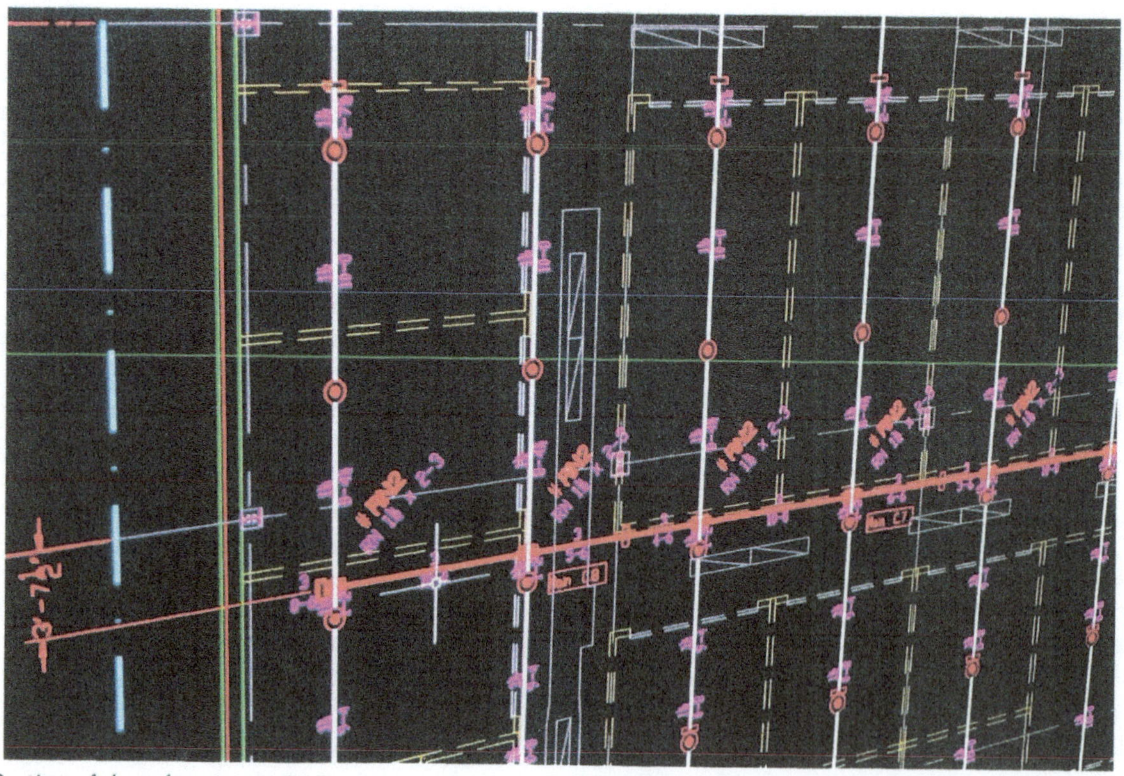

Figure 1 Portion of shop drawing in CAD.

The site plan may be on the same drawing sheet with the plan view, or it may be on a separate sheet, depending upon the size of the project.

1.1.2 Plan Views

Plan views include floor, ceiling, and roof plans. A typical floor plan is shown in *Figure 3*. These drawings are the easiest to understand because they present, in reduced scale, what would be seen if looking down from the top of the structure. On the other hand, these drawings may be the most difficult to read because they contain so many details.

Sprinkler system shop drawings are usually based on the reflected ceiling plan (*Figure 4*). This eliminates much of the clutter seen on floor plans and contains most of the information required for a sprinkler system layout. A reflected ceiling plan shows an overhead plan view of a ceiling. Ceiling plans show tile layouts, light fixtures, sprinkler locations, diffuser locations, speaker locations, etc., all related to the structural reference points and the floor plan.

Figure 5 shows part of a typical shop drawing. Some of the information found on a complete shop drawing includes the following:

- Title block
- North arrow
- General notes
- Keyed notes
- Building walls, exterior and interior, to produce a floor plan
- Structural members to identify support requirements, giving type of roof and ceiling construction
- Superimposed ceiling grid (if applicable), lights, and air conditioning equipment locations to identify ceiling interferences
- Sprinkler locations in all areas spaced according to requirements of the applicable standards
- Location of the riser
- Routing of branch lines, cross mains, and feed mains, giving dimensions of the pipe centerlines from columns, grade beams, permanent walls, and roof decks
- Pipe sizes
- Cut dimensions for the pipe
- Center-to-center dimensions for pipe (where applicable)
- Cut dimensions on riser nipples
- Number designations on branch lines and letter designations on all mains
- Riser detail
- Inspector's test connection on each system including auxiliary dry systems
- Fire Department Connection (FDC) on each building
- Alarm provisions for each system

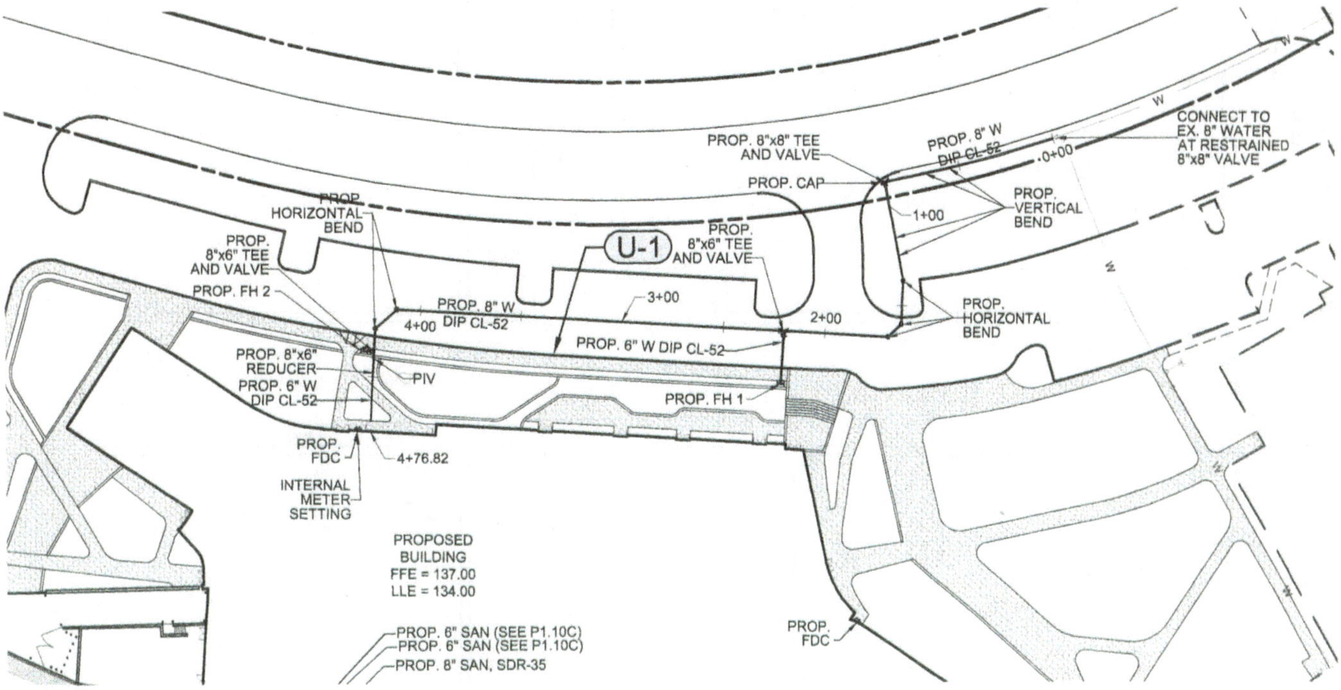

Figure 2 Portion of a typical site plan.

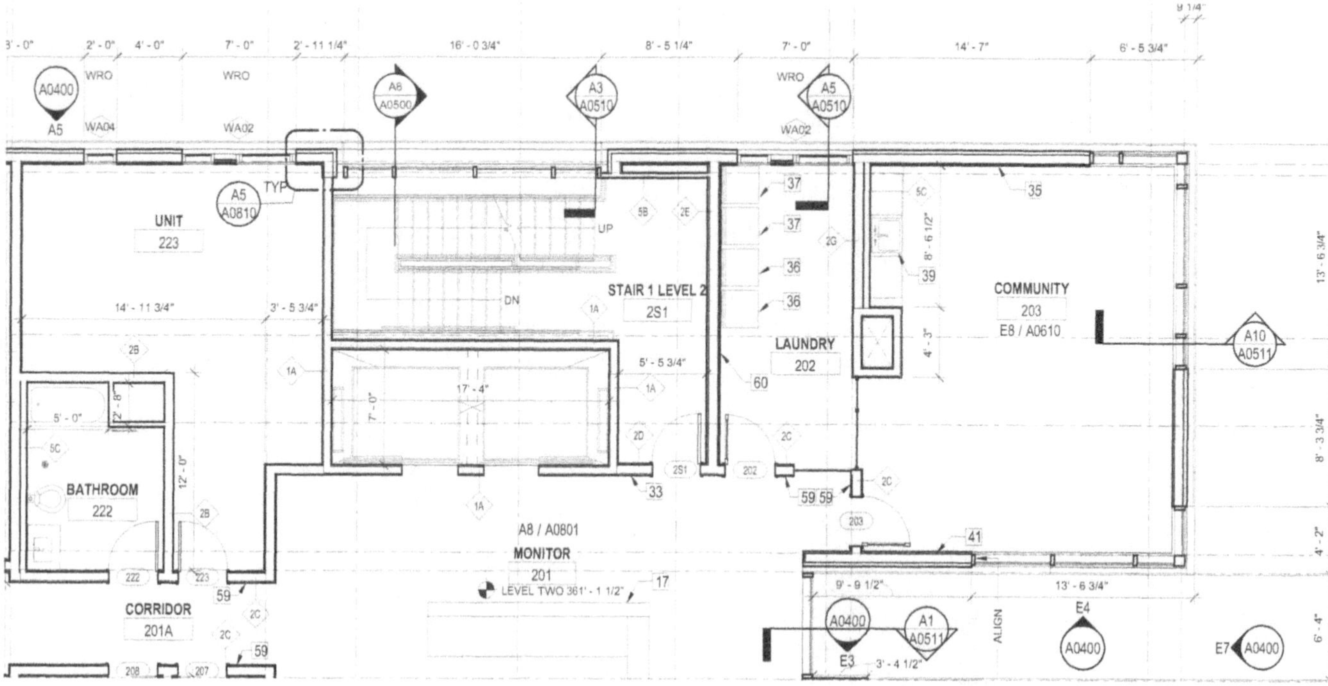

Figure 3 Portion of a typical floor plan.

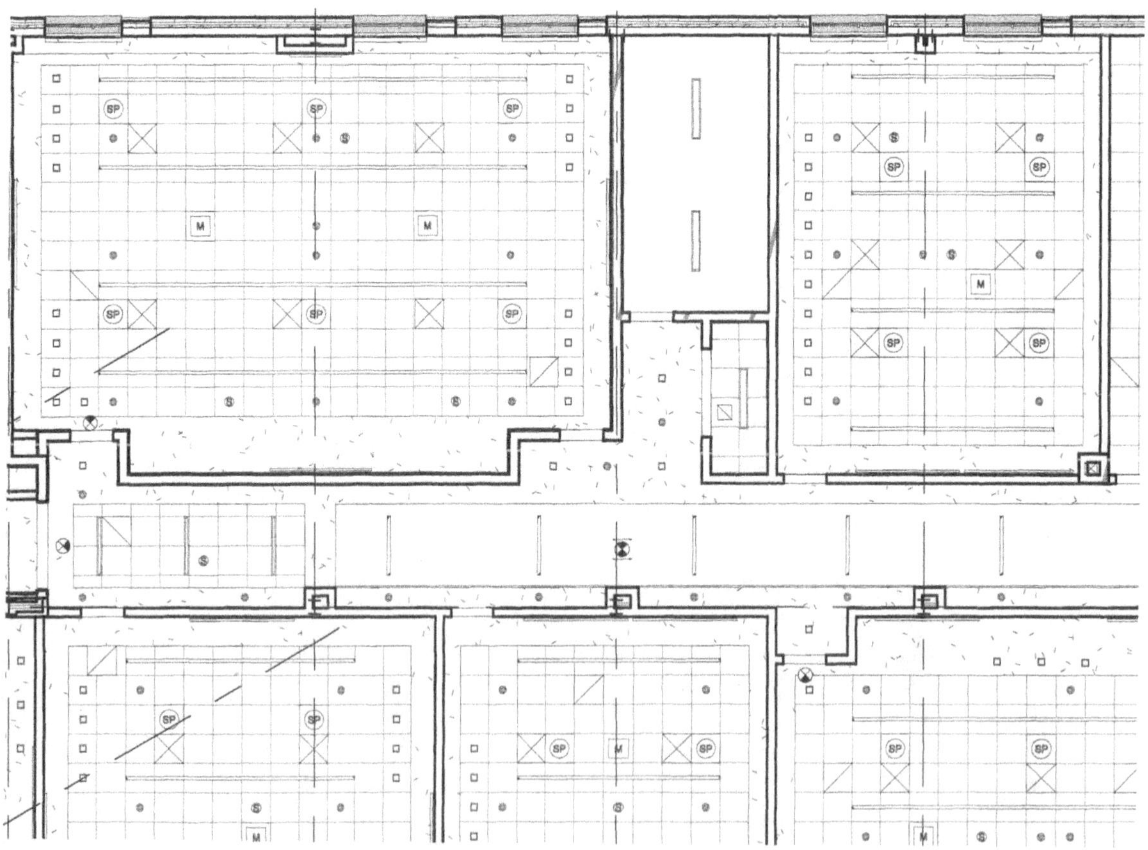

Figure 4 Portion of a reflected ceiling plan.

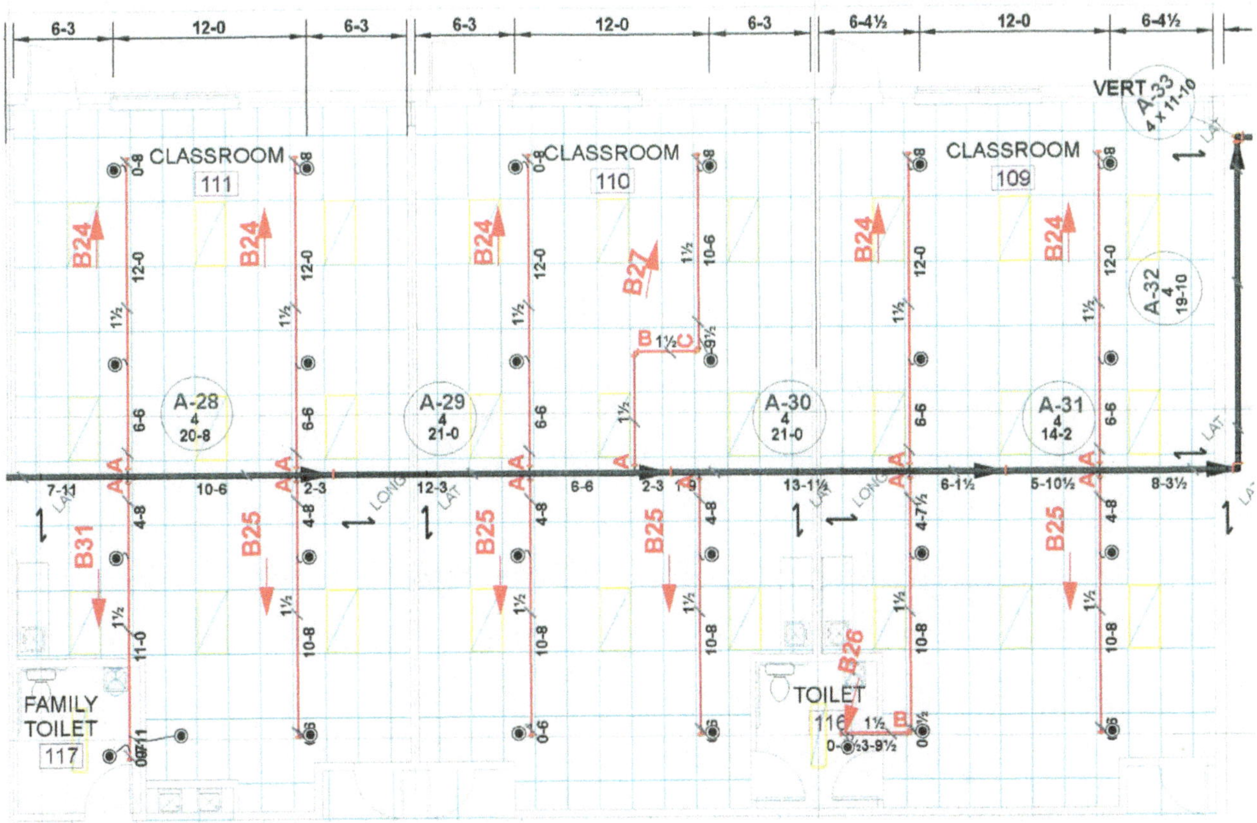

Figure 5 Portion of a typical shop drawing.

1.1.3 Elevation Views

An elevation view is usually a two-dimensional view on a vertical plane showing the building dimensions. This view helps sprinkler fitters determine roof contours and floor levels. An elevation cross section of the building is required in every sprinkler drawing set. Roof contours and floor level information can be conveyed by notations on the plan view. *Figure 6* shows an elevation view of a building section. Note that TOS stands for *top of steel*.

Riser Elevation Details

In addition to an elevation view of the building, elevation views must be supplied to clarify riser assembly details. *Figure 7* shows a typical riser elevation detail.

1.2.0 Title Block, Notes, and Legend

In order to understand and use drawings accurately and effectively, you must know what is included on the drawings. This begins by examining the title block, notes, and drawing legend.

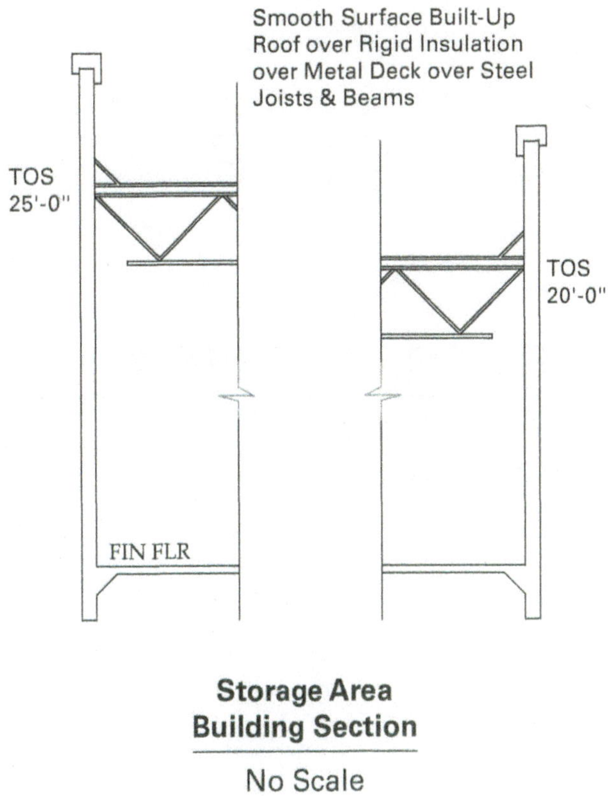

Figure 6 Elevation view.

R1 NEW 6" FIRE SPRINKLER SERVICE UP TO AND INCLUDING A STANDARD FLANGE INSIDE THE BUILDING, (2 HOLED) WILL BE FURNISHED, INSTALLED, BLOCKED, FLUSHED AND TESTED BY OTHERS PER NFPA 24 STANDARD. SEE SITE PLAN FOR CONTINUATION.

R2 (2) 5/8" RESTRAINING RODS REQUIRED FOR 6" UNDERGROUND FIRE SPRINKLER SERVICE

R3 6" STANDARD FLANGE (BY OTHERS)

R4 6" x 4" REDUCING FLANGE

R5 4" AMES MODEL C200BFG DOUBLE CHECK BACKFLOW DEVICE WITH SUPERVISED BUTTERFLY VALVES (WIRING OF SUPERVISORY SWITCHES IS BY OTHERS)

R6 PRESSURE GAUGE

R7 4" FLOW SWITCH (WIRING OF SWITCH IS BY OTHERS)

R8 4" GROOVED VIKING D1 CHECK VALVE

R9 4" TO 4" x 2½" x 2½" STANDARD FINISH FLUSH MOUNT FIRE DEPARTMENT CONNECTION

R10 4" TO WET PIPE SYSTEM

R11 2" MAIN DRAIN TO BE PIPED TO THE EXTERIOR OF THE BUILDING

R12 SPARE SPRINKLER HEAD CABINET STOCKED IN ACCORDANCE WITH NFPA 13.

R13 ALARM BELL WITH WEATHERPROOF BACK ON EXTERIOR (WIRING OF BELL IS BY OTHERS)

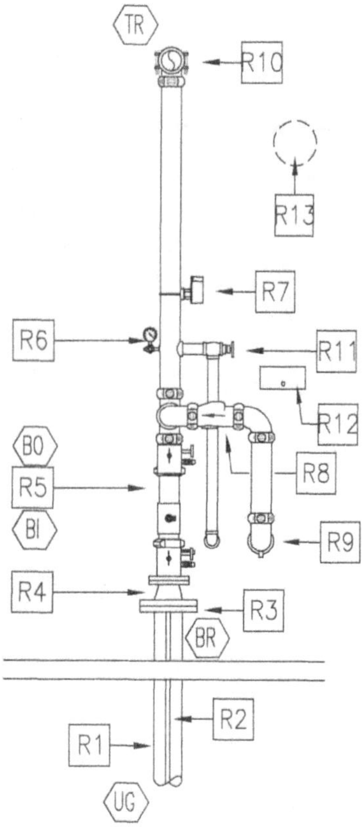

RISER ASSEMBLY DETAIL & NOTES

SCALE: 1/4" = 1'-0"

Figure 7 Typical riser elevation detail.

Using Shop Drawings to Track Time on a Job

In addition to the installation information found on the shop drawings, there are two critical questions the sprinkler fitter must answer on any job:

- Where are the drawings located on the jobsite? They are typically with the supervisor or in the job box, but the location may vary depending on the company. On some jobs, the shop drawings are only available on the job tablet.
- What is the job name and contract/job number for tracking time on this project? This can be found in the drawing's title block. All work on the project is linked to this number (the address alone is not enough). Always note this information at the start of each shift so you are not scrambling to remember when completing your time sheet.

Many fitters highlight completed areas on the shop drawing as the installation progresses. This helps to keep track of your place when working on large or complex projects and can also be used to quickly convey the job status to your supervisor or others.

1.2.1 Title Block

A title block is essential for every drawing. Title blocks vary from source to source. *Figure 8* shows a typical title block. At a minimum, the title block must include the important basic working plan information required by NFPA 13. The working plan information is as follows:

- Name of the owner and occupant
- Location, including street address
- Name, address, and phone number of the contractor

In addition, many title blocks contain these other items:

- Date of creation of the drawing
- Scale of the drawing
- The sheet number and the number of sheets in the package

Construction drawings are usually drawn to scale. The scale indicates the size relationship between an object in the drawing and the object's actual size. The scale used on a drawing depends on the size of the objects being shown, the space available on the drawing, and the type of plan. Scaled drawings show objects such as buildings, rooms, doors, or piping assemblies reduced to a smaller size.

1.2.2 Directional Arrow (North Arrow)

A directional arrow is required showing the north point of the compass. A directional arrow must appear on both the plot plan and the building plan view. *Figure 9* shows various directional arrows.

With most maps it is understood that the top of the map is north. However, this is not always the case on construction drawings. You might see construction drawings where the front of the building faces north and the top of the drawing is south. In other cases, the street directions and building layout do not conform with the main compass points and the directional arrow may be at an angle.

Some structures are too large to show on a single drawing. Structures such as nursing homes and hospitals may have wings in the shape of spokes on a wagon wheel, in which case, each wing drawing may have a different orientation of the north arrow.

1.2.3 General Notes

The general notes section of a drawing package includes information a fitter must know before performing an installation. General notes apply to general things, such as applicable standards, type and size/schedule of pipe and fittings, and how the pipe lengths are dimensioned. *Figure 10* shows an example of general notes from a typical shop drawing. In this example, U.N.O. means *unless noted otherwise*. If you are unsure of a term, ask for a definition. Like symbols, abbreviations and acronyms may vary from drawing to drawing.

1.2.4 Keyed Notes

Keyed notes provide critical information when space does not permit adequate written information in the main field of the drawing. By using a numbered reference and a keyed note section, it is easy for the reader to locate information needed with the drafting details. *Figure 11* shows a keyed note box.

Figure 8 Typical title block.

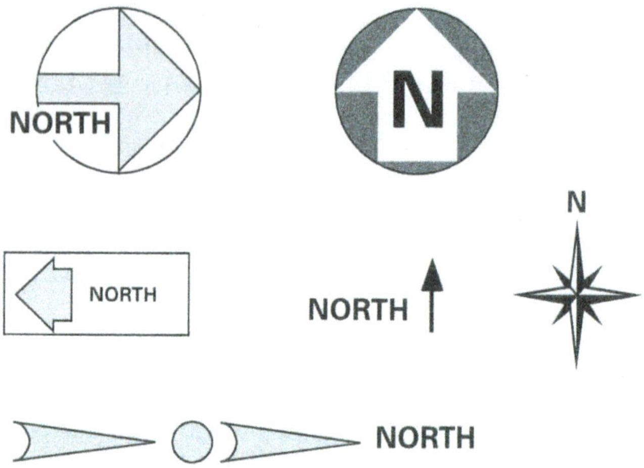

Figure 9 Directional arrows.

Think About It

You are installing the sprinklers in a retail location and notice that a decorative soffit was added that does not appear on the drawings and causes one of the sprinklers to violate the obstruction rules. What should you do and why?

1.2.5 Revisions

The shop drawings used in the field should be the latest dated revision from the company. Those drawings should be based on the most recent numbered and dated revisions of the architect's drawing package. As a result of timing, that may not always be the case.

It is not uncommon for a fitter to discover, as a sprinkler system is being installed, that changes in the building layout have occurred that impact the approved shop drawings. When revisions are required, it is important that the cause of the change be determined. The first thing that should be done is to notify the sprinkler system layout technician of any discrepancies.

Questions relating to revisions include the following:

- Is it a change documented by drawings or change orders that were never given to the sprinkler contractor? If so, the changes should appear in the latest architect's drawing package or in the change orders on file with the general contractor.
- Was the discrepancy caused by a field change in the sprinkler system layout?
- Is this a field change by another contractor that was not provided to the sprinkler contractor?

Verifying that the drawings at the jobsite are current is essential when checking with the office to determine whether the cause is a job change that was not communicated to the sprinkler contractor. A drawing revision block will be present on the sprinkler shop drawings. A similar revision block will be on the architect's drawings at the construction site. The issue date of both must be checked to determine the latest version. *Figure 12* shows a typical revision block.

When changes are discovered in the field, the additional work is documented by the fitter, layout technician, and project manager/estimator before the work is done. Change orders are used to obtain payment for the material and labor involved when the job is not within the scope of work originally bid by the sprinkler contractor.

GENERAL NOTES:

1. NEW FIRE PUMP, NEW SPRINKLER SYSTEMS, AND EXISTING SPRINKLER SYSTEM MODIFICATIONS ARE PER NFPA 13 2019 EDITION AND NFPA 20 2019 EDITION. SEE HAZARD CLASSIFICATION TABLE BELOW.

2. ALL PENDENT SPRINKLERS IN ACOUSTICAL CEILING TILE SHALL BE INSTALLED CENTER OF TILE UNLESS OTHERWISE NOTED.

3. ALL NEW PIPE 2½" AND LARGER IS SCHEDULE 10 BLACK STEEL PIPE WITH ROLL GROOVED ENDS FOR USE WITH GROOVED FITTINGS.

4. ALL NEW PIPE 1" - 2" IS SCHEDULE 40 BLACK STEEL WITH THREADED ENDS FOR USE WITH CAST IRON CLASS 125 THREADED FITTINGS U.N.O.

5. ALL PIPES PASSING THROUGH INTERIOR FLOORS, WALLS, AND PARTITIONS TO BE SLEEVED WITH SCHEDULE 10 BLACK STEEL PIPE OR EQUAL.

6. LISTED SPRINKLER GUARDS TO BE INSTALLED ON ALL SPRINKLERS SUBJECT TO DAMAGE IN ACCORDANCE WITH NFPA 13.

7. ALL NEW MAIN OUTLETS ARE SHOP WELDED OR MECHANICAL TEE TYPE. WHERE MECHANICAL TEE OUTLETS ARE UTILIZED PIPE CUT OUT WILL BE TIED TO THE MAIN.

8. UNDERGROUND FLUSH CERTIFICATE TO BE PROVIDED AND VERIFIED PRIOR TO SYSTEM TIE-IN.

9. HYDROSTATIC TESTING ON SPRINKLER SYSTEMS TO BE PERFORMED IN ACCORDANCE WITH 28.2.1 OF NFPA 13 - 2019 EDITION.

Figure 10 Example of general notes.

The project manager/estimator is responsible for obtaining clearance to perform the work. Do not perform this work unless authorized to do so.

1.2.6 As-Builts

As-builts are revised drawings that show the final installation after any changes have been made. Only fitters can supply the information to create as-built drawings. Every change made by the fitters must be noted on a copy of the drawing and given to the layout technician to document the actual installation. This is known as *redlining* as it is often done using a red pen. It is important to mark changes neatly and carefully so they can be easily understood by the layout technician. Be sure to provide as much information as possible, including all dimensions, part numbers, etc. Photos of the completed installation can be taken using a phone or tablet and sent along with the redlines. They are helpful to clarify positioning details for the layout technician.

Some construction contracts require as-built drawings to be supplied to the AHJ and/or the building owner upon completion of the installation. As-builts are essential for the sprinkler contractor's files when performing warranty work, modifications, and additions.

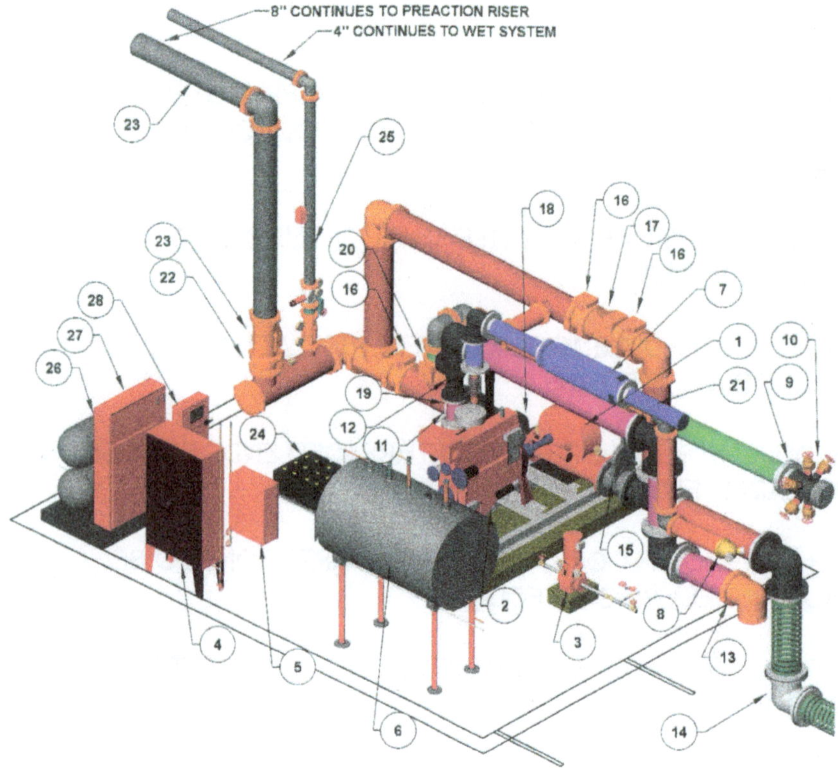

FIRE PUMP LEGEND			
Key Note	Size	Description	Count
1	10 x 6	DIESEL FIRE PUMP, 2000 GPM	1
2	-	DIESEL ENGINE	1
3	1¼	JOCKEY PUMP, 20 GPM	1
4	-	FIRE PUMP CONTROLLER	1
5	-	JOCKEY PUMP CONTROLLER	1
6	-	DIESEL TANK, DOUBLE-WALL	1
7	6	DIESEL ENGINER MUFFLER	1
8	6 x 2½	SIAMESE FDC w/ SIGNAGE	1
9	8	HOSE VALVE TEST HEADER	1
10	2½	TEST HOSE VALVE	6
11	6	PRESSURE RELIEF VALVE	1
12	6 x 10	WASTE CONE	1
13	10	RELIEF DISCHARGE	1
14	10	D.I. UNDERGROUND SUPPLY	1
15	10	OS&Y VALVE	1
16	10	BUTTERFLY VALVE w/ TAMPER	3
17	10	CHECK VALVE	1
18	10 x 6	CONCENTRIC REDUCER	1
19	10	WAFER CHECK VALVE	1
20	8	BUTTERFLY VALVE w/ TAMPER	1
21	6	CHECK VALVE	1
22	8	BUTTERFLY VALVE w/ TAMPER	1
23	8	PREACTION VALVE	1
24	-	BATTERY RACK	1
25	4	EXISTING WET SYSTEM RISER	--
26	-	NITROGEN GENERATOR	1
27	-	NITROGEN GENERATOR CONTROLLER	1
28	-	ADDRESSABLE PREACTION RELEASE PANEL	1

Figure 11 3D CAD drawing with keyed note box and references.

REVISIONS			
NO.	DATE	BY	DESCRIPTION
1	7/8/2020	RH	REVISED PER ASI #1 ADDITION OF NEW WALLS
2	8/3/2020	RH	REVISED PER ASI #2 RELOCATION OF DUCT WORK IN MECHANICAL ROOM #2
3	8/26/2020	RH	REVISED PER ASI #3 NEW CEILING LAYOUT IN CONFERENCE ROOM

Figure 12 Typical revision block.

1.2.7 Dimensions

Drawing dimensions can be expressed in several ways. Dimensions under one foot are stated in inches (e.g., 5½" or 7¼"). Occasionally the inch mark in the fraction may be omitted, as in 5½ or 7¼. Dimensions under one foot can also be expressed using a zero as a foot placeholder, such as: 0'-5½".

Dimensions over one foot are stated in feet and inches (e.g., 1'-0" or 9'-6½"). *Figure 13* shows typical dimensions on a drawing.

The red lines shown in *Figure 13* are sprinkler piping. Notice that there is a dimension above the pipe and a dimension below the pipe. The top dimension is stated in inches. This figure indicates the nominal pipe size. The bottom dimension is the length of that section of pipe and is stated in feet and inches.

Center-to-Center and Cut Lengths

Pipe lengths may be specified as the center-to-center length between fittings and/or the cut length. In some instances, the center-to-center length and the cut length will both be given. *Figure 14* shows center-to-center and cut lengths.

There may be occasions when a dimension is missing, such as when drawings have portions labeled not-to-scale (N.T.S). This is used to show details that would otherwise be too small to illustrate. In addition, last-minute changes are sometimes made to the layout by modifying numerical dimensions without redrawing to scale prior to printing. If dimensions are unclear or missing, consult with the layout technician for clarification.

Beams and Columns

The grade beam is the ultimate reference point for all construction dimensions. This is the structural beam forming the perimeter of the building. Sometimes this may be the only accurate point of reference for fitters from which to work. *Figure 15* shows the grade beam as shown on a shop drawing.

Each type of building construction has a symbol that identifies it with a particular structural type. You are probably most familiar with I-beam construction. Today's construction practice identifies these as wide flange beams, noted as *WF*. *Figure 16* illustrates the most frequently used steel shapes.

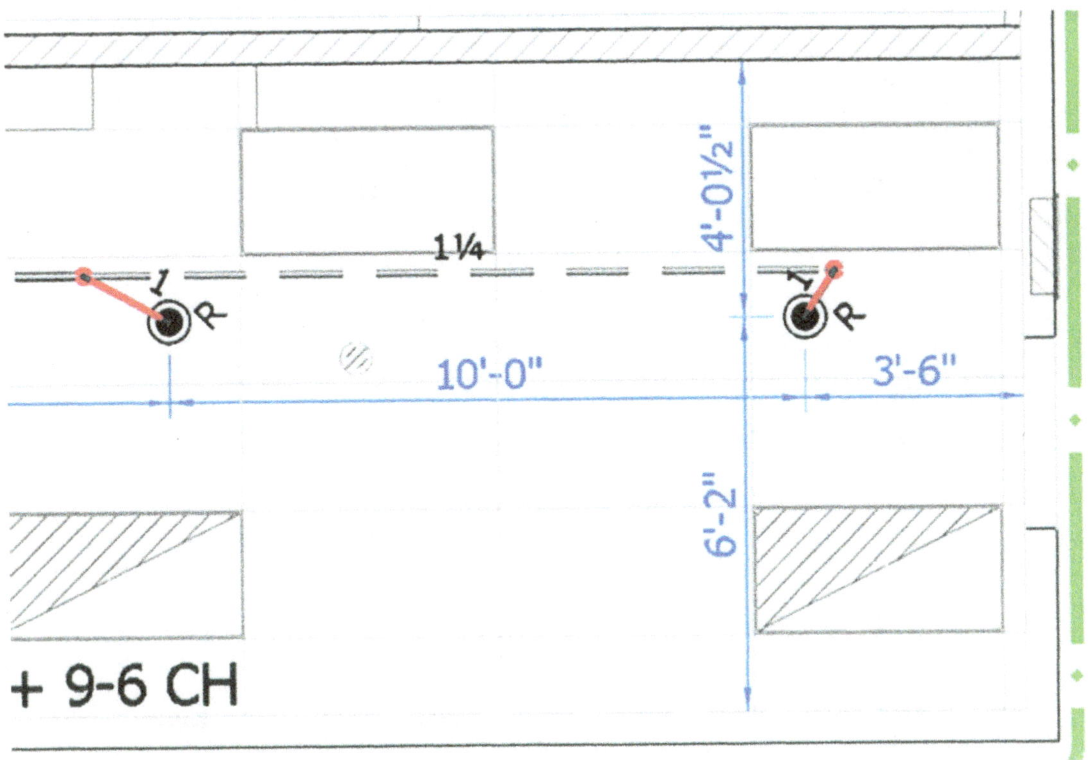

Figure 13 Dimensions on a drawing.

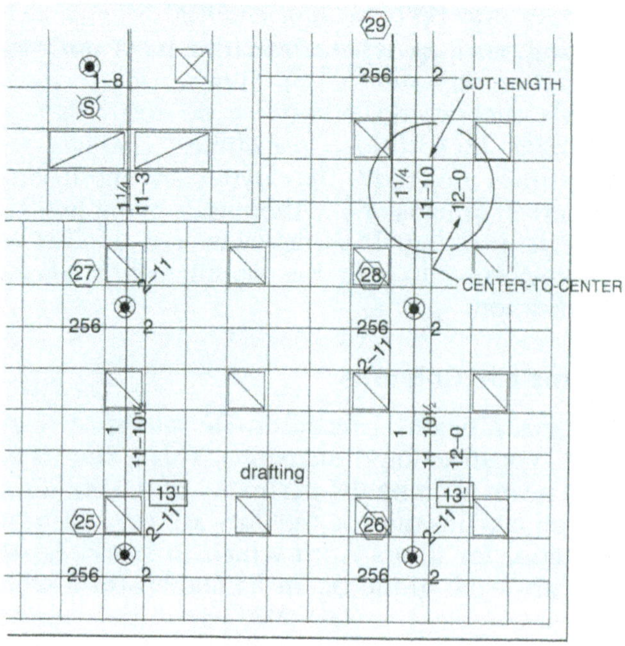

Figure 14 Center-to-center and cut lengths.

A round column is used to support beams and girders when loadbearing walls do not exist. Columns are made of steel pipe and are usually filled with concrete. *Figure 17* shows the elevation and plan symbols for a column.

Columns of square steel tubing, concrete, and wide flange beams are also common in commercial construction. Frequently, the wide flange column will be encased in concrete or some other fire protective material. *Figure 18* shows various column symbols.

Note that all beams, regardless of type, are shown as a line in a plan view. If the beams are bar joists, for example, this information would only be identified in the section by a note describing the construction type.

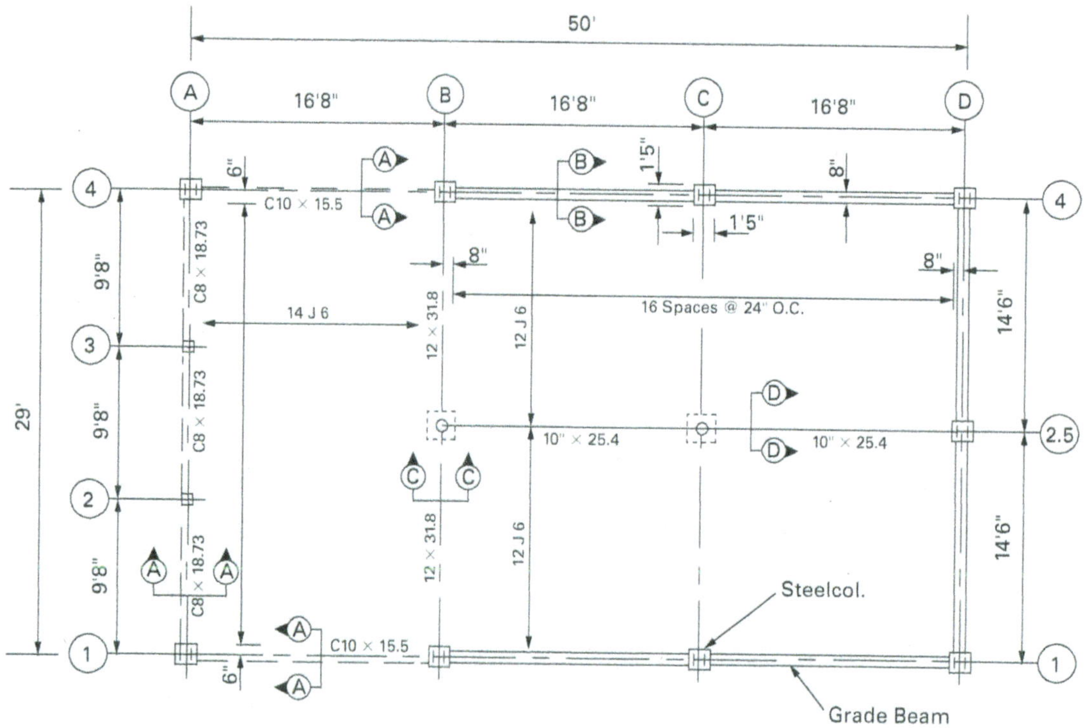

Figure 15 Typical grade beam.

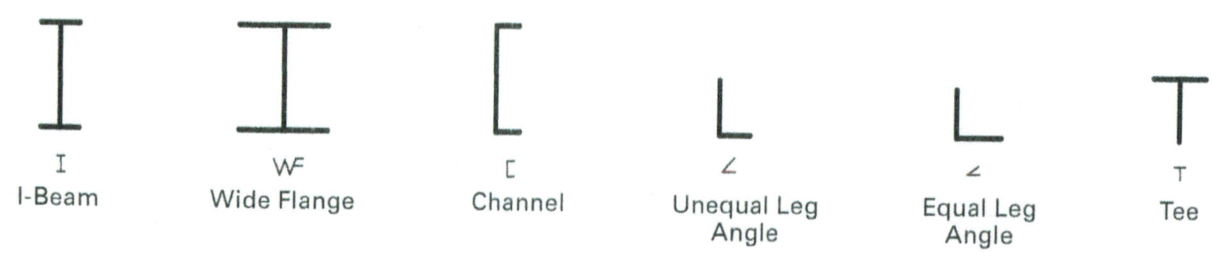

Figure 16 Structural symbols for steel shapes.

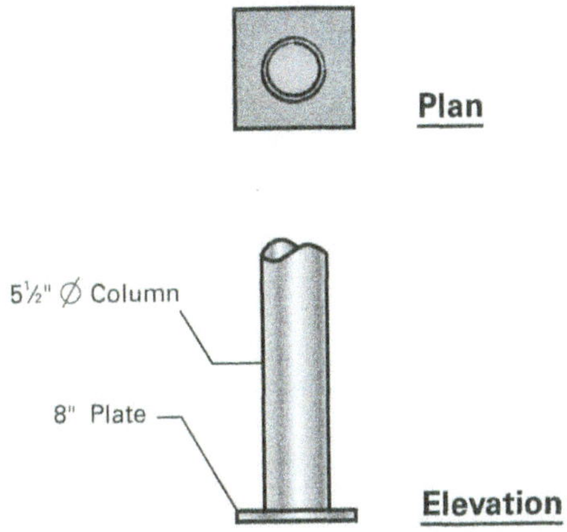

Figure 17 Elevation and plan symbols for a column.

1.2.8 Drawing Legends

One of the key factors in reading a drawing correctly is the ability to recognize symbols. The specific symbols used on a drawing are generally defined in a legend. *Figure 19* shows a typical drawing legend. Sprinkler shop drawings must also include a sprinkler legend that details all of the information required to identify the types and locations of the sprinklers in the system. Sprinkler legends are discussed in detail later in this module.

1.3.0 Specifications

Specifications include a complete listing of approved materials and fixtures along with their manufacturers, part/model numbers, and dimensions, as well as installation procedures and required system tests. On smaller projects, the specifications can be a note on a construction drawing; on larger projects, they can be a separate document that accompanies the set of drawings. Specifications are written by the project's architect or engineer.

Specifications are often not job-specific, though plans always are. In other words, the same basic specifications may be used for several jobs, while only those elements in the specifications that apply to a specific project will appear in that project's construction drawings. In a sense, specifications are similar to a written version of the schedules provided in many types of construction drawings.

Since there is not enough space on the drawings to include all the required technical information about the project, the layout technician develops a set of specifications to provide important technical information about the work and quality of materials. The drawing set shows what to install, while the specifications explain what materials to use and how to do it.

Building Structure in a Plan View

It can be difficult to picture a two-dimensional plan as a three-dimensional building in your mind. When looking at plans, note the beam depth shown in the section next to the beam, ceiling heights, elevations of the piping, and any other information regarding elevation. This will provide a better mental picture of the actual structure.

While drawings show exact dimensions and where objects are located, specifications describe the materials to be used, hardware requirements, and other performance guidelines for the project.

Contractors and their employees are responsible for adhering to project specifications and meeting standards of quality. Specifications serve three general purposes:

- Define the responsibilities of the layout technician, owner, contractor, and subcontractors.
- Supplement the working drawings with detailed technical information regarding the quality of work and materials to be used.
- Support the drawings as part of the contract documents.

In general, if there is a discrepancy between the drawing set and the specifications, defer to the specifications. Inform your supervisor if you spot inconsistencies in construction documents.

For ease of use, specifications are formatted into a series of sections dealing with the construction requirements for the various areas of work on a construction project. People who use the specifications must be able to find all the information they need without spending too much time looking for it. The most common specification writing format used in North America is MasterFormat®. The 2016 version of MasterFormat® is shown in *Figure 20*. Because the specifications on some jobs can be very large documents, it is helpful to know where to find the information on the sprinkler system and its components. Sprinkler fitting falls under the Facility Services Subgroup under Fire Suppression (Division 21). Division 21 is further divided into sections and subsections for different types of sprinkler systems and components. For example, 21 13 00 covers the different types of sprinkler systems, and 21 13 13 covers wet pipe sprinkler systems. Information on hangers and supports can be found in 21 05 29.

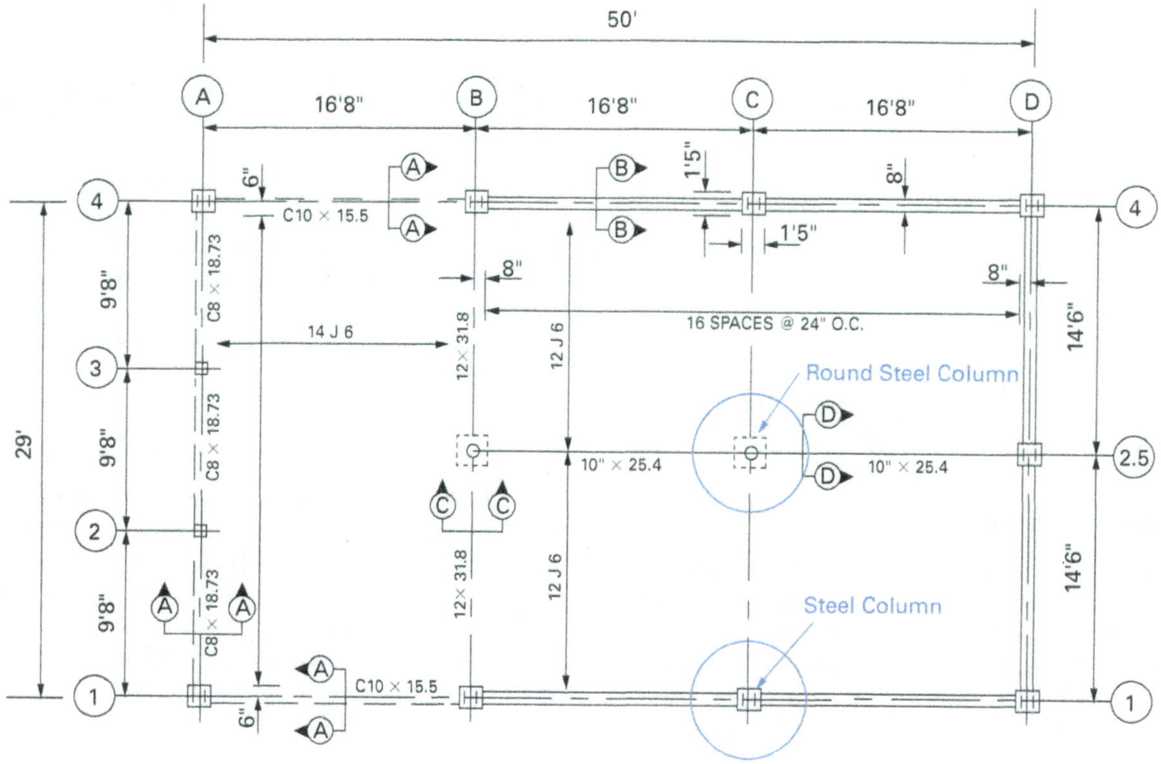

Figure 18 Column symbols.

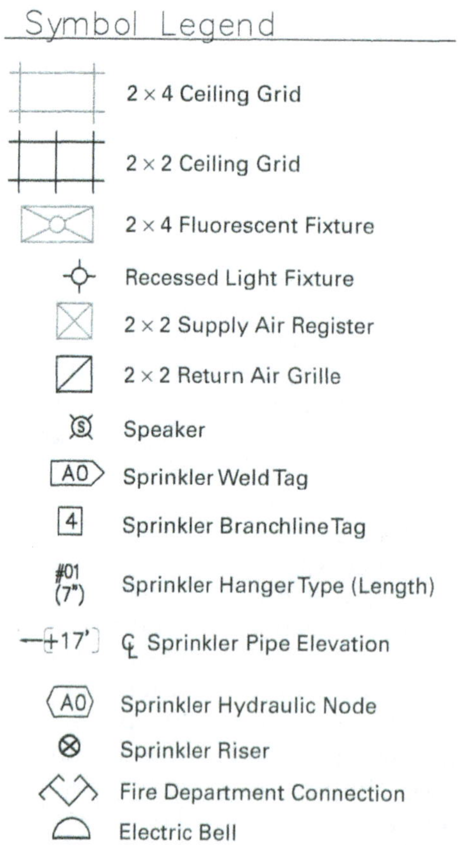

Figure 19 Typical drawing legend.

Applications Guide

MasterFormat Groups, Subgroups, and Divisions

PROCUREMENT AND CONTRACTING REQUIREMENTS GROUP

Division 00 – Procurement and Contracting Requirements
Introductory Information
Procurement Requirements
Contracting Requirements

SPECIFICATIONS GROUP

GENERAL REQUIREMENTS

Division 01 – General Requirements

FACILITY CONSTRUCTION SUBGROUP

Division 02 – Existing Conditions
Division 03 – Concrete
Division 04 – Masonry
Division 05 – Metals
Division 06 – Wood, Plastics, and Composites
Division 07 – Thermal and Moisture Protection
Division 08 – Openings
Division 09 – Finishes
Division 10 – Specialties
Division 11 – Equipment
Division 12 – Furnishings
Division 13 – Special Construction
Division 14 – Conveying Equipment
Division 15 – Reserved for Future Expansion
Division 16 – Reserved for Future Expansion
Division 17 – Reserved for Future Expansion
Division 18 – Reserved for Future Expansion
Division 19 – Reserved for Future Expansion

FACILITY SERVICES SUBGROUP

Division 20 – Reserved for Future Expansion
Division 21 – Fire Suppression

Division 22 – Plumbing
Division 23 – Heatin, Ventilating, and Air-Conditioning (HVAC)

Division 24 – Reserved for Future Expansion
Division 25 – Integrated Automation
Division 26 – Electrical
Division 27 – Communications
Division 28 – Electronic Safety and Security
Division 29 – Reserved for Future Expansion

SITE AND INFRASTRUCTURE SUBGROUP

Division 30 – Reserved for Future Expansion
Division 31 – Earthwork
Division 32 – Exterior Improvements
Division 33 – Utilities
Division 34 – Transportation
Division 35 – Waterway and Marine Construction
Division 36 – Reserved for Future Expansion
Division 37 – Reserved for Future Expansion
Division 38 – Reserved for Future Expansion
Division 39 – Reserved for Future Expansion

PROCESS EQUIPMENT SUBGROUP

Division 40 – Process Interconnections
Division 41 – Material Processing and Handling Equipment
Division 42 – Process Heating, Cooling, and Drying Equipment
Division 43 – Process Gas and Liquid Handling, Purification, and Storage Equipment
Division 44 – Pollution and Waste Control Equipment
Division 45 – Industry-Specific Manufacturing Equipment
Division 46 – Water and Wastewater Equipment
Division 47 – Reserved for Future Expansion
Division 48 – Electrical Power Generation
Division 49 – Reserved for Future Expansion

19 Introduction and Applications Guide

Figure 20 2016 MasterFormat®.

1.4.0 NFPA Requirements for Shop Drawings

NFPA 13 defines which information is required on a shop drawing. It is important to note that not all information on the plan is for the fitter. Shop drawings are created for many entities, including the fitter, inspector, architect, engineer, and insurance company. You will notice additional information on the plan and must be able to determine if it affects the installation. The minimum data required by NFPA 13 on all plans includes the following:

- *Name of owner and occupant* — This information is found in the title block in the lower right-hand corner of the plan. It identifies the name of the project, the AHJ, who owns the building, and who is going to occupy the building.
- *Location, including street address* — This information is also found in the title block. It is used to help locate the building, as well as the city and insurance records. In some cases, the address is not yet shown on the street, and the building (especially in multiple building projects), may be hard to find. If so, there may be a site plan or reference plan on the same sheet of the layout that will help in locating the correct building.
- *Point of compass* — The point of compass, commonly referred to as the *directional arrow* or *north arrow,* can be located anywhere on the drawing, but the preferred location is near the title block. That way when the drawing is folded, the title block and north arrow can both be seen at once.
- *Full height cross section, or schematic diagram, including structural member information, if required for clarity; including ceiling construction and method of protection for nonmetallic piping* — The full height cross section is used to visualize what the building looks like for a reviewer that does not have a full set of plans. It can be located anywhere it fits on the plan and is generally one of the last items added. A schematic is used less often but shows more information than a building section. It can also be located anywhere on the plan. Two important parts of the section or schematic are the ceiling construction and the protection of nonmetallic piping. The ceiling construction is important because it impacts the sprinkler spacing and deflector distances. This can change from one area to another, so all areas should be identified. The method of protecting nonmetallic piping is critical as the Listing of nonmetallic pipe requires that it be protected in most installations.
- *Ceiling/roof heights and slopes* — This information must be listed if not shown in the full height cross section.
- *Location of partitions* — Any partition that affects the sprinkler location or interferes with the sprinkler spray pattern must be shown on the shop drawings. Full-height partitions are typically shown as solid lines and partial-height partitions are shown as dashed lines. There are some cases in which the sprinklers will throw over partitions such as short pony walls. These partitions are usually not shown on the shop drawings.
- *Location of fire walls* — Fire walls are sometimes shown shaded or simply called out on the drawing. Openings in fire walls must be sealed to the correct fire rating.
- *Occupancy class of each area or room* — This information is sometimes difficult to find on a plan; however, it has a considerable impact as many system requirements are based on the type of occupancy. It should be located in each area unless all areas are the same.
- *Location and size of concealed spaces, closets, attics, and bathrooms* — The locations of these spaces are shown in the plan view and are important, as are all areas that potentially need to be sprinklered.
- *Small enclosures in which no sprinklers are to be installed* — In most cases, not only will the area be shown on the plan, but it will also be called out as having no sprinklers. This alerts the AHJ and the fitter that sprinklers are not required.
- *Size of city main in street and whether dead-end or circulating* — If dead-end, the drawing must indicate direction and distance to nearest circulating main and city main test results and system elevation relative to the test hydrant. This information could be shown on the site plan or could be found in the notes. It is important for verifying the water supply in the field.
- *Other sources of water supply, with pressure or elevation* — Other sources of water supply would include tanks or ponds that supplement the main water supply. These are shown in numerous ways and could be covered by general notes. The important actions for the fitter are to verify that all sources are connected correctly to the system, and that there have not been any sources omitted in the field that are called for on the plan.

- *Sprinkler information* — Sprinkler information includes the make, type, model, nominal K-factor, and sprinkler identification number (SIN). It is found in the sprinkler legend. The fitter must verify that the sprinklers are available and installed in the correct locations.
- *High-temperature sprinklers* — The temperature rating and location of high-temperature sprinklers will be noted in the sprinkler legend and/or identified on the shop drawing. Always verify the temperature of every sprinkler and check the drawing for any high-temperature sprinklers that must be installed due to the location of heaters and other heat sources.
- *Total area protected by each system on each floor* — This is to ensure that the system size is not too large. It is sometimes found as a general note or on the site plan.
- *Number of sprinklers on each riser per floor* — This is found in the sprinkler legend or a general note.
- *Total number of sprinklers on each dry pipe system, preaction system, combined dry pipe-preaction system, or deluge system* — This is found in the sprinkler legend or a general note.
- *Approximate capacity in gallons of each dry pipe system* — This shows compliance with the size limitations of dry systems. It is usually found in a general note.
- *Pipe type and schedule of wall thickness* — The pipe type and schedule is listed in a general note. Always check the type of pipe listed on the drawings against the pipe sent to the site. Specifications and hydraulic calculations are both affected if the type of pipe is incorrect.
- *Nominal pipe size and cut lengths of pipe (or center-to-center dimensions)* — Each piece of piping will be marked with its size and length or center-to-center dimension in the case of field-fabricated pipe. The only exception is on typical branch lines in which only one of the branch lines is marked.
- *Location and size of riser nipples* — These are typically marked at each location.
- *Type of fittings and joints and location of all welds and bends* — Fittings and welds are either shown in place on the drawing or called out in a numbered note that corresponds to a location on the drawing.
- *Type and locations of hangers, sleeves, braces, and methods of securing piping* — Any sleeving must be precisely dimensioned as the future piping is fabricated to run through the sleeve. Hangers are usually called out using a note or diagram showing the type and length. Braces are shown with a length and are usually located close to where they are to be installed. Braces are indicated as lateral or longitudinal by the way they are attached to the piping and may include the abbreviations of LAT and LONG.
- *Control valves, check valves, drainpipes, and test connections* — Usually, the size as well as the manufacturer will be identified. (The manufacturer will not be indicated for drain valves.)
- *Backflow prevention assembly* — Information on backflow preventers includes the make, type, model, and size, and means to forward flow test at system demand, including the hose allowance, if required.
- *Dry pipe valves* — The make, type, model, and size of alarm or dry pipe valve must be indicated.
- *Preaction or deluge valves* — This includes the make, type, model, and size and is usually shown in the riser section. Note that identifying the type of valve is often the easiest way to find the type of system being installed.
- *Type and location of alarm bells* — Many times, determining the location of the alarm bell (horn/strobe) on the outside of the building has to be performed on-site. Always consult with the building owner before selecting the location of alarm bells. The type will usually be either a water motor gong or an electric bell, and usually is shown on the riser section or on the shop drawing at the location of the riser.
- *Size and location of standpipe risers, hose outlets, hand hose, monitor nozzles, and related equipment* — This information is important for the fitter because it shows additional piping that must be installed in the building. Note that some equipment may be shown that is not part of the fitter's work but is necessary for the AHJ when reviewing the plan.
- *Private fire service mains* — The drawing must include private service main sizes, lengths, locations, weights, materials, and points of connection to the city main. It must also show the sizes, types, and locations of valves, valve indicators, regulators, meters, and valve pits. The depth that the top of the pipe is laid below grade must also be indicated.
- *Piping provisions for flushing* — Provisions for system flushing are sometimes difficult to locate on a drawing as they may be as simple as a grooved cap at the end of the main. Always verify that the flushing requirement has been satisfied.

- *Equipment installed as an addition to an existing system* — Where the equipment is to be installed as an addition to an existing system, enough of the existing system indicated on the plans is necessary to make all conditions clear.
- *Scale* — A graphic representation of the drawing scale must be included and usually found in the title block. It can also be used to verify if the drawing has been reduced or increased in size when it was reproduced.
- *Name, address, and phone number of contractor* — This is found in the title block.
- *Pressure-reducing valve settings* — The fitter must be aware of the correct setting and location of each pressure-reducing valve. This information can be found in the form of notes or shown as a detail on the shop drawing.
- *Antifreeze solution (if used)* — The type and amount of Listed antifreeze solution is usually shown in the notes.
- *Size and location of hydrants* — The shop drawing shows the size and number of outlets, and if outlets are to be equipped with independent gate valves. It also notes whether hose houses and equipment are to be provided, and by whom. Static and residual hydrants used in flow tests are also shown.
- *FDCs* — The shop drawing must show the size, location, and piping arrangement of FDCs.
- *Edition of NFPA 13* — The year or edition of NFPA 13 to which the sprinkler system is designed is typically found in the notes. It is important as it establishes the requirements in place at the time of design.

Other types of information are required by NFPA for hydraulic calculations or the AHJ but are not critical to the sprinkler fitter. These include the hydraulic data nameplate, hydraulic reference points, minimum rate of water application, total quantity of water and pressure, relative sprinkler and supply elevations, unprotected wall openings, load calculations for sizing sway bracing, and zones of influence used in calculations for seismic bracing. These will be discussed in more detail in NCCER Module 18401, *System Layout*.

Submitting an RFI

One of the documents used to make changes to shop drawings is a Request for Information (RFI). When RFIs are delivered to the appropriate authority, all work in that area must stop until a correction is issued. Any work completed prior to a solution of the problem or error is the responsibility of the contractor. If you know how to solve the problem identified in an RFI, make the solution known to the authority when you submit the RFI. This may save time on the project.

1.0.0 Section Review

1. Sprinkler system shop drawings are usually based on the _____.
   a. floor plan
   b. reflected ceiling plan
   c. roofing plan
   d. site plan

2. What was traditionally known as an *I-beam* is now referred to as a(n) _____.
   a. 2T-beam
   b. WF beam
   c. H-beam
   d. C-beam

3. Where in Division 21 can you find the specifications for the different types of sprinkler systems?
   a. 21 10 00
   b. 21 11 00
   c. 21 12 00
   d. 21 13 00

4. The term *SIN* refers to the manufacturer identification number of each _____.
   a. single-inlet valve
   b. seismic brace assembly
   c. screwed indicating valve
   d. sprinkler type

SECTION TWO

2.0.0 IDENTIFYING SPRINKLER SYSTEM SYMBOLS

Objective

Identify common symbols used in sprinkler system shop drawings.

a. Identify typical sprinkler symbols.
b. Identify common valve and device symbols.
c. Describe the symbols used to show fittings, hangers, and bracing.
d. Identify the symbols for rises, drops, plugs, and hydraulic calculation references.
e. Point out various components and dimensions on a sample shop drawing.

Performance Task

1. Use an instructor-supplied shop drawing to identify various types of information, including the sprinkler legend, title block, north arrow, sprinkler type and temperature, and pipe size.

Trade Term

Sprig: A pipe that rises vertically and supplies a single sprinkler.

Each craft has its own set of symbols. NFPA 170, *Standard for Fire Safety and Emergency Symbols*, publishes a set of fire safety symbols unique to the sprinkler industry. However, it is not a requirement and symbol standardization has not occurred in the sprinkler industry. Some layout technicians follow NFPA 170, and others do not. The symbols presented in this module represent shop drawings from various contractors. Each layout technician may develop a unique set of symbols over time, so it is not unusual to see a diversity of symbols in use within one company. Always refer to the legend on each drawing before beginning an installation. *Figure 21* shows a typical legend. If a drawing has no legend, communicate with the layout technician before proceeding.

2.1.0 Sprinkler Symbols

There are many symbols used to describe sprinkler types and configurations. *Figure 22* shows typical sprinkler symbols. Upright and pendent sprinklers are usually indicated on a drawing as a circle, while sidewall sprinklers are often shown as triangles. Nozzles or open sprinklers must be different from closed sprinklers and may be shown as a diamond or other symbol as indicated in the sprinkler legend. Note that while there are multiple symbols for each type of sprinkler, they are normally used to show different types. For example, an open circle may be used for a chrome upright, and an open circle with an X for a brass upright. These symbols are based on NFPA 170, but there are variations. For example, the custom sprinkler symbol shown in *Figure 22* was created for a specific sprinkler type.

CAUTION

Drawing symbols may vary depending on the system layout technician or company. It is critical to check the symbols on every drawing to make sure the correct device is installed.

Since there are a number of symbols that can be used for each type of sprinkler, it can be confusing to match symbols to multiple sprinkler types. That is why the sprinkler legend is so important (*Figure 23*). The sprinkler legend lists the characteristics of each sprinkler used on a job, including the type, K-factor, SIN, and finish. It can also be used to determine the number of spare sprinklers required in the spare sprinkler box. (Remember, you must leave a minimum of two sprinklers of each type with a total of 6 sprinklers on systems with less than 300 sprinklers, 12 sprinklers on systems with between 300 and 1,000 sprinklers, and 24 sprinklers on systems with over 1,000 sprinklers.) A list of the spare sprinklers included must be left in the spare sprinkler box, along with sprinkler wrenches for each sprinkler in the system.

NOTE

Some layout technicians may include the temperature in the sprinkler symbol, but only when the sprinkler departs from ordinary temperature.

SYMBOL LEGEND			
SYMBOL	DESCRIPTION	SYMBOL	DESCRIPTION
[10'–0"]	PIPE ELEVATION AFF		DENOTES HANGER LOCATION
[*12"]	ELEVATION BELOW TOP OF STEEL		SWAY BRACE
TOS	ELEVATION TOP OF STEEL		FIRE DEPT CONNECTION
	RISE UP / DOWN		ELECTRIC BELL
	RISER / STANDPIPE	FS	FLOW SWITCH
	NEW PIPE		FIRE HYDRANT
	EXISTING PIPE	8'–0"	CEILING HEIGHT
1	HYDRAULIC REFERENCE POINT	RISER RM	ROOM NAME

Figure 21 Typical shop drawing legend.

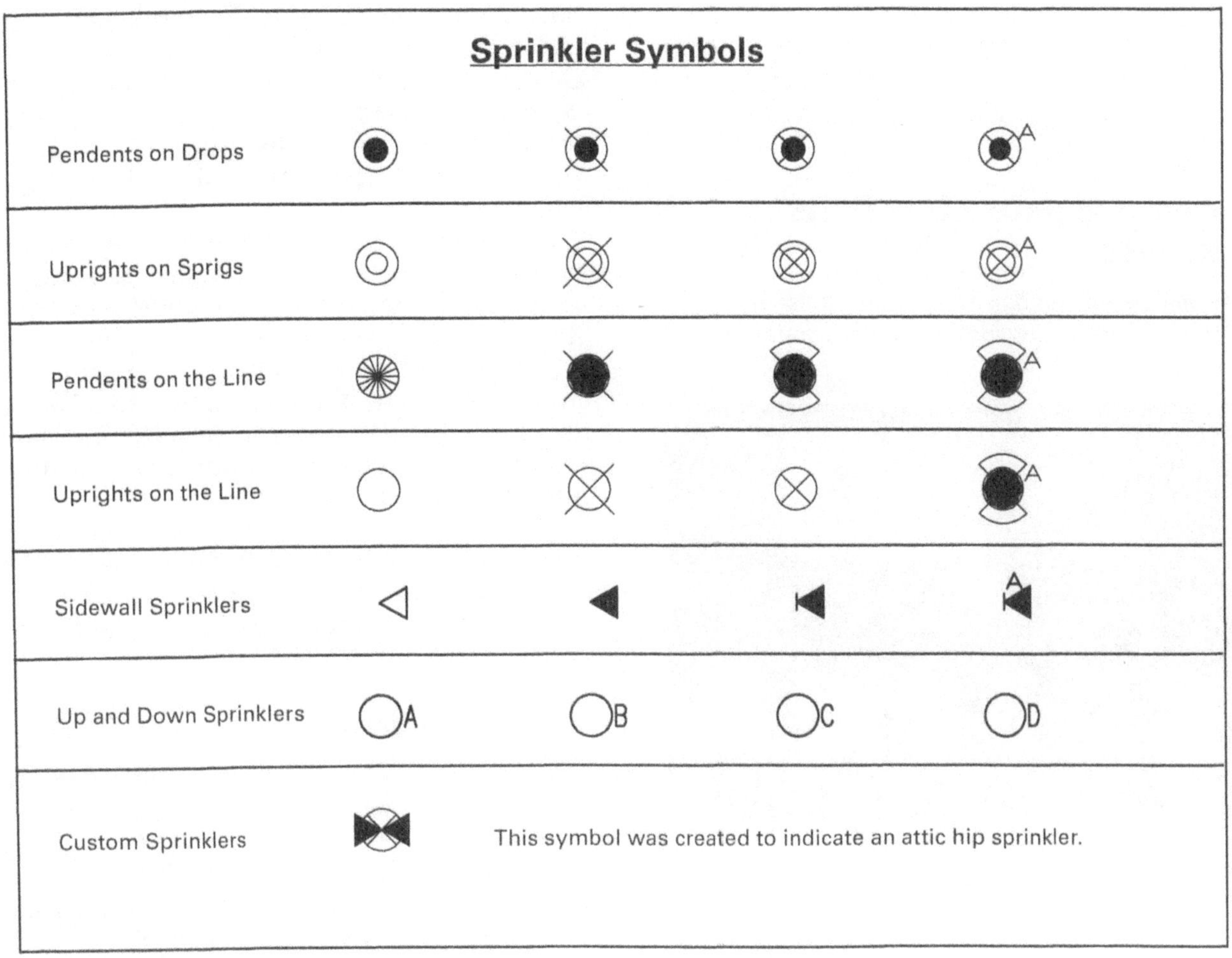

Figure 22 Typical sprinkler symbols.

Symbol	Manufacturer	SIN	Quantity	K-Factor	Type	Size (NPT)	Response	Finish	Temperature	Escutcheon	Note
⊗	Reliable	RA1425	75	5.6	Upright	½	Quick	Brass	155°F	N/A	
⦸	Reliable	RA1425	90	5.6	Upright	½	Quick	Brass	200°F	N/A	
◉	Reliable	RA3415	62	5.6	Pendent	½	Quick	White	165°F	Concealed	
◆	Reliable	RA3415	6	5.6	Pendent	½	Quick	White	212°F	Concealed	
▲	Reliable	R5734	3	5.6	Pendent	1	Quick	White	155°F	Concealed	Dry
●	Reliable	R5734	2	5.6	Pendent	1	Quick	White	200°F	Concealed	Dry
◎	Reliable	RA1414	10	5.6	Pendent	½	Quick	Brass	155°F	N/A	
			Total = 248								

Figure 23 Example sprinkler legend.

Symbols for Other Trades

As you work with construction drawings, you will see many types of building symbols. In addition to sprinkler symbols, you may see architectural symbols, engineering symbols, mechanical symbols, and electrical symbols. It will be helpful to become familiar with these symbols so that when you see them on drawings, you will understand how they impact the sprinkler installation.

Spare Sprinkler Box Labels

The spare sprinkler box must include a list or label showing the quantity and type of each sprinkler in the box, as shown here.

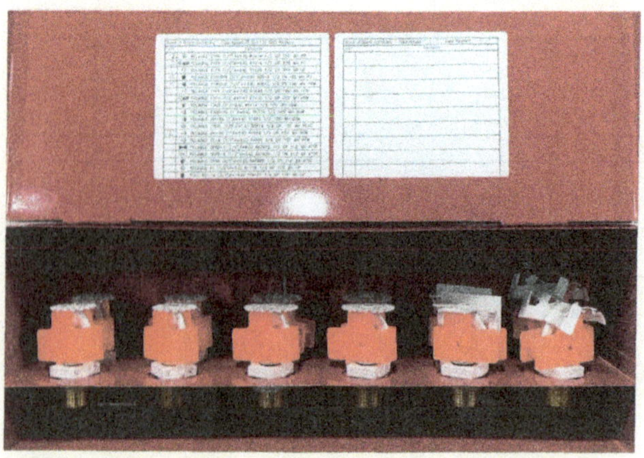

Figure Credit: Taylor Schumacher

2.2.0 Valve and Device Symbols

There are many kinds of valves and devices. It isn't possible to establish a consistent symbol for each device. Always refer to the drawing legend to identify the valves and devices to be installed. *Figure 24* shows some of the standard symbols in use.

2.2.1 Riser Valves

No. 1 in *Figure 24* illustrates two symbols used for the plan view of a riser with an alarm check valve. No. 2 shows three elevation view symbols, reflecting the NFPA 13 pure schematic method first, then a modification to indicate a flanged-by-flanged (F×F) valve, and a third possibility indicating a grooved-by-grooved (G×G) valve. Grooved-by-flanged or G×F valves are also available.

No. 3 illustrates a symbol for a dry pipe valve in a plan view. No. 4 illustrates the same valve in elevation view. No. 5 illustrates a method for indicating the presence of a quick opening device.

No. 6 illustrates a form for a preaction valve in the plan view. No. 7 illustrates, first the NFPA 13 pure schematic form for a preaction valve in elevation, followed by adaptations for flanged and grooved configurations.

No. 8 illustrates a plan view configuration for a deluge valve. No. 9 shows the NFPA 13 convention for the elevation view of a deluge valve, and adaptations for flanged and grooved conventions.

2.2.2 Check Valves

Nos. 10 and 11 in *Figure 24* illustrate forms suitable for a check valve. The simple arrowhead is used in NFPA 13. The arrow form suggests one-way flow. Some drawings will depict a check valve with both a zigzag and an arrow, as shown in *Figure 25*. For installations in pipe, the same conventions as used for an alarm check can be used but are placed in a horizontal format. Since

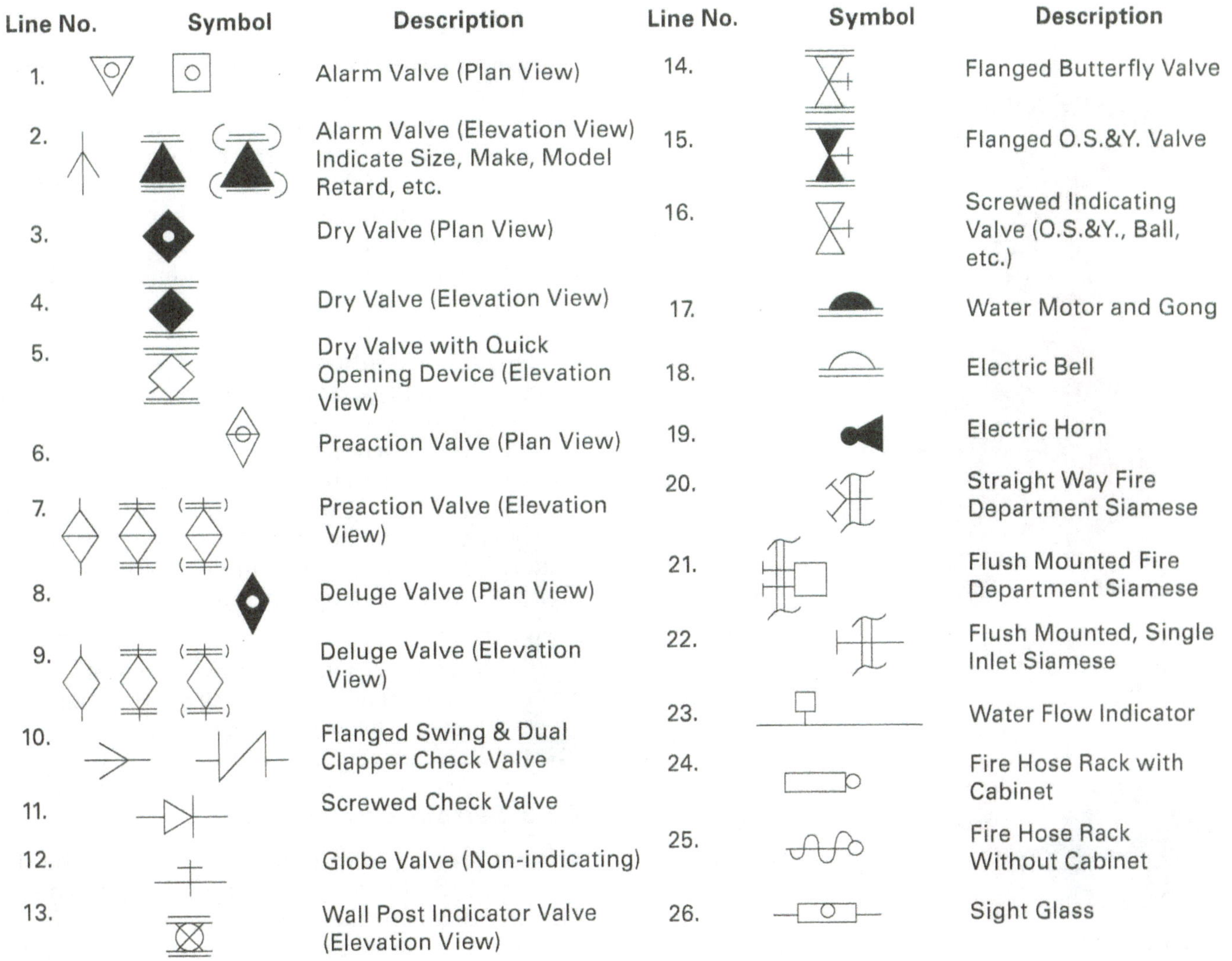

Figure 24 Typical valve and other device symbols.

Figure 25 Check valve and symbol.

these come in screwed, flanged, grooved, wafer, and swing clapper forms, review the notes that define what is being indicated by the symbol.

> **NOTE**
> An arrow on the valve should indicate the direction of flow. However, the fitter should check direction of flow before any installation.

2.2.3 Hand-Operated Valves

No. 12 in *Figure 24* illustrates the most common form for a simple globe valve. No. 16 illustrates a symbol that may be used for an OS&Y gate valve. In addition to indicating whether the valve is a rising stem or not, drawings may also indicate the type of connection. The OS&Y valve shown in *Figure 26* is a grooved-by-flanged or G×F type.

Figure 27 shows an indicating butterfly valve and symbol.

Figure 26 G×F rising stem OS&Y valve and symbol.

Figure 27 Indicating butterfly valve and symbol.

2.2.4 Underground Valves and Piping

Figure 28 shows typical symbols used to indicate underground valves and associated piping. It also shows the common symbols for hydrants and FDCs.

2.3.0 Fittings, Hangers, and Bracing

All shop drawings must identify the types of fittings and hangers required for the installation. In areas of seismic activity, the types and general locations of sway bracing must also be shown.

2.3.1 Fitting Symbols

A variety of fittings are used in the sprinkler industry. Fitting symbols are defined in general notes, keyed notes, and special notes on the shop drawings. Where all of the fittings are the same on a drawing, the general notes may be the only place they are identified. *Figure 29* summarizes some of the symbols used for fittings.

2.3.2 Hangers

Hangers required for the system may be shown on the piping in their proper locations or indicated in the notes. For buildings in which the piping is hung from beams, the hanger is shown on the intersection of the beam and piping. Buildings with more structure available to hang to, such as a flat concrete slab, generally do not include specific locations and the fitter can adjust the hanger slightly for field conditions but not beyond the limits stated in NFPA 13.

Hangers are often shown as small slashes across the piping and the type is shown by a number reference or symbol. The length of the rod is shown next to the hanger type or covered by a general note if the hangers are mostly typical. *Figure 30* shows some typical symbols used for hanger and support material, underground piping, and underground devices. *Figure 31* shows a hanger detail.

> **NOTE**
> Some drawings may not show individual hanger locations. However, all drawings should indicate the hanger spacing and type.

2.3.3 Seismic Bracing

Seismic bracing information may include the diameter of the brace and the pipe length. This is illustrated much like a pipe size and length for a piece of system piping. The locations where bracing is required are also indicated on the plan. However, only the general location is shown, not the exact location of bracing attachment to the structure and the system piping. Consult with the layout technician if questions arise regarding the location of the bracing. A note or table of brace fasteners is located on the drawing to indicate what size fastener to use to attach the brace to the structure. *Figure 32* shows the standard symbols for two-way and four-way braces. Braces are indicated as lateral (LAT) or longitudinal (LONG) by the way they are attached to the piping. Remember that longitudinal bracing runs along the pipe, while lateral bracing is attached at an angle to the pipe.

2.4.0 Miscellaneous Sprinkler Notations

System notations are symbols used as descriptors on a drawing. Before assuming that a particular symbol has a certain meaning, read the notes, and check the legend. *Figure 33* shows some typical system notations in a drawing legend. Like other symbols, they may vary from drawing to drawing.

No.	Symbol	Description
1.		Post Indicator Valve
2.		Gate Valve with Roadway Box
3.		Gate Valve in Pit
4.		2-Way Fire Hydrant
5.		3-Way Fire Hydrant
6.		Housed Hydrant (Specify model and manufacturer of house, and equipment furnished.)
7.		Hydrant With House (Specify model and manufacturer of house, and equipment furnished.)
8.		Angle Body F.D.C Siamese (Specify maunfacturer, model, finish, and size.)
9.		Public Main (Indicate circulating or dead end, material, and size.)
10.		New Private Main (Indicate size and material.)
11.		Existing Private Main to be Abandoned
12.		Existing Private Main
13.		Pump Suction Main

Figure 28 Underground valves and associated piping.

No.	Symbol	Description
1.		Companion Flange
2.		Grooved Coupling
3.		Threaded Coupling
4.		Flanged Union
5.		Grooved Reducer with Takeout
6.		Grooved Concentric Reducer
7.		Threaded Reducer
8.		Flanged Threaded Reducer
9.		Welded Nipple
10.		Weld-o-Let
11.		Flanged Tee
12.		Grooved Tee
13.		Flanged ELL
14.		Grooved 90° ELL
15.		Grooved 45° ELL
16.		Threaded 90° ELL
17.		Threaded 45° ELL
18.		Threaded Tee

Figure 29 Typical fitting symbols.

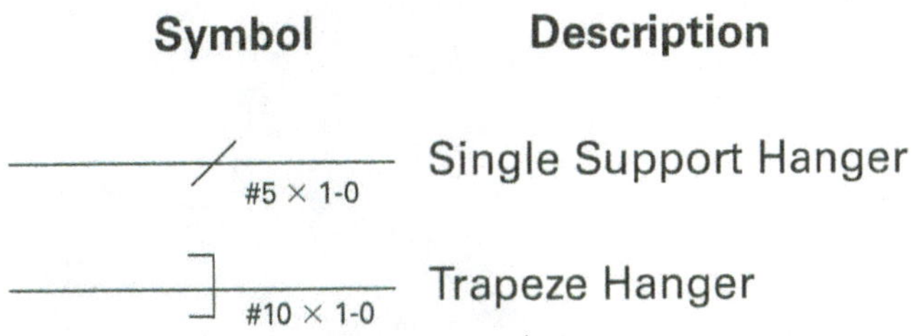

Figure 30 Hanger and underground material symbols.

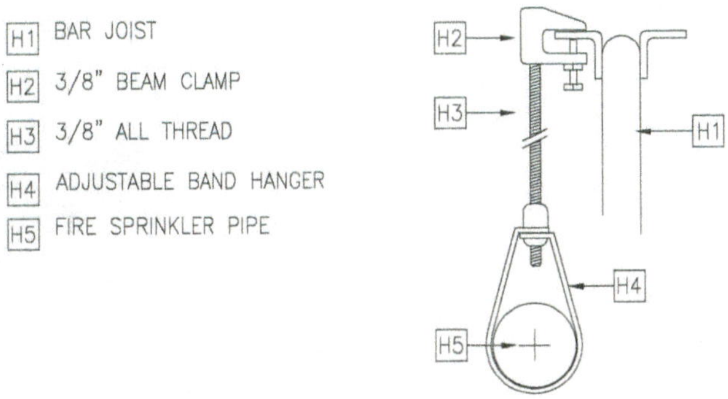

Figure 31 Typical hanger detail and notes.

2.4.1 Rises and Drops

There are several ways in which rises and drops in the sprinkler piping are shown on drawings. *Figure 34* shows rises and drops. Note that the label for vertical pipe length is shown on a 45-degree angle. This calls attention to it and is often the way rises and drops are shown on shop drawings.

2.4.2 Plugs for Future Work

There are several situations in which plugs are installed in systems to accommodate future work. *Figure 35* shows several ways plugs are installed for future use. One way to install a future plug is on a sprig. A nipple rises from the branch line to a tee with a 1" plug to the side. Another way to install a future plug is on an armover to a tee with a plug below the sprinkler.

2.4.3 Sprigs and Drops

Sprigs and drops are depicted separately from sprinklers shown on branch lines. An example of this is shown in *Figure 36*.

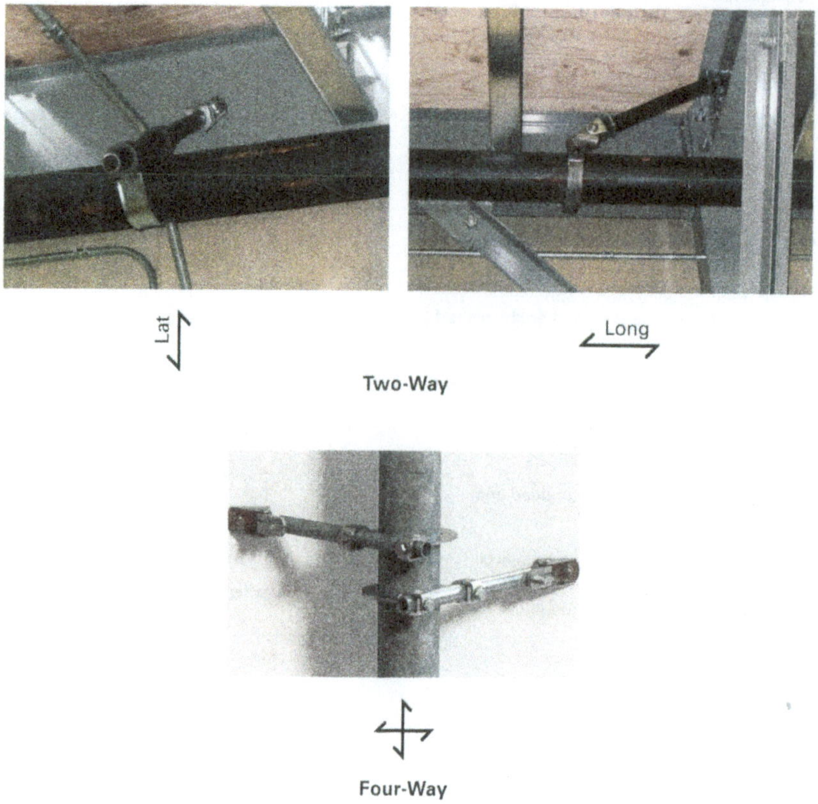

Figure 32 Two-way and four-way braces.

No.	Symbol	Description
1.	#2	Fabrication Branch Line Number
2.	[2]	Hydraulic Branch Line Number
3.	7 7 D12/C12	Hydraulic Reference Point
4.	(2-6) ℄ 30" T.O.J.	Centerline Elevation Below Reference (Identify reference on each drawing sheet.)
5.	25-0 25-0	Building Elevation Note reference on all plans as: –Bottom of deck to finished floor. –Bottom of z-purlin to finished floor. –Bottom of joist to finished floor.
6.	(8-6)	Height of Suspended Ceiling Above Finished Floor
7.	═════	Walls to Roof Deck
8.	= = = = =	Walls to Ceiling Height

Figure 33 Typical system notations.

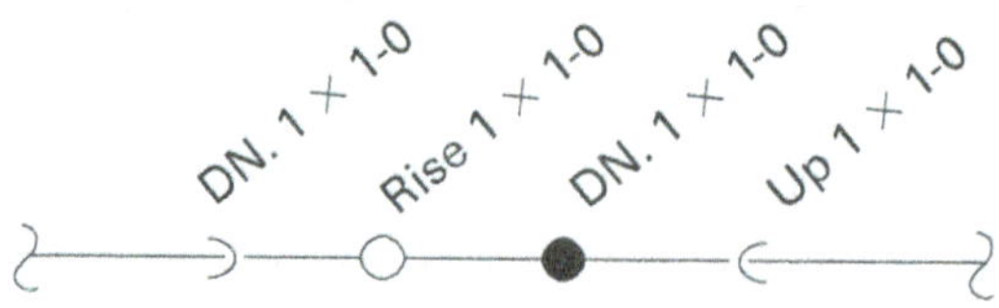

Figure 34 Rises and drops.

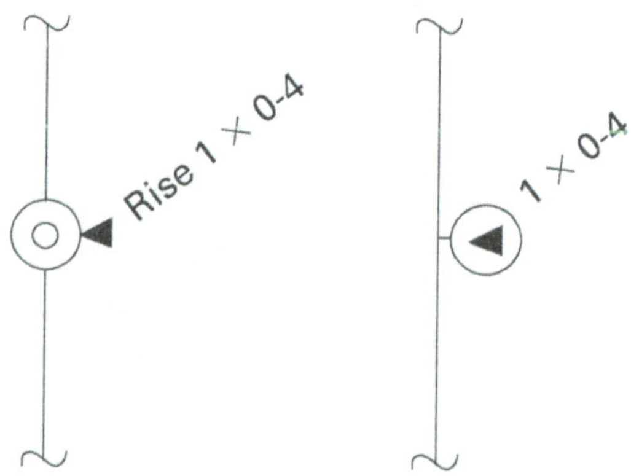

Figure 35 Plug installation.

2.4.4 Sprinklers Above and Below Ceilings

In some circumstances, sprinklers are required to be both above and below the ceiling and be directly over each other. When this is the case, it may be indicated as shown in *Figure 37*.

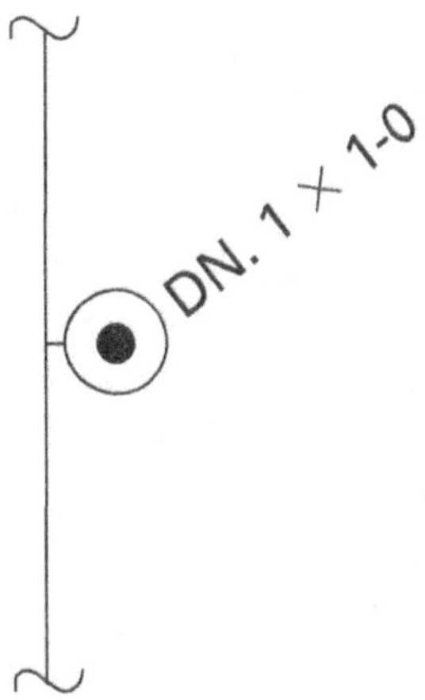

Figure 37 Sprinklers above and below a ceiling combined in a single symbol.

2.4.5 Calculation References

Reference points for use in checking hydraulic calculations are typically shown on shop drawings as hexagons (*Figure 38*). The number or letter inside the hexagon refers to a point in the system referenced in the hydraulic calculation sheets submitted to the AHJ. They are not of any specific use to the installer; however, they will show where the most remote area of sprinkler operation is located.

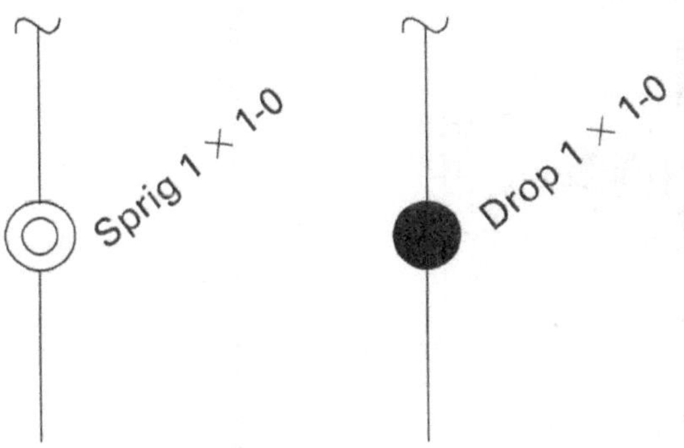

Figure 36 Sprigs and drops.

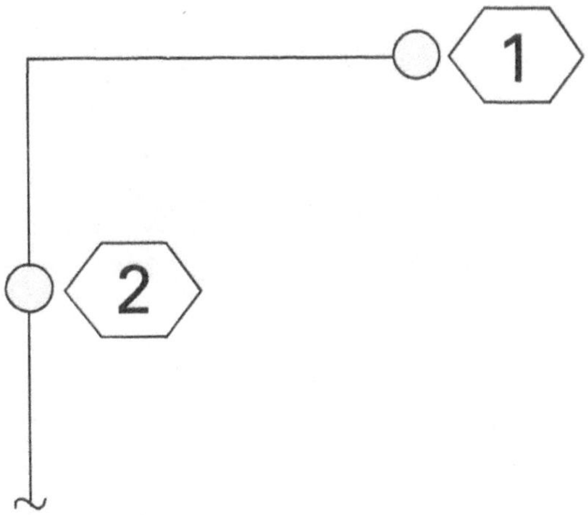

Figure 38 Reference points.

2.5.0 Reading a Sample Shop Drawing Exercise

Now that you are familiar with sprinkler shop drawings and symbols, you can apply them to interpreting an actual shop drawing in order to select and install components. Refer to the portion of a sample shop drawing shown in *Figure 39* and the associated sprinkler legend in *Figure 40* to answer the following questions:

1. What size pipe will be installed?
2. Assuming north is at the top of the drawing, what is the distance between the west wall and the first sprinkler line?
3. How many sprinklers will be installed in this area?
4. What type of sprinklers will be installed?
5. What is the cut length between fittings?

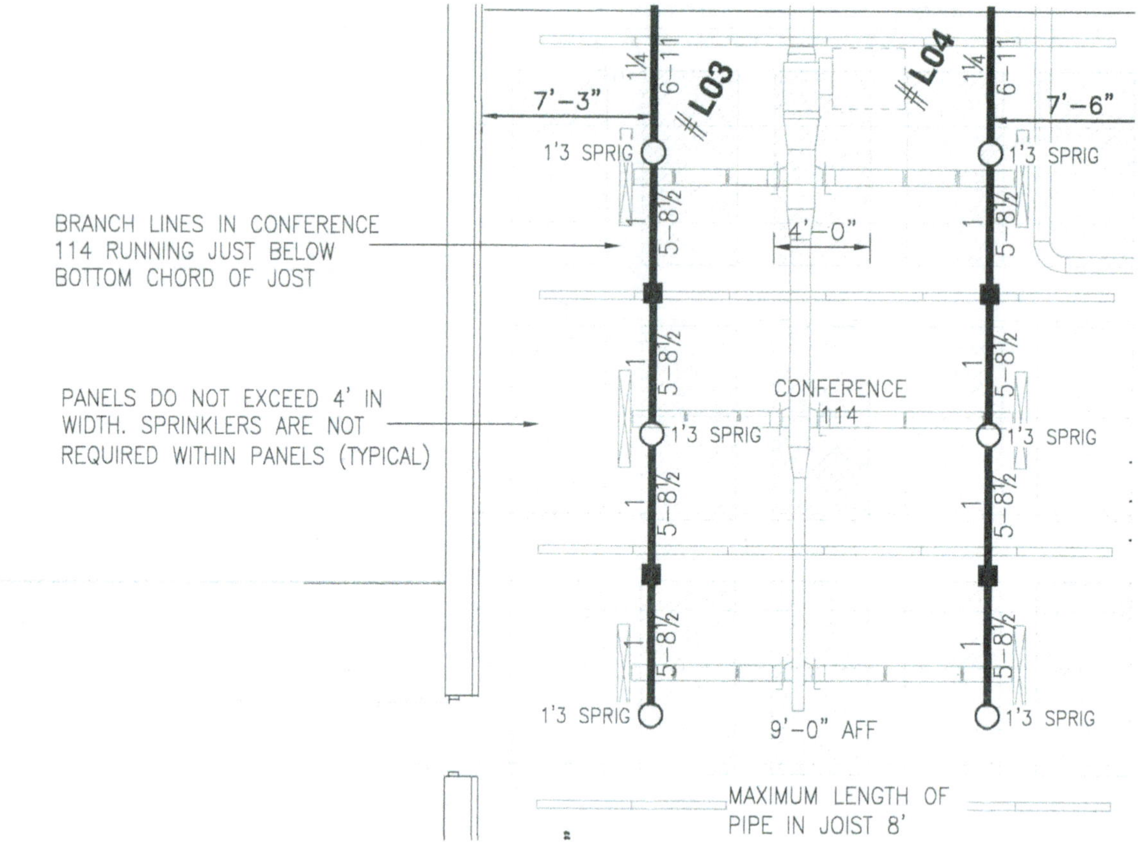

Figure 39 Shop drawing exercise.

SPRINKLER	FINISH	MFG	MODEL	K	NPT	ID #	TEMP	SYM.	ESCUTCHEON/OTHER	QTY.
QUICK RESPONSE UPRIGHT	BRASS	VIKING	QR	5.6	1/2	VK3001	155	○	NONE	24
STANDARD RESPONSE UPRIGHT	BRASS	VIKING	SR	5.6	1/2	VK1001	200	⊘	HEAD CAGE	12
QUICK RESPONSE PENDENT	BRASS	VIKING	MIRAGE	5.6	1/2	VK462	135 155	●	WHITE COVER PLATE	46
QUICK RESP EXTENDED CVRG PEND	BRASS	VIKING	MIRAGE QR EC	8.0	3/4	VK634	135 135	◓	WHITE COVER PLATE	10
QUICK RESPONSE SIDEWALL	WHITE	VIKING	MICROFAST	5.6	1/2	VK305	155	◀	SURFACE ESCUTCH	1
TOTAL SPRINKLERS THIS PROJECT										93

Figure 40 Sprinkler legend for shop drawing exercise.

The following answers can be found in *Figure 41*:

1. **The pipe size is 1" as indicated by the numeral 1 above the pipe on the drawing.**
2. **The distance between the west wall and the first line is 7'-3".**
3. **Six sprinklers will be installed in this conference room.**
4. **According to the sprinkler legend in *Figure 40*, the sprinklers in this area are quick response upright sprinklers with a sprinkler identification number of VK3001.**
5. **The cut length between fittings is 5'-8½".**

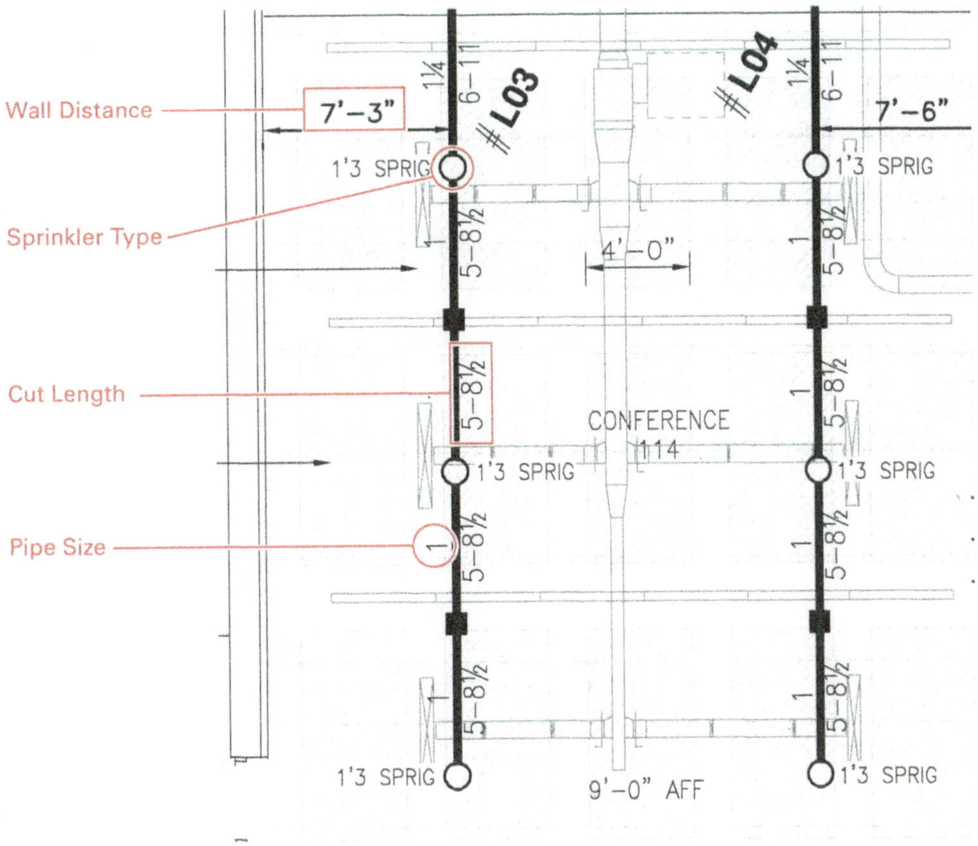

Figure 41 Shop drawing exercise solution.

2.0.0 Section Review

1. The symbols for nozzles or open sprinklers *must* be different from closed sprinklers.
   a. True
   b. False

2. The valve that may be shown as an arrow or a zigzag with or without an arrow is a _____.
   a. check valve
   b. butterfly valve
   c. gate valve
   d. deluge valve

3. With regard to sway bracing attachments, shop drawings _____.
   a. show the exact locations of bracing attachments
   b. show the general locations of bracing attachments
   c. list the type of attachments required but do not show their locations
   d. never show bracing attachments

4. A hydraulic reference point is often shown as a number within a _____.
   a. circle
   b. hexagon
   c. square
   d. triangle

5. What are the lengths of all sprigs shown in *Figure 39*?
   a. 8"
   b. 12"
   c. 1'-3"
   d. 1'-6"

Section Three

3.0.0 Building Information Modeling (BIM)

Objective

Identify the advantages of Building Information Modeling (BIM).

a. Define BIM 4D through 7D.
b. Describe how 3D scanning is used in renovation work.

Trade Term

Building Information Modeling (BIM): An interactive building modeling and design methodology that uses special software to design and manage the construction and operation of systems and structures.

Building Information Modeling (BIM) is an interactive approach to building design and construction. While CAD is used to create static 2D and 3D drawings, BIM can be used to create dynamic 3D models that can link each component to a wide variety of information, including cost, scheduling, and product data (*Figure 42*).

BIM uses special software, such as Revit, to show the building through its entire lifecycle, from the design phase through each stage of construction, operation, and maintenance. This allows layout technicians to present different design and equipment options, including factors such as cost and build time. It also helps identify and resolve design and scheduling issues before construction begins. In addition, each component can be linked to product data, such as installation and maintenance instructions.

BIM software can be used to prepare material takeoffs, shop drawings, and as-built drawings. Fitters can use mobile devices to access 3D models that can be viewed from any angle and the models can also be seen as plan views (*Figure 43*). After construction, BIM models can also be used to prepare schedules for inspection and maintenance.

3.1.0 BIM 4D and Beyond

A 3D model shows three dimensions: length, width, and height. But what does it mean to go to BIM 4D and beyond? This is where BIM links each major system component to data such as scheduling, cost, sustainability, and maintenance. BIM provides a living record of each component within the structure over its life cycle. BIM dimensions include the following:

- 4D BIM adds the dimension of time. This may include the lead time when ordering, time to install, and the time required to be fully operational (including inspection and testing). 4D BIM also allows the user to see the project at each stage of construction. This is helpful when planning. For example, firestopping inserts must go in before the concrete is poured, and supervisory switches cannot be connected until the wiring is in place. Using BIM to track the building's progress helps to avoid conflicts, delays, and rework.
- 5D BIM adds the dimension of cost. This includes the costs of purchasing and installing a component, ongoing operation and maintenance costs, and eventual replacement. At the planning stage, BIM allows layout technicians to switch out objects on the model to see how they impact both the initial and lifecycle costs of a system or structure. For example, suppose Pump A costs $10,000 but requires replacement every five years, and Pump B costs $20,000 but can be expected to last 20 years or more. Which pump is the better value over time? In 20 years, you will have replaced the cheaper pump four times, spending $40,000 (not including labor and down time). This makes it much more costly than the more expensive pump. Lifecycle cost comparisons can help the building owner see the value in using high-quality products. 5D BIM also tracks the cost of any change orders in real time, which are then communicated to all stakeholders. This can be helpful when explaining why a small change can have a big cost impact.
- 6D BIM adds sustainability information. This allows the layout technician to present various energy-saving scenarios, such as the addition of solar power. Stakeholders can then make informed decisions about the long-term benefits of each design.
- 7D BIM adds facility management information. All of the warranty and maintenance information is stored within the building model. This allows fitters to click on a component and access data such as installation instructions and inspection schedules. Schedules can be exported to a separate document or used to trigger reminders. This is helpful when planning shutdowns.

Figure 42 BIM 3D model.

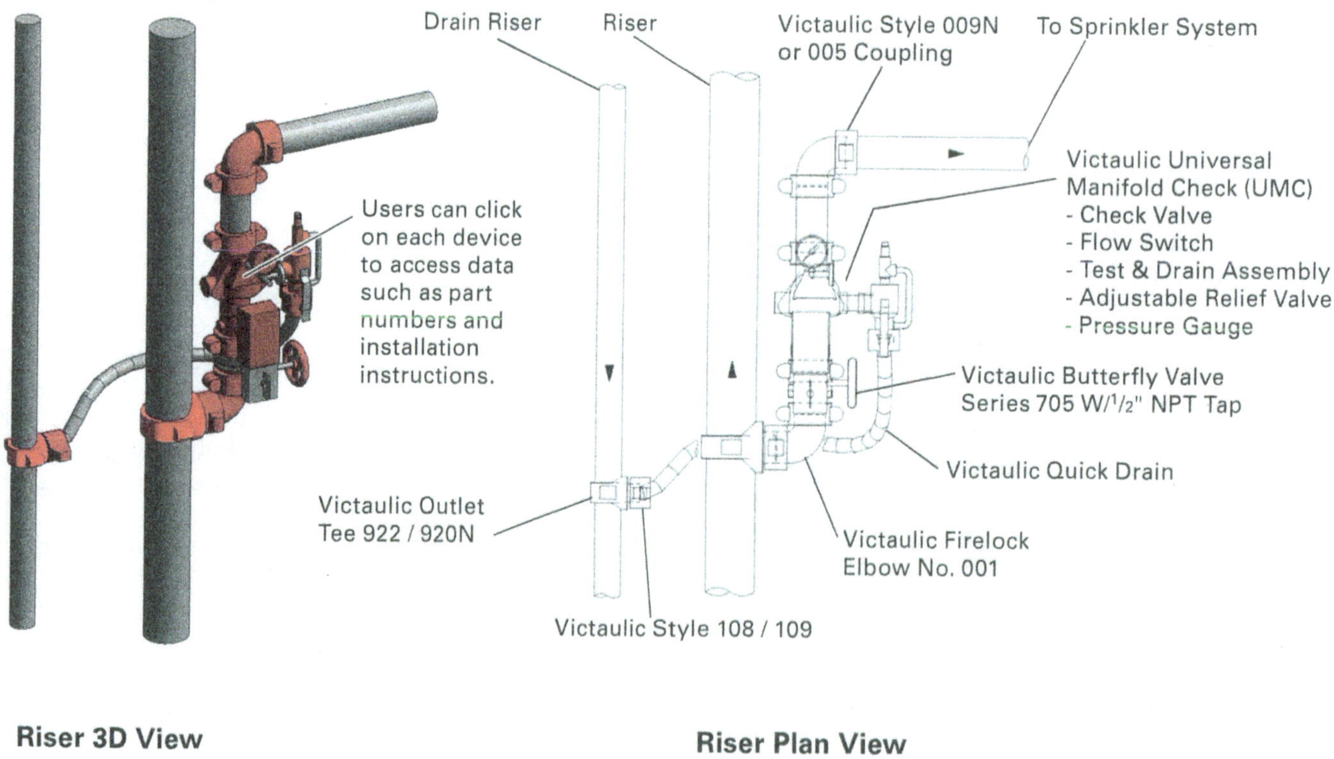

Figure 43 3D and plan views of a stair riser.

BIM can also be accessed using virtual reality (VR) headsets, enabling users to walk through 3D models at any point in the construction process. This provides the ability to review and interact with the model from any location. This is helpful to avoid surprises for the building owners and can also be used to identify potential problems. For example, does a system drain tie into a public sewer system, violating NFPA requirements? Will a door swing into a narrow walkway, presenting a safety hazard? Is the equipment too close to an obstruction, making it difficult to install or service? VR allows users to test drive an installation and make corrections before the building is constructed.

3.2.0 3D Scanning in Renovation Work

Laser scanning, also known as *high-definition surveying (HDS)*, can be used to create a 3D model of an existing installation for use by a layout technician in another location. Drones can be used to carry scanners where human access is difficult, such as on rooftops or underground. A layout technician then uses the 3D model to add new components and prepare both drawings and estimates, saving travel time and expenses. *Figure 44* shows a 3D scan of an existing installation with new components added using exact locations and dimensions.

BIM Mandates

BIM offers many advantages, including saving time, cost, energy, and rework. While it has not taken the place of construction drawings, it is rapidly gaining traction across the globe. In the United States, BIM is now required on all federal Government Services Administration (GSA) new construction and renovation projects.

Figure 44 3D scan with new installation overlay.

3.0.0 Section Review

1. Information on project costs is covered under BIM _____.
   a. 4D
   b. 5D
   c. 6D
   d. 7D

2. 3D laser scanning can be used to map an existing installation with precise dimensions.
   a. True
   b. False

Review Questions

1. The drawings that typically contain the *most* information for the sprinkler system layout are the _____.
   a. site plans
   b. plan views
   c. elevation views
   d. section views

2. Diffuser locations can be found in the _____.
   a. elevation view
   b. reflected ceiling plan
   c. site plan
   d. keyed notes

3. To determine roof contours and floor levels, refer to the _____.
   a. site plan
   b. plan view
   c. elevation view
   d. riser view

4. A drawing title block typically contains the _____.
   a. occupant name
   b. north arrow
   c. sprinkler types
   d. notes

5. In a drawing package, information on how pipe lengths are dimensioned is found in the _____.
   a. title block
   b. revision block
   c. legend block
   d. general notes

6. When numbered notes are assigned to associated numbers on a drawing, they are called _____.
   a. general notes
   b. reference notes
   c. revisions
   d. keyed notes

7. As-built drawings are marked up on the original drawings by the _____.
   a. architect
   b. engineer
   c. AHJ
   d. fitter

8. When a necessary dimension is missing from a shop drawing, the fitter should _____.
   a. use the drawing scale to determine the missing dimension
   b. consult with the layout technician for clarification
   c. estimate the dimension using nearby objects for reference
   d. assume the dimension is not important

9. The ultimate reference point for all construction dimensions is the _____.
   a. column
   b. grade beam
   c. main floor slab
   d. second floor slab

10. When sprigs and drops are depicted on a plan, the length of vertical pipe is typically shown _____.
    a. at a 45-degree angle
    b. in a box
    c. using horizontal text
    d. vertically

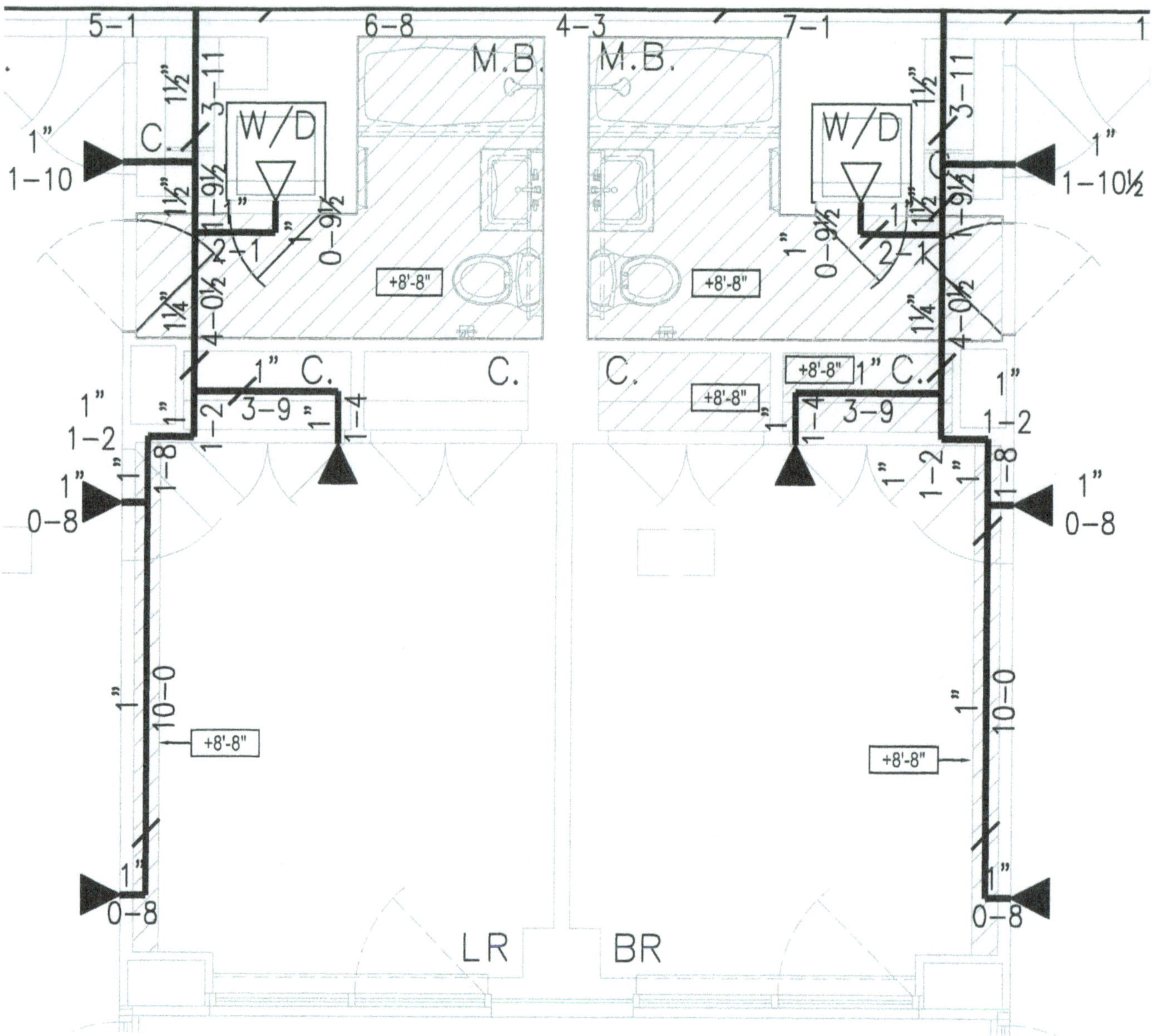

Figure Credit: Lora Deeds NICET III WBSL Bamford, Inc.

Figure RQ01

11. How many sprinklers will be installed in the area shown in *Figure RQ01*?
    a. 6
    b. 8
    c. 10
    d. 12

12. All of the sprinklers shown in *Figure RQ01* are *most* likely to be _____.
    a. sidewall sprinklers
    b. upright sprinklers on drops
    c. pendent sprinklers
    d. upright sprinklers on the line

13. What is the size of the sprinkler piping shown in the bedroom in *Figure RQ01*?
    a. 1"
    b. $1\frac{1}{4}$"
    c. $1\frac{1}{2}$"
    d. 2"

14. In *Figure RQ01*, what is the distance between sprinklers in the living room?
    a. 3'-11"
    b. 6'-8"
    c. 10'-0"
    d. 12'-6"

15. The lead time required when ordering system components can be accounted for in BIM _____.
    a. 4D
    b. 5D
    c. 6D
    d. 7D

Trade Terms Introduced in This Module

Building Information Modeling (BIM): An interactive building modeling and design methodology that uses special software to design and manage the construction and operation of systems and structures.

Column: A round structural member used to support beams and girders where loadbearing walls do not exist.

Elevation view: Usually a two-dimensional view on a vertical plane showing internal and external dimensions of buildings. This view helps sprinkler fitters determine roof contours and floor levels. This is sometimes called a *sectional view*.

General notes: A written list on a drawing that provides information such as applicable standards, types of pipe, and fittings.

Grade beam: The structural beam forming the perimeter of the building and used as the ultimate reference point for all construction dimensions.

Keyed notes: A system by which numbered notes are assigned to associated numbers on a drawing to provide detailed information without crowding the drawing itself.

Legend: A list of symbols and associated definitions used on a drawing.

Plan view: A two-dimensional layout or drawing of a building.

Reflected ceiling plan: An architectural term used to describe a plan view of a ceiling that typically shows the lights, fixtures, diffusers, and sprinklers within a given area.

Revision block: A box or table on a drawing that lists the date of every revision to the drawing.

Site plan: A bird's-eye view of a building in small scale.

Sprig: A pipe that rises vertically and supplies a single sprinkler.

Sprinkler identification number (SIN): A unique manufacturer identification number for every sprinkler type, change in K-factor or shape, deflector characteristic, pressure rating, or thermal sensitivity.

Symbol: A written or printed character or sign representing a drawing element.

Title block: A boxed area on a drawing that shows the installation location, address, contractor, owner, occupant, and other identifying information as required by NFPA 13.

Appendix

Construction Drawing

The QR code below will provide access to an example shop drawing that can be used with Performance Task 1.

SCAN ME

Additional Resources

This module presents thorough resources for task training. The following reference material is suggested for further study.

International Building Code, Latest Edition. Falls Church, VA: International Code Council.

NFPA 13, *Standard for the Installation of Sprinkler Systems*, Latest Edition. Quincy, MA: National Fire Protection Association.

NFPA 170, *Standard for Fire Safety and Emergency Symbols*, Latest Edition. Quincy, MA: National Fire Protection Association.

Figure Credits

Robert Hsieh NICET III WBSL Strickland Fire Protection, Module Opener, Figures 2–4, 8, 10, 12–13, 23

Lora Deeds NICET III WBSL Bamford, Inc., Figures 5, 7, 21, 31, 39–41

Steven Helkin NICET II WBSL Piper Fire Protection, Figure 11

Used with permission of Johnson Controls, Figures 14–15, 17–18, 27, 34–38

© 2018 The Construction Specifications Institute, Inc. (CSI). MasterFormat® excerpt and trademarks used under permission from CSI., Figure 20

AFSA Technical Department, Figures 24, 28–30, 33

Victaulic, Figures 25, 42–44

Reliable Automatic Sprinkler Co., Inc., Figure 26

Chuck Bamford, Figure 32

Courtesy of Eaton, Figure 32

Section Review Answer Key

Section 1.0.0

Answer	Section Reference	Objective
1. b	1.1.2	1a
2. b	1.2.7	1b
3. d	1.3.0	1c
4. d	1.4.0	1d

Section 2.0.0

Answer	Section Reference	Objective
1. a	2.1.0	2a
2. a	2.2.2	2b
3. b	2.3.3	2c
4. b	2.4.5	2d
5. c	2.5.0	2e

Section 3.0.0

Answer	Section Reference	Objective
1. b	3.1.0	3a
2. a	3.2.0	3b

Answer to Think About It

Section 1.2.5

In addition to being a life safety issue that will prevent this sprinkler from operating as expected in the event of a fire, violating the NFPA 13 obstruction rules could also result in the system failing inspection. Stop and inform your supervisor, who will coordinate with the layout technician to update the drawings and determine a solution.

User Update

NCCER makes every effort to keep its textbooks up-to-date and free of technical errors. We appreciate your help in this process. If you find an error, a typographical mistake, or an inaccuracy in NCCER's curricula, please submit a User Update form by visiting **https://www.nccer.org/olf**. You can also scan the QR code using the camera on your phone or mobile device to access the form.

Standard Spray Sprinklers

Overview

This module describes the characteristics of standard spray sprinklers. It also covers types of construction, NFPA 13 light and ordinary hazard occupancy classifications, and installation considerations.

Module 18205

18205

Standard Spray Sprinklers

Objectives

Successful completion of this module prepares you to do the following:

1. Identify the characteristics of standard spray sprinklers.
   a. Identify sprinkler temperature ratings and their associated color codes.
   b. Define K-factor.
   c. Describe sprinkler location, spacing, and positioning requirements.
2. Identify NFPA 13 occupancy classifications.
   a. Define light hazard occupancies.
   b. Define ordinary hazard occupancies.
3. Identify the types of construction.
   a. Describe the types of unobstructed construction.
   b. Describe the types of obstructed construction.
4. Describe the installation considerations for standard spray sprinklers.
   a. Explain the spacing requirements for standard spray pendent (SSP) and standard spray upright (SSU) sprinklers.
   b. Explain the spacing requirements for standard spray sidewall (SSW) sprinklers.

Performance Tasks

Under supervision, you should be able to do the following:

1. Given an instructor-supplied obstruction scenario, space a standard coverage spray sprinkler to avoid the obstruction.
2. Determine the maximum sprinkler spacing based on different types of construction.

Trade Terms

Cold soldering
Combustible
Fusing (operating) temperature
Joist
Limited-combustible
Noncombustible
Obstructed construction
Pendent
Sidewall
Small room
Unobstructed construction
Upright

Industry Recognized Credentials

If you are training through an NCCER-accredited sponsor, you may be eligible for credentials from NCCER's Registry. The ID number for this module is 18205. Note that this module may have been used in other NCCER curricula and may apply to other level completions. Contact NCCER's Registry at 1.888.622.3720 or go to **www.nccer.org** for more information.

You can also show off your industry-recognized credentials online with NCCER's digital badges. Transform your knowledge, skills, and achievements into badges that you can share across social media platforms, send to your network, and add to your resume. For more information, visit **www.nccer.org**.

NOTE

Unless otherwise specified, references in parentheses following figure and table numbers refer to NFPA 13.

Contents

Figures

Tables

Section One

1.0.0 Standard Spray Sprinkler Characteristics

Objective

Identify the characteristics of standard spray sprinklers.

a. Identify sprinkler temperature ratings and their associated color codes.
b. Define K-factor.
c. Describe sprinkler location, spacing, and positioning requirements.

Trade Terms

Pendent: Refers to a sprinkler installed so that the deflector is below the fusible element and designed to direct the water stream downward against the deflector.

Sidewall: Refers to a sprinkler installed so that the deflector discharges most of the water away from the nearby wall in a pattern resembling one quarter of a sphere, with a small portion of the discharge directed at the wall behind the sprinkler.

Upright: Refers to a sprinkler installed so that water distribution is directed upward against a deflector.

Standard spray sprinklers have been used extensively over the last fifty years. The most common type of standard spray sprinkler is the standard coverage sprinkler. These sprinklers are available in three configurations: upright sprinklers, which are typically installed on the top of pipe; pendent sprinklers, which are typically installed on the bottom of pipe; and sidewall sprinklers, which are typically installed on the side of pipe.

Both upright and pendent sprinklers have water distribution or spray patterns that resemble an umbrella shape. One of the characteristics of their spray pattern is that the diameter of the pattern measures 16' at a point 4' below the sprinkler when it is discharging at 15 gpm. A pendent sprinkler may be referred to as a *standard spray pendent (SSP)* and an upright sprinkler may be referred to as a *standard spray upright (SSU)*. Pendent sprinklers are commonly used where pipe is concealed. Upright sprinklers are used where pipe is run exposed. The pipe serves as built-in protection to limit physical damage to the upright sprinklers.

Sidewall sprinklers have water distribution or spray patterns that resemble half of the umbrella shape. They are designed to discharge most of the water away from the nearby wall with a small portion of the discharge directed at the wall behind the sprinkler.

Sprinklers consist of a fusible element (link or bulb), orifice, deflector, and frame (*Figure 1*). All sprinklers have these basic components. However, they all do not perform the same. Many different types of sprinklers are available. It is critical that sprinkler fitters know the type, function, and limitations of each sprinkler before installing it.

Standard coverage sprinklers use a fusing element to release the seat at a given time delay and temperature. Standard coverage sprinklers are available in both standard-response and quick-response types. Often the intended use or occupancy classification of the area will dictate sensitivity requirements. Light hazard occupancies, which are generally places where people gather, must have a quick-response sensitivity, while ordinary hazard occupancies, such as manufacturing facilities, may have a standard-response sensitivity. It is important to ensure that all sprinklers within a given area have the same sensitivity. This ensures that the sprinklers nearest the fire activate first, avoiding the activation of unneeded sprinklers, which might overwhelm the water supply.

Listing agencies require that quick-response sprinklers activate significantly faster than standard-response sprinklers. It is therefore necessary to ensure that the correct type of sprinkler is installed.

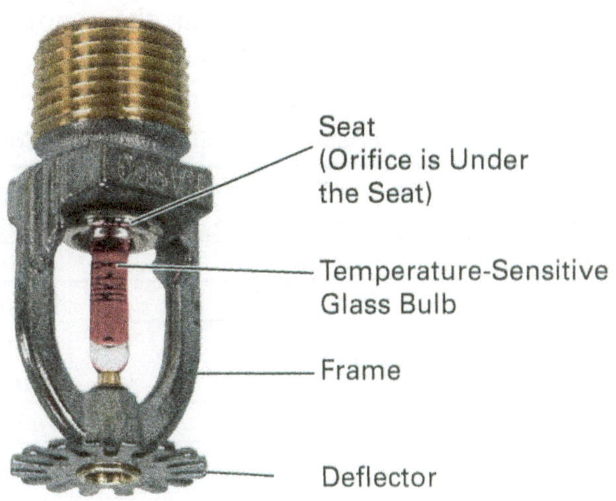

Figure 1 Sprinkler construction.

1.1.0 Temperature Ratings

The temperature ratings of sprinklers are categorized into temperature classifications. *Table 1* identifies these temperature classifications and the associated temperature range for each type. The most common temperature classifications are ordinary, intermediate, and high.

Normally, ordinary temperature-rated sprinklers are used throughout a building. The decision to use one temperature classification over another is influenced by the maximum ambient temperature of the room, special conditions such as proximity to heat sources, or design considerations. *Table 1* shows that the maximum ambient ceiling temperature for an ordinary temperature-rated sprinkler is 100°F. Exceeding the maximum ambient temperature rating of a sprinkler can place stress on the fusible element, potentially resulting in an inadvertent activation. Take care to ensure sprinklers are not exposed to high temperatures prior to installation.

It is sometimes necessary to install sprinklers near heat sources, such as the unit heaters installed in large spaces including big box stores, garages, and warehouses (*Figure 2*). Unit heaters are also installed above commercial doorways in cooler climates to provide a blanket of warm air that prevents cold air from rushing in whenever the door opens. Follow NFPA 13 requirements for sprinklers installed near these heat sources. *Figure 3* and *Table 2* summarize the minimum allowable temperature ratings of sprinklers near common heat sources.

Unit heaters should be indicated on the approved shop drawings. When conditions do not match the approved shop drawings, notify your supervisor, who will consult with the sprinkler system layout technician so that the conditions can be evaluated and corrective action taken.

Figure 2 Unit heater.

Fire protection engineers or layout technicians may specify a specific temperature rating for a sprinkler. The ratings specified are based on design considerations, such as the Listing limitations of a sprinkler at certain spacings or reductions in the hydraulic calculation used to size the sprinkler system piping.

Minimum temperatures are also a concern. Most fire protection systems are water filled. As conditions approach freezing temperatures,

Table 1 Temperature Ratings, Classifications, and Color Codings (NFPA 13, *Table 7.2.4.1*)

Maximum Ceiling Temperature (°F)	Temperature Rating (°F)	Temperature Classification	Color Code	Glass Bulb Colors
100	135–170	Ordinary	Uncolored or black	Orange or red
150	175–225	Intermediate	White	Yellow or green
225	250–300	High	Blue	Blue
300	325–375	Extra high	Red	Purple
375	400–475	Very extra high	Green	Black
475	500–575	Ultra high	Orange	Black
625	650	Ultra high	Orange	Black

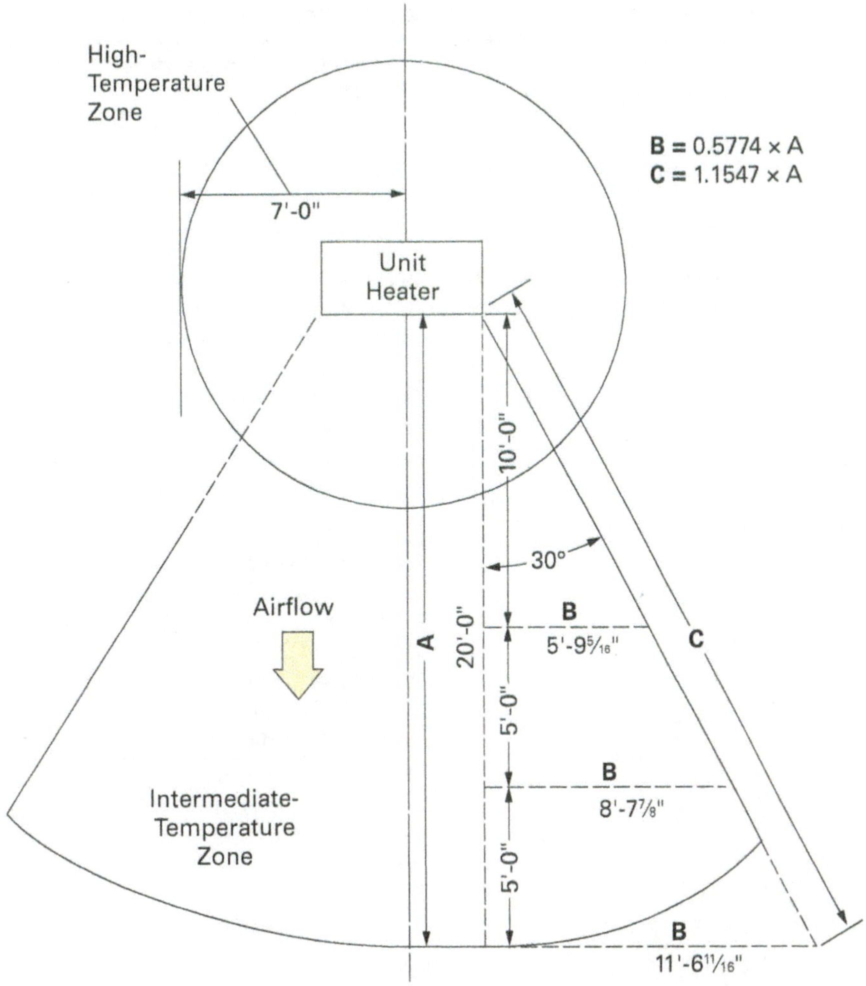

Note: This training material is not affiliated with nor has it been reviewed or approved by the NFPA. Reproduced with permission of NFPA from NFPA 13, *Standard for the Installation of Sprinkler Systems*, 2019 edition. © 2018, National Fire Protection Association. For a full copy of NFPA 13, please go to www.nfpa.org.

Figure 3 High-temperature and intermediate-temperature zones at unit heaters and radiant heaters (NFPA 13, *Figure 9.4.2.5*).

consideration to the type of sprinkler selected must be given. Systems installed in areas subject to freezing conditions may be filled with compressed gas. In these situations, upright sprinklers or dry pendent sprinklers are preferred. Pendent sprinklers may be used when the system is an antifreeze system, or the piping and sprinkler are installed on a return bend in a heated area.

1.2.0 Sprinkler K-Factors

Standard coverage sprinklers are available in a wide variety of orifice sizes or *K-factors*. A sprinkler's K-factor represents the water flow from the sprinkler at a given pressure. Historically, NFPA 13 referenced orifice sizes, such as ½" or ¹⁷⁄₃₂", but now references all sprinklers by their nominal K-factor (*Table* 3). K-factors for standard coverage sprinklers range from 1.4 to 8 or higher, depending on the manufacturer.

All new sprinklers are permanently marked with an alphanumeric *sprinkler identification number (SIN)* (*Figure* 4). These markings provide a unique identification for every change in K-factor, orientation, pressure rating, and thermal sensitivity. Every manufacturer has its own marking system. Before installing any sprinklers, check that the SIN, K-factor, and temperature match the sprinkler legend on the approved shop drawings.

Table 2 Temperature Ratings of Sprinklers Based on Distance from Heat Sources [NFPA 13, *Table 9.4.2.5(a)*]

Type of Heat Condition	Ordinary-Temperature Rating	Intermediate-Temperature Rating	High-Temperature Rating
(1) Heating ducts			
(a) Above	More than 2'-6"	2'-6" or less	—
(b) Side and below	More than 1'-0"	1'-0" or less	—
(c) Diffuser	Any distance except as shown under *Intermediate-Temperature Rating* column	*Downward discharge*: Cylinder with 1'-0" radius from edge extending 1'-0" below and 2'-6" above *Horizontal discharge*: Semicylinder or cylinder with 2'-6" radius in direction of flow extending 1'-0" below and 2'-6" above	—
(2) Unit heater and radiant heater			
(a) Horizontal discharge		*Discharge side*: 7'-0" radius to 20'-0" radius pie-shaped cylinder extending 7'-0" above and 2'-0" below heater, see *Figure 3*; also 7'-0" radius cylinder more than 7'-0" above unit heater	7'-0" radius cylinder extending 7'-0" above and 2'-0" below unit heater
(b) Vertical downward discharge (for sprinklers below unit heater, see *Figure 3*)		7'-0" radius cylinder extending upward from an elevation 7'-0" above unit heater	7'-0" radius cylinder extending from top of unit heater to an elevation 7'-0" above unit heater
(3) Steam mains (uncovered)			
(a) Above	More than 2'-6"	2'-6" or less	—
(b) Side and below	More than 1'-0"	1'-0" or less	—
(c) Blowoff valve	More than 7'-0"	—	7'-0" or less

1.3.0 Sprinkler Location, Spacing, and Position

The location, spacing, and position requirements of a sprinkler depend on the occupancy in which the sprinkler is to be used, the type of construction in which it is installed, and the particular type of sprinkler. Consult all available information to verify sprinkler specifications before any installation.

The location, spacing, and position of sprinklers are indicated on the approved shop drawings. The location shows the sprinkler fitter where to install the sprinklers, the spacing indicates the distance between the sprinklers, and the position refers to how they are installed to meet code compliance and be cost-effective.

NFPA 13 states that sprinklers must be installed throughout the premises. The sprinklers must be located so that the discharge does not exceed the maximum protection area for each sprinkler and positioned to optimize the activation time and water distribution. Improper location of a sprinkler can delay its activation and result in the distribution pattern being obstructed by structural and architectural features.

The best source of information about a sprinkler is the manufacturer's data sheet. See the *Appendix* for an example. Data sheets contain all Listing parameters for the sprinkler as well as specific instructions for installation, including tightening torques, use of Teflon® tape, etc. Some of these parameters may differ from NFPA 13 standards. If this occurs, follow the product Listing for specific requirements.

Table 3 Sprinkler Discharge Characteristics (NFPA 13, *Table 7.2.2.1*)

Nominal K-factor [gpm/(psi)$^{1/2}$]	K-factor Range [gpm/(psi)$^{1/2}$]	Percent of Nominal K-5.6 Discharge	Thread Type
1.4	1.3–1.5	25	½" NPT
1.9	1.8–2.0	33.3	½" NPT
2.8	2.6–2.9	50	½" NPT
4.2	4.0–4.4	75	½" NPT
5.6	5.3–5.8	100	½" NPT
8.0	7.4–8.2	140	¾" NPT or ½" NPT
11.2	10.7–11.7	200	½" NPT or ¾" NPT
14.0	13.5–14.5	250	¾" NPT
16.8	16.0–17.6	300	¾" NPT
19.6	18.6–20.6	350	1" NPT
22.4	21.3–23.5	400	1" NPT
25.2	23.9–26.5	450	1" NPT
28.0	26.6–29.4	500	1" NPT

Notes:
The nominal K-factor for dry-type sprinkler is used for sprinkler selection. See NFPA 13, *Section 27.2.4.10.3* for use of adjusted dry-type sprinklers for hydraulic calculation purposes.

Figure 4 Sprinkler identification number.

Changes sometimes occur on construction projects that are not corrected on the approved shop drawings before work begins. At the start of each job, inspect all areas to ensure that the sprinklers can be located as specified on the approved shop drawings. This inspection provides an opportunity to resolve any problems before installing the sprinklers. For example, if a beam or lighting fixture creates an obstruction, armovers may be required to complete the installation. Even the most experienced layout technician can make a location error or may have incomplete information from other contractors working on the project. The sprinkler fitter must recognize when something is not right and consult with the layout technician to find and document a solution.

Always check with your supervisor before making system modifications. Any modification to the original contract should be submitted as a change order and clearly marked on the field drawings to be incorporated into the as-built drawings.

1.3.1 Sprinkler Storage and Handling

Sprinklers are precision life-safety devices and must be treated with care to prevent damage. Keep the following in mind when storing and handling sprinklers:

- Sprinklers are equipped with a temperature-sensitive glass bulb or fusible link, both of which must be protected against temperature extremes. Store sprinklers in a cool, dry environment. Never store sprinklers outdoors or in unconditioned spaces. Always check to make sure you have not left a box of sprinklers on the truck. The temperature of a vehicle parked in the sun can quickly exceed the sprinkler's temperature rating.
- Sprinklers are shipped with a protective plastic cap or strap to prevent damage during construction (*Figure 5*). Do not remove the cap/strap until construction is complete.
- Handle sprinklers with care and keep them in the manufacturer's packaging until they are installed. Sprinklers are easily damaged by rough handling. Never throw or drop boxes of sprinklers.

CAUTION

In the past, it was common to throw sprinklers into a bag or bucket to carry them up a ladder and around a jobsite. This is no longer done as it can result in damage, including bent deflectors and broken bulbs.

1.3.2 Sprinkler Installation

Sprinklers must be installed properly to prevent damage and ensure proper operation. Keep the following in mind when installing sprinklers:

- Before installation, verify that the sprinkler matches the model, orifice size, and temperature rating shown on the sprinkler legend.

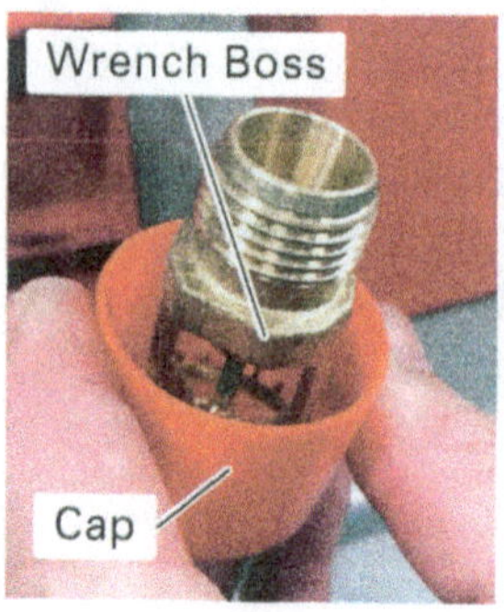

Cap

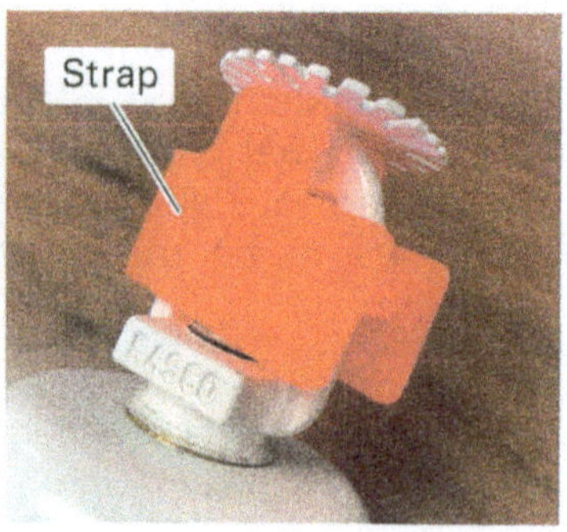

Strap

Figure 5 Sprinkler cap and strap.

- Always read and follow the manufacturer's instructions for the installation of each sprinkler type. Never assume anything. Instructions often change as products are improved.
- Do not install sprinklers until the piping system is fully installed and in place. Never attach sprinklers to a fitting and then install the fitting. For example, do not install a sprinkler into a CPVC reducing coupling and then solvent-cement the coupling into place. Always connect the coupling first.
- Inspect each sprinkler before it is installed. Do not install sprinklers that have been dropped or damaged. Make sure that glass bulbs are not cracked or missing fluid. Check for damaged fusible links. Look for bent or damaged frames, deflectors, and threads. Discard damaged sprinklers.
- Install the sprinkler in its intended orientation only. Sprinkler deflectors are designed to discharge in a specific orientation.
- If an obstruction prevents the installation of a sprinkler as indicated on the approved shop drawings, stop and inform your supervisor, who will consult with the layout technician to determine and document a solution.
- Do not handle sprinklers by the deflector, which may be easily damaged.
- Thread the sprinkler into the fitting by hand, then tighten the sprinkler using the manufacturer-approved sprinkler wrench. Sprinkler wrenches are not interchangeable and vary by manufacturer, frame style, and sprinkler type. Make sure the sprinkler wrench is positioned against the wrench boss and not the frame or bulb. Tighten the sprinkler to the maximum torque indicated in the manufacturer's instructions. Do not over-tighten.

WARNING!

Never use a standard wrench, such as a crescent wrench, to install sprinklers. Using the incorrect wrench can damage the sprinkler, resulting in a failed inspection and preventing the sprinkler from operating in the event of a fire.

- Carefully remove sprinkler straps/caps after construction is complete and the sprinklers are ready to be put into service. Follow the manufacturer's instructions for strap/cap removal.
- Do not paint or alter sprinklers. This will violate the product Listing and prevent the sprinkler from operating as expected in the event of a fire.
- Before leaving the job, make sure to include at least two of every sprinkler type in the spare sprinkler box, along with their matching wrenches.

Did You Know?

The first standard spray sprinklers were installed in the United States in 1953. Old-style sprinklers sprayed 40 to 60 percent of the water upward toward the ceiling. Today, standard spray sprinklers spray all their water downward at the fire. This change was based on research showing that fire control is improved when sprinklers spray downward.

1.0.0 Section Review

1. A sprinkler with a blue glass bulb has a temperature classification of _____.
   a. ordinary
   b. high
   c. extra high
   d. very extra high

2. A sprinkler with a K-factor of 5.6 has an NPT type of _____.
   a. ½"
   b. ⅝"
   c. ¾"
   d. 1"

3. If a sprinkler's Listing differs from NFPA 13 requirements, follow the Listing requirements.
   a. True
   b. False

SECTION TWO

2.0.0 TYPES OF OCCUPANCIES

Objective

Identify NFPA 13 occupancy classifications.

a. Define light hazard occupancies.
b. Define ordinary hazard occupancies.

Trade Term

Combustible: Refers to materials that can burn.

Occupancies in NFPA 13 are general classifications of the contents of a structure and do not correlate with any building code. Occupancy classifications are used to determine which types of sprinklers are required, how much water these sprinklers need, and at what location they must be installed based on the contents and potential hazards in a building. The sprinkler location, spacing, and position cannot be determined until the occupancy is defined.

Occupancies covered by NFPA 13 standards include:

- Light hazard, including residential
- Ordinary hazard
- Extra hazard
- Storage areas

This module focuses on light and ordinary hazard occupancies.

2.1.0 Light Hazard Occupancies

A light hazard occupancy is a place where there are people but there is no heavy machinery operating and there is no storage. It is essentially a place where people assemble.

Light hazards also include occupancies or portions of other occupancies in which the quantity and/or combustibility of contents are low and fires with relatively low rates of heat release are expected. Typical light hazard occupancies include the following:

- Animal shelters
- Churches
- Clubs
- Eaves and overhangs (if combustible construction with no combustibles beneath)
- Educational
- Hospitals, including animal hospitals and veterinary facilities
- Institutional
- Kennels
- Libraries, except large stack rooms
- Museums
- Nursing or convalescent homes
- Offices, including data processing
- Residential
- Restaurant seating areas
- Theaters and auditoriums, excluding stages and prosceniums
- Unused attics

2.1.1 Residential Occupancies

Residential occupancies are covered by three different NFPA standards: NFPA 13, NFPA 13R, and NFPA 13D. The NFPA requirements for residential sprinklers differ from those for standard coverage sprinklers.

NFPA 13 allows the use of residential sprinklers in dwelling units and their associated corridors. Dwelling units are defined as one or more rooms arranged for the use of one or more individuals living together, as in a single housekeeping unit normally having cooking, living, sanitary, and sleeping facilities. For the purposes of NFPA 13, dwelling units include hotel rooms, dormitory rooms, apartments, condominiums, sleeping rooms in nursing homes, and similar units. Therefore, a building such as a six-story hotel could use residential sprinklers in the dwelling units and corridors without applying the light hazard installation criteria, but the rest of the building would have to meet NFPA 13 criteria for that specific occupancy. For example, a hotel may have a mechanical room, parking area, and retail area that are all ordinary hazard occupancies. The hotel may also have lobby and restaurant areas that are light hazard occupancies. Even so, the hotel may still use residential sprinklers in the dwelling units and associated corridors.

Residential sprinklers must provide a fast response to ensure life safety. These sprinklers are designed to throw water higher than standard spray sprinklers. High wall wetting protects walls that may be covered in combustible materials, such as curtains or pictures, and any furniture positioned against the walls.

Did You Know?

Residential sprinklers have proven to be so critical to life safety that some municipalities now mandate their use in new residential construction.

2.2.0 Ordinary Hazard Occupancies

Ordinary hazard occupancies include areas with machinery and/or storage of combustible materials. There are two groups of ordinary hazard occupancies.

Group 1 hazards (OH1) include occupancies or portions of other occupancies in which combustibility is low, quantity of combustibles is moderate, stockpiles of combustibles do not exceed 8', and fires with moderate heat-release rates are expected. Typical Group 1 hazards include the following:

- Automobile showrooms
- Bakeries
- Beverage manufacturing
- Canneries
- Dairy products manufacturing and processing
- Electronic plants
- Glass and glass products manufacturing
- Laundries
- Restaurant service areas
- Porte cocheres (covered driveway areas, as in hotel entrances)
- Mechanical rooms

Group 2 hazards (OH2) include occupancies or portions of other occupancies in which quantity and combustibility of contents is moderate to high, stockpiles do not exceed 12', and fires with moderate to high rates of heat release are expected. Typical Group 2 hazards include the following:

- Agricultural facilities
- Barns and stables
- Cereal mills
- Chemical plants (ordinary)
- Confectionery products
- Distilleries
- Dry cleaners
- Exterior loading docks (Note that exterior loading docks used only for loading and unloading of ordinary combustibles should be classified as OH2. For the handling of flammable and combustible liquids, hazardous materials, or where utilized for storage, exterior loading docks should be protected based upon the actual occupancy and the materials handled on the dock, as if the materials were actually stored in that configuration.)
- Feed mills
- Horse stables
- Leather goods manufacturing
- Libraries (large stack room areas)
- Machine shops
- Metal working
- Mercantile
- Paper and pulp mills
- Paper processing plants
- Piers and wharves
- Plastics fabrication, including blow molding, extruding, and machining; excluding operations using combustible hydraulic fluids
- Post offices
- Printing and publishing
- Racetrack stable/kennel areas, including those stable/kennel areas, barns, and associated buildings at state, county, and local fairgrounds
- Repair garages
- Resin application areas
- Stages
- Textile manufacturing
- Tire manufacturing
- Tobacco products manufacturing
- Wood machining
- Wood product assembly

2.2.1 Multiple Occupancies

Buildings can contain multiple occupancies. For example, an office building may have retail floors at ground level, offices above, and a mechanical room on the roof. The retail and mechanical room would be ordinary hazard occupancies that are separated by office space, which is a light hazard occupancy. Another example would be light hazard office space in a portion of a high-piled storage warehouse. In addition, the attic over any space, as long as it is not used for storage, is typically a light hazard occupancy. For example, a machine shop or processing area would typically be an ordinary hazard occupancy, but the attic above it would be a light hazard occupancy. Each type of occupancy requires different amounts of water and different sprinkler types.

The occupancy examples described in the various hazard classifications represent the norm for those occupancy types. Unusual fuel loading, combustible characteristics, and the potential for change must be considered in the selection of a particular occupancy.

2.0.0 Section Review

1. A light hazard occupancy is *best* described as a(n) _____.
   a. storage area
   b. place of assembly
   c. agricultural facility
   d. manufacturing facility

2. Which of the following would be classified as an OH2 occupancy?
   a. Repair garage
   b. Church
   c. Restaurant service area
   d. Museum

Section Three

3.0.0 Types of Construction

Objective

Identify the types of construction.

a. Describe the types of unobstructed construction.
b. Describe the types of obstructed construction.

Trade Terms

Joist: A structural member that provides support.

Noncombustible: A material that, in the form in which it is used, and under the conditions anticipated, will not ignite, burn, support combustion, or release flammable vapors when subjected to fire or heat.

Obstructed construction: Beams, trusses, or other structural members that impede heat flow or water distribution and affect the ability of sprinklers to control or suppress a fire.

Unobstructed construction: Construction where beams, trusses, or other members do not impede heat flow or water distribution in a manner that materially affects the ability of sprinklers to control or suppress a fire.

Construction types are defined by NFPA 13. In some cases, they will affect the sprinkler installation, depending on the deflector distance and spacing requirements. Just as the occupancy puts limitations on the spacing location and types of sprinklers that can be used, construction types further narrow the available sprinklers and limit the installation rules.

The two types of construction are unobstructed and obstructed. Correctly identifying the construction type is critical to determining the sprinkler spacing, location, and position.

3.1.0 Unobstructed Construction

Unobstructed construction is construction where beams, trusses, or other members do not impede heat flow or water distribution in a manner that materially affects the ability of sprinklers to control or suppress a fire. Unobstructed construction has horizontal structural members that are not solid, where openings are at least 70 percent of the cross-sectional area and member depth does not exceed the least dimension of the openings. This type represents all construction types where the spacing of structural members exceeds $7\frac{1}{2}$' on center.

The following are examples of unobstructed construction. Definitions from NFPA 13 are provided as guidelines to assist in determining the construction type.

3.1.1 Bar Joist Construction

Bar joist construction (*Figure 6*) refers to construction employing joists that have steel truss-shaped members. Wood truss-shaped members that consist of wood top and bottom chord members with steel tube or bar webs, are also defined as *bar joists*. Bar joist construction includes noncombustible or combustible roof or floor decks.

3.1.2 Open-Grid Ceiling Construction

Open-grid ceilings have openings that are $\frac{1}{4}$" or larger in the least dimension, the thickness of the ceiling material does not exceed the least dimension of the openings, and such openings constitute at least 70 percent of the ceiling area.

Figure 6 Bar joist construction.

3.1.3 Smooth Ceiling Construction

As defined in NFPA 13, smooth ceiling construction includes the following:

- Flat-slab, pan-type reinforced concrete
- Continuous, smooth bays formed by wood, concrete, or steel beams spaced more than $7\frac{1}{2}$' on center with beams supported by columns, girders, or trusses
- Smooth roof or floor decks supported directly on girders or trusses spaced more than $7\frac{1}{2}$' on center
- Smooth monolithic ceilings of at least $\frac{3}{4}$" of plaster on metal lath or a combination of materials of equivalent fire-resistive rating attached to the underside of wood joists, wood trusses, and bar joists
- Open-web steel beams, regardless of spacing
- Smooth shell-type roofs, such as folded plates, hyperbolic paraboloids, saddles, domes, and long barrel shells
- Suspended ceilings of combustible or noncombustible construction
- Smooth, monolithic ceilings with fire resistance less than that specified above attached to the underside of wood joists, wood trusses, and bar joists

One type of smooth ceiling construction is shown in *Figure 7*.

3.1.4 Wood Truss Construction

Wood truss construction refers to parallel or pitched wood chord members connected by open wood members (webbing) supporting a roof or floor deck. Trusses that have steel webbing (similar to bar joist construction), and top and bottom wood chords exceeding 4" in depth, should also be considered wood truss construction.

3.2.0 Obstructed Construction

Obstructed construction is construction where beams, trusses, or other members impede heat flow or water distribution in a manner that affects the ability of sprinklers to control or suppress a fire. It is important to note that this refers to both the obstruction of the spray pattern of the sprinkler and the heat flow across a ceiling. A common error is to believe that locating a sprinkler below a beam makes the construction type unobstructed since the spray pattern would be below the beam. Because the beam would interfere with the heat flow across the ceiling, the construction would be considered obstructed. The following are examples of obstructed construction.

Figure 7 Smooth ceiling construction.

3.2.1 Beam-and-Girder Construction

Beam-and-girder construction as used in NFPA 13 includes noncombustible and combustible roof or floor decks supported by wood beams of 4" or larger, concrete or steel beams spaced 3' to $7\frac{1}{2}$' on center, and either supported on or framed into girders. According to NFPA 13, where supporting a wood plank deck, this includes semi-mill and panel construction, and where supporting (with steel framing) gypsum plank, steel deck, concrete, tile, or similar material, this would include much of the so-called noncombustible construction.

3.2.2 Concrete Tee Construction

Concrete tee construction (*Figure 8*) refers to solid concrete members with stems (legs) having a nominal thickness less than the nominal height.

3.2.3 Composite Wood Joist Construction

Composite wood joist construction (*Figure 9*) refers to wood I-beams constructed of wood flanges and solid wood web that support a floor or roof deck. Composite wood joists may vary in depth up to 48", may be spaced up to 48" on centers, and may span up to 60' between supports. Joist channels should be firestopped to the full depth of the joists, using material equal to the web construction so that individual channel areas do not exceed 300 ft^2.

3.2.4 Panel Construction

Panel construction includes ceiling panels formed by members capable of trapping heat to aid the operation of sprinklers and limited to a maximum of 300 ft^2 in area.

Figure 8 Concrete tee construction.

There should be no unfilled penetrations in the cross-sectional area of the bounding structural members, including the interface at the roof. Beams spaced more than $7\frac{1}{2}$' apart and framed into girders qualify as panel construction, provided the 300 ft^2 area limitation is met.

3.2.5 Wood Joist Construction

Wood joist construction refers to solid wood members of a rectangular cross section, which may vary from 2" to 4" nominal width and up to 14" nominal depth. They are spaced up to 3' on centers, span up to 40' between supports, and support a floor or roof deck. Solid wood members less than 4" nominal width and up to 14" nominal depth, spaced more than 3' on center, are also considered wood joist construction.

Figure 9 Composite wood joist construction.

3.0.0 Section Review

1. A type of unobstructed construction is _____.
   a. beam-and-girder construction
   b. concrete tee construction
   c. bar joist construction
   d. wood joist construction

2. Because a beam would interfere with the heat flow across a ceiling, the construction would be considered obstructed.
   a. True
   b. False

SECTION FOUR

4.0.0 SPRINKLER SPACING REQUIREMENTS

Objective

Describe the installation considerations for standard spray sprinklers.

a. Explain the spacing requirements for standard spray pendent (SSP) and standard spray upright (SSU) sprinklers.
b. Explain the spacing requirements for standard spray sidewall (SSW) sprinklers.

Performance Tasks

1. Given an instructor-supplied obstruction scenario, space a standard coverage spray sprinkler to avoid the obstruction.
2. Determine the maximum sprinkler spacing based on different types of construction.

Trade Terms

Cold soldering: A condition that occurs when sprinklers are positioned in close proximity to one another. It results when one sprinkler distributes water on another sprinkler, preventing or delaying the adjacent sprinkler's activation by cooling it below the fusing temperature.

Fusing (operating) temperature: The temperature at which a sprinkler activates.

Limited-combustible: A material with a structural base of noncombustible material; an example is gypsum wallboard.

Small room: A compartment of light hazard occupancy classification having unobstructed construction and a floor area not exceeding 800 ft^2.

Sprinkler spacing is based on numerous factors, such as occupancy, construction type, sprinkler type, and hydraulic calculations. Sprinklers must be installed throughout the premises, they must not exceed their maximum allowable spacing, and they must be located to provide satisfactory activation time and water distribution.

4.1.0 Standard Spray Pendent (SSP) and Standard Spray Upright (SSU) Sprinklers

The most common types of sprinklers are standard pendent and upright spray sprinklers. These sprinklers are typically installed in light hazard occupancies with smooth, horizontal ceilings or sloped, flat ceilings.

4.1.1 Area of Coverage and Maximum Spacing

Once the occupancy type and the requirement for sprinklers have been determined, the maximum distance between standard spray sprinklers and maximum area of coverage can be determined. Maximum coverage is the floor area a sprinkler is capable of covering in the event of a fire. The maximum coverage area refers to the maximum protection area per sprinkler. It is based on response time, type of fire expected (occupancy), and type of sprinkler used. *Table 4* and *Table 5* show the maximum protection areas of sprinklers for light and ordinary hazard occupancies. Note the table is the maximum value but may not represent the design spacing, which may be limited by other factors and can only be found in the approved shop drawings.

Additional guidelines for spacing include the following:

- The distance from sprinklers to walls must not exceed one-half of the allowable distance between sprinklers, except when using the small room rule that allows 9' off of a single wall.
- Sprinklers must be located a minimum of 4" from a wall.
- Where standard spray sprinklers are spaced less than 6' on center, baffles sized not less than 8" long and 6" high must be installed. The tops of baffles must extend 2" to 3" above the deflectors of upright sprinklers. The bottoms of baffles must extend downward to a level at least even with the deflectors of pendent sprinklers. This is to prevent cold soldering, which results when one sprinkler distributes water on another sprinkler, preventing or delaying the adjacent sprinkler's activation by cooling it below its fusing (operating) temperature.

Table 4 Protection Areas and Maximum Spacing of Standard Pendent and Upright Spray Sprinklers for Light Hazard [NFPA 13, *Table 10.2.4.2.1(a)*]

Construction Type	System Type	Maximum Protection Area (ft^2)	Maximum Spacing (ft)
Noncombustible unobstructed	Hydraulically calculated	225	15
Noncombustible unobstructed	Pipe schedule	200	15
Noncombustible obstructed	Hydraulically calculated	225	15
Noncombustible obstructed	Pipe schedule	200	15
Combustible unobstructed with no exposed members	Hydraulically calculated	225	15
Combustible unobstructed with no exposed members	Pipe schedule	200	15
Combustible unobstructed with exposed members 3' or more on center	Hydraulically calculated	225	15
Combustible unobstructed with exposed members 3' or more on center	Pipe schedule	200	15
Combustible unobstructed with exposed members less than 3' on center	All	130	15
Combustible obstructed with exposed members 3' or more on center	All	168	15
Combustible obstructed with exposed members less than 3' on center	All	130	15
Combustible concealed spaces inaccordance with NFPA 13, *Section 10.2.6.1.4*	All	120	15 parallel to the slope, 10 perpendicular to the slope*

Notes:
*See NFPA 13, *Section 10.2.6.1.4.4*.

Table 5 Protection Areas and Maximum Spacing of Standard Pendent and Upright Spray Sprinklers for Ordinary Hazard [NFPA 13, *Table 10.2.4.2.1(b)*]

Construction Type	System Type	Maximum Protection Area (ft^2)	Maximum Spacing (ft)
All	All	130	15

S × L Method

NFPA 13 determines the protection area of coverage for a sprinkler (A_s) using the S × L method. This method adheres to the following procedure:

Step 1 Measure the distance between the sprinklers on the branch line or twice the distance to the wall in the case of the end sprinkler on a branch line (*Figure 10*). Whichever is greater, call this *S*.

Step 2 Measure the distance between branch lines or twice the distance to the wall in the case of the end branch line. Whichever is greater, call this *L*.

Step 3 Multiply *S* and *L* to determine the area the sprinkler is covering.

Step 4 For example, if the spacing off the wall is 5' (*Figure 11*), multiply this measurement by 2 for twice the distance from the wall, or 10'. If the distance between the sprinklers on the branch line is 10', *S* is equal to 10'. If the distance to the wall from the end sprinkler of a branch line is 6' (6' × 2 = 12') and the distance between branch lines is 12', *L* is 12'. The maximum protection area of coverage of this sprinkler is 120 ft^2, since 10' × 12' = 120 ft^2.

Although it seems very simple, there are a few other important design considerations. For example, the maximum coverage per sprinkler cannot be exceeded. Therefore, the maximum distance between branch lines in ordinary hazard occupancies cannot be used in both directions (15' × 15' = 225 ft^2, not 130 ft^2). A reduction in the spacing would be required—such as 15' × 8' = 120 ft^2, which is less than 130 ft^2.

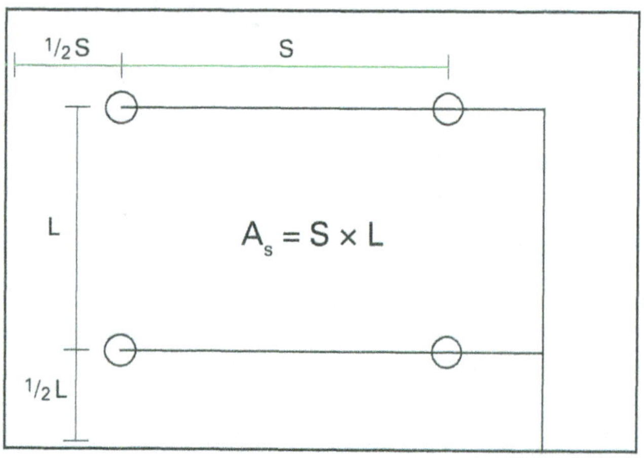

Figure 10 Determining the protection area of coverage for a sprinkler.

Therefore, more information is needed than just the rules. Also, the maximum distance off of a wall is one-half the maximum distance between the sprinklers.

WARNING!

It is a common misconception that a distance of 7'-6" off the wall can be applied in all instances. However, the maximum distance between sprinklers can be limited by the hydraulic calculations, which may be less than the maximum allowed by NFPA 13. Always follow the approved shop drawings, sprinkler Listing, and manufacturer's installation instructions. Over-spacing can result in a safety hazard.

Hydraulic calculations also play a role in the maximum spacing of a sprinkler. The layout technician uses hydraulic calculations to determine the pipe size. These hydraulic calculations are based on the spacing of the sprinklers. The layout technician has the option to space sprinklers at less than the maximum and when doing so, limits the maximum area of coverage of the sprinklers to a square footage less than the maximum allowed by NFPA 13. This could be done for a number of reasons that may not be obvious to the sprinkler fitter. What is important to understand is that this is very common. If the layout technician limits an ordinary hazard occupancy to 120 ft^2 instead of 130 ft^2 by the hydraulic calculations, the system is said to provide just enough water to control the anticipated fire if the sprinklers do not exceed 120 ft^2, not 130 ft^2. This would affect the maximum square footage that the two S × L dimensions could cover.

Occasionally, sprinkler fitters must reroute sprinklers in the field or determine if a change in the building affects the spacing. When doing this, keep in mind the S × L rule and do not think of spacing in general terms, such as 15' between sprinklers or 7'-6" off of the wall. Also, it is critical to know if the sprinklers are standard spray sprinklers or special extended coverage sprinklers. Extended coverage sprinklers have different NFPA 13 requirements.

Small Room Rule

The small room rule is an allowance in the standard to go beyond the maximum distance up to 9' off of a single wall for a standard pendent or upright sprinkler. This rule can only be applied in rooms that are 800 ft^2 or less, have unobstructed construction, and are classified as

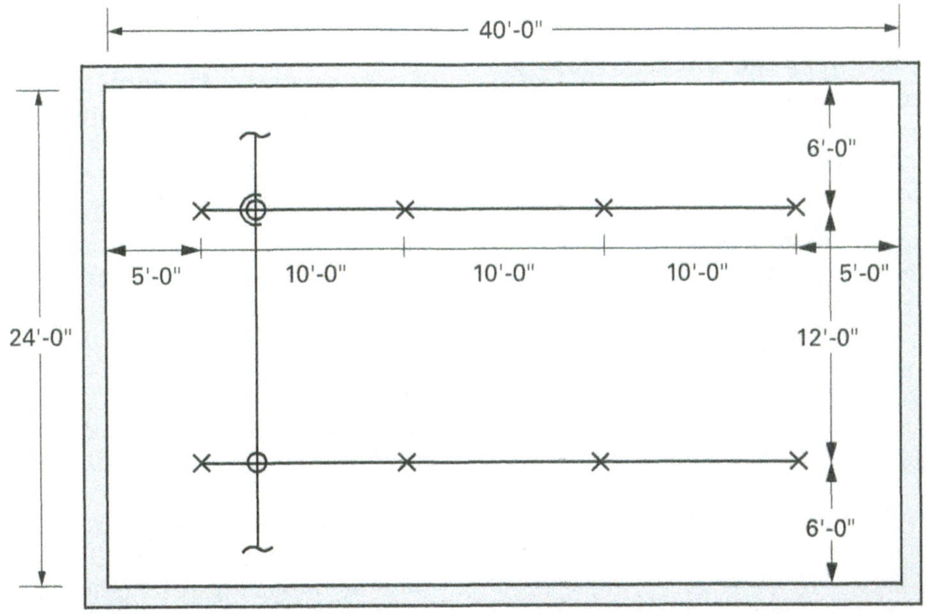

Figure 11 Sprinkler maximum protection area of coverage example.

light hazard. The spacing in the room cannot exceed the maximum spacing for a hydraulically calculated system, or 225 ft^2, or a pipe schedule system at 200 ft^2.

The spacing is determined by dividing the square footage of the room by the number of sprinklers in the room. This is the only use of this type of spacing in any NFPA standard. The main reason for this is to allow for lights and diffusers in small rooms, but not increase the amount of sprinklers in the room excessively. If the lights were in just the right location, the number of sprinklers required would go up with little or no benefit to the level of protection. Although the small room rule is used to determine the proper number of sprinklers for a small area with lights and diffusers, there is no prohibition against using the small room rule at any time. In other words, it is not a requirement to only use it when it is necessary to avoid lights and diffusers.

Where walls are angled or irregular, the maximum horizontal distance between a sprinkler and any point of floor area protected by that sprinkler should not exceed 0.75 times the allowable distance permitted between sprinklers, with the provision that the maximum perpendicular distance is not exceeded. For 15' spacing, this would be 15' × 0.75 = 11.25'.

Baffles

Contrary to widespread belief, there is no minimum spacing for standard spray sprinklers in the NFPA standards. However, there is a rule that requires baffles. Baffles stop one sprinkler from spraying another sprinkler and delaying it or preventing it from activating. Baffles must be installed wherever sprinklers are less than 6' apart. Baffles must be located midway between sprinklers and arranged to protect the actuating elements. Baffles must be of noncombustible or limited-combustible material that will stay in place before and during sprinkler operation. The baffles must be not less than 8" long and 6" high. The tops of baffles must extend 2" to 3" above the deflectors of upright sprinklers. The bottoms of baffles must extend downward to a level at least even with the deflectors of pendent sprinklers.

CAUTION

A baffle is not a heat collector. In fact, there is no such thing as a heat collector. In some cases, a piece of metal installed over a sprinkler located too far from the ceiling or deck has been referred to as a *heat collector* or *baffle*. There is no Listing for such a device nor has one ever been proven to be effective. In fact, there is evidence that this device will inhibit performance and should be avoided.

The US Fire Problem

According to the National Fire Protection Association, a Fire Department in the United States responds to a fire every 24 seconds. Local Fire Departments responded to 1,291,500 fires in 2019. These fires caused roughly 3,700 civilian deaths, 16,600 civilian injuries, and $14.8 billion in property damage.

Spacing Under Pitched Roofs

Spacing of sprinklers under pitched roofs requires a different and somewhat more complex set of rules. The position of deflectors must be parallel to ceilings, roofs, or the incline of stairs. Exceptions include the following:

- Sprinklers that are installed in the peak of a pitched roof must be installed with the deflector in the horizontal position.
- Pitched roofs having slopes not exceeding 2" per foot are considered horizontal in the application of this rule, and sprinklers must, according to NFPA 13, be permitted to be installed with deflectors horizontal.

It is important to measure the distance between sprinklers either on the branch lines or between the branch lines, running up or down the slope of a pitched surface, and along the slope.

Also, sprinklers under or near the peak must have deflectors located not more than 3' vertically down from the peak. The combination of these rules allows the sprinklers to be spaced in pitched roofs. *Figure 12* shows sprinklers in pitched roofs.

There are exceptions for noncombustible or limited-combustible, non-accessible vertical duct, electrical, or mechanical shafts. Attic sprinklers also have different requirements, as described in NFPA 13.

Sprinklers under a roof or ceiling in combustible concealed spaces of wood joist or wood truss construction, with members 3' or less on center and a slope having a pitch of 4 in 12 or greater, must be quick response. They must also be installed so that a row of sprinklers is located within 12" horizontally of the peak (*Figure 13*). Sprinklers installed along the eave must be located not less than 5' from the outer line of the concealed space.

4.1.2 Positioning

Sprinkler positioning is critical to system performance. The position is determined by the type of sprinkler, occupancy, and type of construction. Proper positioning includes determining the distance between the deflector and structures as well as avoiding obstructions and locations near heat sources. It also includes protection against damage, as in gymnasiums or storage areas. Any sprinkler subject to physical damage must be protected by a Listed guard or protective device.

An improperly positioned sprinkler may cause the sprinkler to activate too slowly or too quickly, or the spray pattern may be blocked by an obstruction and not provide full coverage.

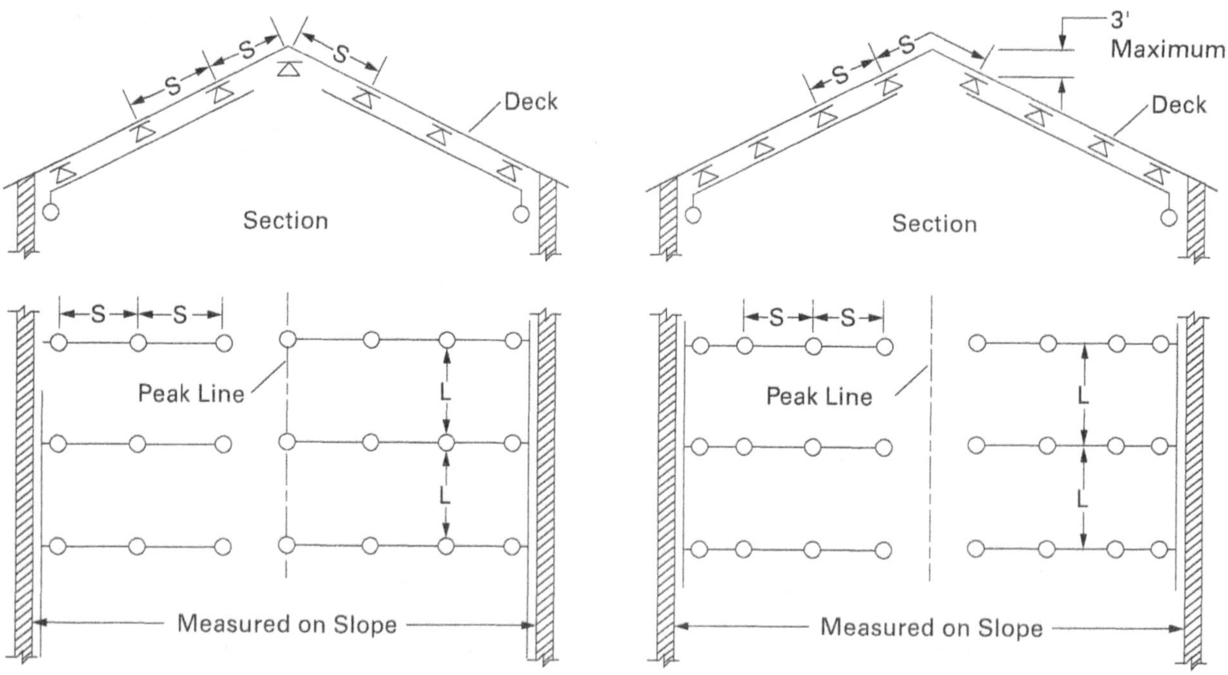

Figure 12 Sprinklers under pitched roofs [NFPA 13, *Figures 10.2.6.1.3.1(a) and (b)*].

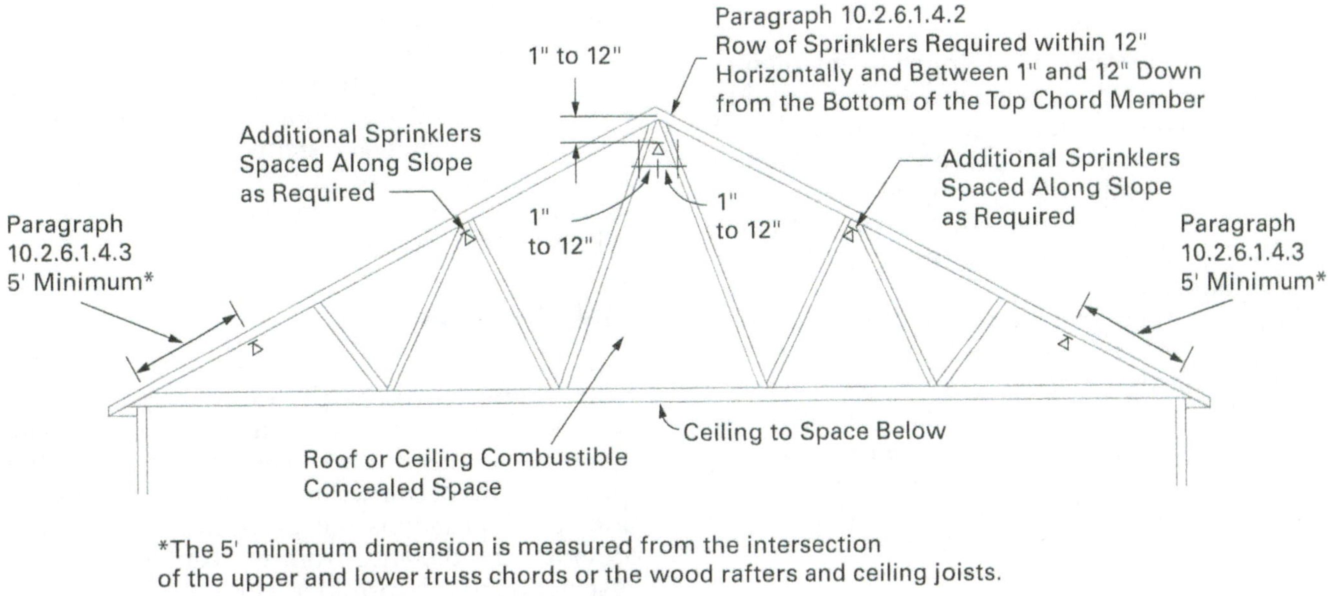

Figure 13 Sprinklers under a roof or ceiling in combustible concealed spaces (NFPA 13, *Figure 10.2.6.1.4*).

The deflector distance of a sprinkler is always measured from the deflector regardless of whether it is a pendent or an upright. The rules are based on the types of construction.

Obstructed Construction

Under obstructed construction, the sprinkler deflector must be installed in one of the following arrangements:

- The sprinkler deflector must be located 1" to 6" below the structural members and a maximum distance of 22" below the ceiling/roof deck.
- Sprinklers installed with the deflector at or above the bottom of the structural member to a maximum of 22" below the ceiling/roof deck where the sprinkler is installed per NFPA 13, *Section 10.2.7.1.2*.
- Deflectors must be a minimum of 1" and a maximum of 12" below the ceiling where sprinklers are installed in each bay of obstructed construction.
- Sprinklers installed with deflectors within the horizontal planes 1" to 6" below composite wood joists to a maximum of 22" below the ceiling/roof deck where joist channels are firestopped to the full depth of the joists, using materials equivalent to the web construction so that individual channel areas do not exceed 300 ft^2.
- Deflectors of sprinklers under concrete tee construction with stems spaced less than 7$\frac{1}{2}$' but more than 3' on centers can be located at or above a horizontal plane 1" below the bottom of the stems of the tees and must comply with NFPA 13, *Table 10.2.7.1.2*.

Unobstructed Construction

Under unobstructed construction, the distance between the sprinkler deflector and the ceiling must be a minimum of 1" and a maximum of 12" with the exception of special ceiling sprinklers (concealed, recessed, and flush types). Special ceiling sprinklers are permitted to have the operating element above the ceiling and the deflector located nearer to the ceiling where it is installed in accordance with their Listing.

Where there is a vertical change in ceiling elevation within the area of coverage of the sprinkler creating a distance of more than 36" between the upper ceiling and the sprinkler deflector, a vertical plane extending down from the ceiling at the change in elevation is considered a wall for the purpose of sprinkler spacing (*Figure 14*). Where the distance between the upper ceiling and the sprinkler deflector is less than or equal to 36", the sprinklers are permitted to be spaced as though the ceiling was flat, provided the obstruction rules are observed.

4.1.3 Clearance

A minimum clearance of 18" must be maintained between the top of storage and ceiling sprinkler deflectors. This is important to the sprinkler fitter since he or she is the person most capable of determining if this rule is being followed. If the clearance is not maintained, the spray pattern will not be allowed to form, resulting in unprotected floor space.

The exceptions include where other standards specify greater minimums. In these cases, the other standards must be followed and a minimum clearance of 36" must be permitted for special sprinklers. The 18" dimension is not intended to limit the height of shelving on a wall or shelving against a wall in accordance with NFPA 13.

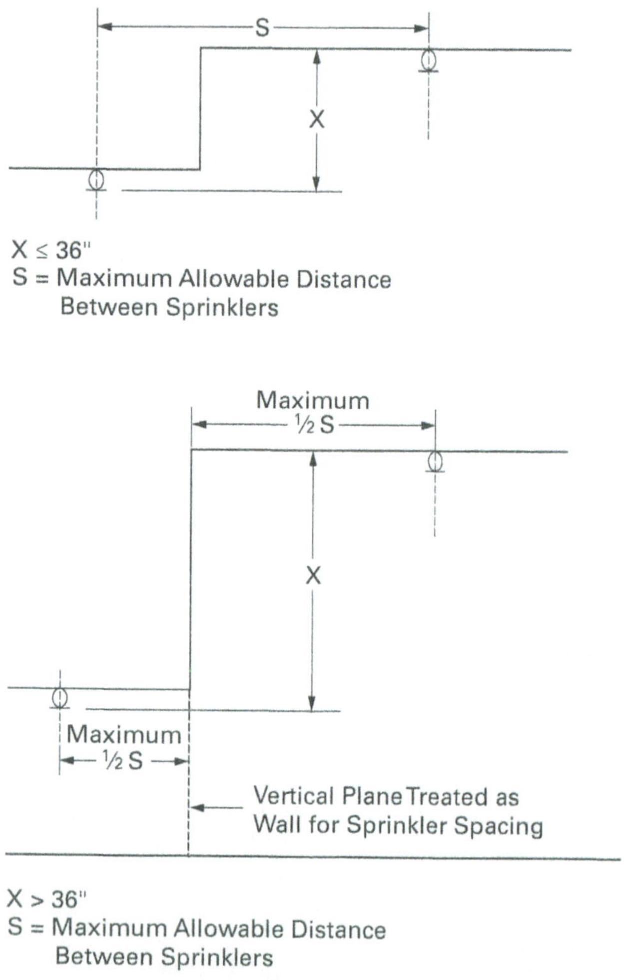

Figure 14 Vertical changes in ceiling elevations [NFPA 13, *Figures 10.2.6.1.1.3(A) and (B)*].

Where shelving is installed on a wall and is not directly below sprinklers, the shelves, including what is on them, may extend above the level of a plane located 18" below the ceiling sprinkler deflectors. Shelving, and items stored on shelving that is placed directly below the sprinklers, may not extend above the plane located 18" below the ceiling sprinkler deflectors.

4.1.4 Obstructions

Sprinklers should be located so as to minimize obstructions to discharge, or additional sprinklers should be provided to ensure adequate coverage of the hazard. The exception is that sprinklers are permitted to be spaced on opposite sides of obstructions not exceeding 4' in width, provided the distance from the centerline of the obstruction to the sprinklers does not exceed one-half the allowable distance permitted between sprinklers. The sprinklers should be arranged to comply with *Figure 15* and *Table 6*.

Sprinklers must be installed under fixed obstructions over 4' in width, such as large ductwork. Note that this dimension includes the insulation, which may not be installed until after the HVAC system has been tested. Sprinklers located under obstructions must either be installed below or within 3" of the outside edge of the obstruction. The deflector of any sprinkler installed beneath a fixed obstruction must be located no more than 12" below the bottom of the obstruction.

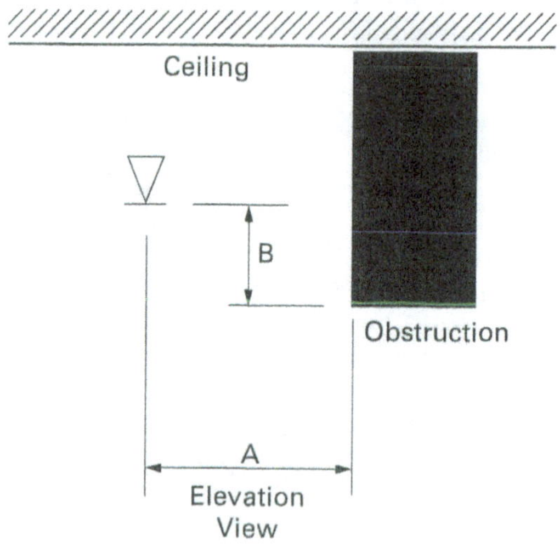

Figure 15 Positioning sprinklers to avoid obstructions to discharge (SSU/SSP) [NFPA 13, *Figure 10.2.7.1.2(a)*].

Table 6 Positioning of Sprinklers to Avoid Obstructions to Discharge (SSU/SSP) (NFPA 13, *Table 10.2.7.1.2*)

Distance from Sprinklers to Side of Obstruction (A)	Maximum Allowable Distance of Deflector Above Bottom of Obstruction (in) (B)
Less than 1'	0
1' to less than 1'-6"	2½
1'-6" to less than 2'	3½
2' to less than 2'-6"	5½
2'-6" to less than 3'	7½
3' to less than 3'-6"	9½
3'-6" to less than 4'	12
4' to less than 4'-6"	14
4'-6" to less than 5'	16½
5' to less than 5'-6"	18
5'-6" to less than 6'	20
6' to less than 6'-6"	24
6'-6" to less than 7'	30
7' to less than 7'-6"	35

Notes:
For A and B, refer to *Figure 15*.

Obstructions Less than 18" Below the Sprinkler Deflector

With obstructions less than 18" below the sprinkler deflector that prevent the pattern from fully developing, the sprinkler should be positioned such that it is located at a distance three times greater than the maximum dimension of the obstruction up to a maximum of 24". *Figure 16* shows the minimum distance from obstructions in the vertical orientation. *Figure 17* shows the minimum distance from obstructions in the horizontal orientation.

Exceptions to these obstruction rules include the following:

- For light and ordinary hazard occupancies, only structural elements must be considered.

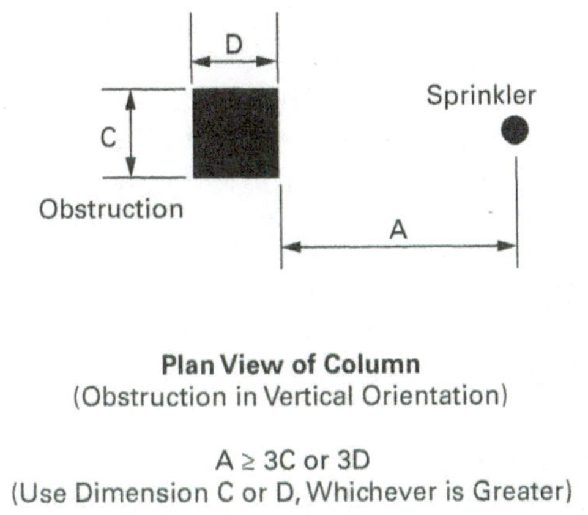

Figure 16 Minimum distance from obstruction in the vertical orientation (SSU/SSP) [NFPA 13, *Figure 10.2.7.2.1.3(a)*].

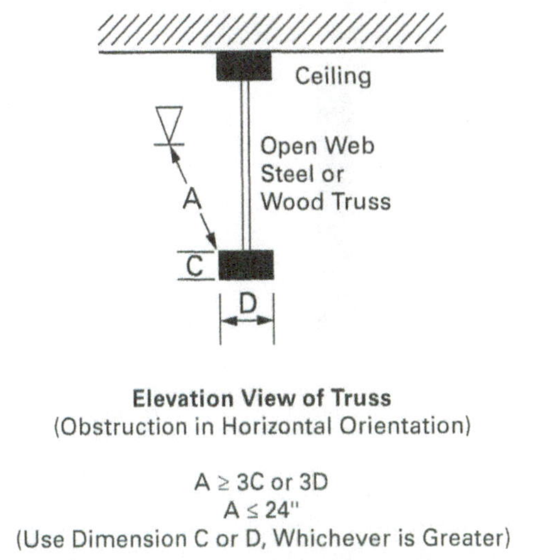

Figure 17 Minimum distance from obstruction in the horizontal orientation (SSU/SSP) [NFPA 13, *Figure 10.2.7.2.1.3(b)*].

- Sprinklers are permitted to be spaced on opposite sides of an obstruction not exceeding 4' in width, provided the distance from the centerline of the obstruction to the sprinklers does not exceed one-half the allowable distance between sprinklers.
- Where the obstruction consists of open trusses 20" or greater apart (24" on center), sprinklers are permitted to be located one-half the distance between the obstruction created by the truss, provided that all truss members are not greater than 4" (nominal) in width.
- These obstruction requirements do not apply to sprinklers installed in the centerline of bar joists or trusses.
- These obstruction requirements do not apply to sprinkler system piping less than 3" in diameter.

Suspended or Floor-Mounted Vertical Obstructions

The distance from sprinklers to privacy curtains, freestanding partitions, room dividers, and similar obstructions in light hazard occupancies should be in accordance with *Figure 18* and *Table 7*.

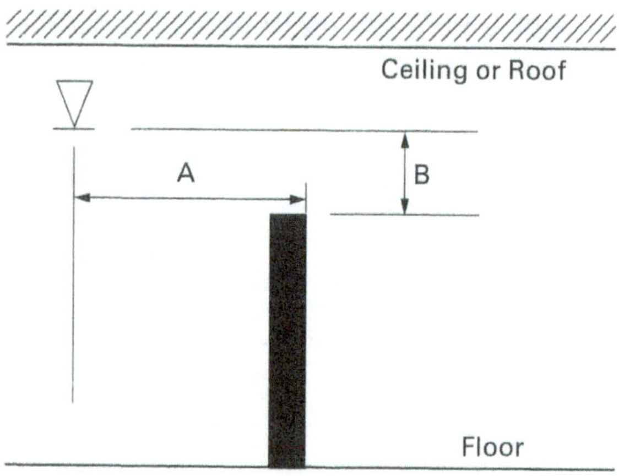

Note: This training material is not affiliated with nor has it been reviewed or approved by the NFPA. Reproduced with permission of NFPA from NFPA 13, *Standard for the Installation of Sprinkler Systems*, 2019 edition.

Figure 18 Suspended or floor-mounted obstruction (standard sidewall spray sprinklers) in light hazard occupancy only (NFPA 13, *Figure 10.2.7.2.2*).

Other Obstruction Considerations

Obstructions that interrupt the water discharge in a horizontal plane more than 18" below the sprinkler deflector in a manner to limit the distribution from reaching the protected hazard should comply with the following:

- Sprinklers should be installed under fixed obstructions over 4' wide, such as ducts, decks, cutting tables, and overhead doors, with the exception of obstructions that are not fixed in place, such as conference tables.
- Sprinklers installed under open gratings should be of the intermediate level/rack storage type or otherwise shielded from the discharge of overhead sprinklers.

4.2.0 Standard Spray Sidewall (SSW) Sprinklers

Sidewall sprinklers are typically installed in light hazard occupancies with smooth, horizontal ceilings or sloped, flat ceilings. Sidewall sprinklers may be used in ordinary hazard occupancies but must be specifically Listed for such use.

4.2.1 Area of Coverage and Maximum Spacing

The maximum coverage area refers to the maximum protection area per sprinkler. It is based on response time, occupancy, and sprinkler type. *Table 8* shows the maximum protection area of sprinklers for various occupancies.

The protection area for each sprinkler should be determined using the protection area rule S × L, where *S* is the distance to the next sprinkler (or to the wall, in the case of an end sprinkler on a branch line) upstream and downstream. Choose the larger of either twice the distance to the wall or the distance to the next sprinkler. *L* is the distance to the opposite side of the room. (Note that *S* and *L* are not interchangeable.) Where sprinklers are provided on both sides of the room, *L* should be half the distance between the walls. This calculation is similar to upright and pendents, except one dimension is along a wall.

4.2.2 Positioning

Sidewall sprinkler deflectors (vertical type) are typically positioned not more than 6" or less than 4" from walls and ceilings.

Table 7 Suspended or Floor-Mounted Obstructions in Light Hazard Occupancies Only (SSU/SSP) (NFPA 13, *Table 10.2.7.2.2*)

Horizontal Distance (A)	Minimum Vertical Distance Below Deflector (in) (B)
6" or less	3
More than 6" up to 9"	4
More than 9" up to 12"	6
More than 12" up to 15"	8
More than 15" up to 18"	9½
More than 18" up to 24"	12½
More than 24" up to 30"	15½
More than 30"	18

Notes:
For A and B, refer to *Figure 18*.

Table 8 Protection Areas and Maximum Spacing (Standard Sidewall Spray Sprinklers) (NFPA 13, *Table 10.3.3.2.1*)

	Light Hazard		Ordinary Hazard	
	Combustible Ceiling Finish	Noncombustible or Limited-Combustible Ceiling Finish	Combustible Ceiling Finish	Noncombustible or Limited-Combustible Ceiling Finish
Maximum distance along the wall (S) (ft)	14	14	10	10
Maximum room width (L) (ft)	12	14	10	10
Maximum protection area (ft^2)	120	196	80	100

Horizontal sidewall sprinklers may be positioned 6" to 12" or 12" to 18" below noncombustible ceilings, where Listed for these positions. Horizontal and vertical sidewall sprinklers are permitted to be positioned with their deflectors less than 4" from the wall where they are mounted.

Sidewall sprinklers that are installed under a sloped ceiling must be located at the high point of the slope and positioned to discharge downward. The deflector must be parallel to the sloped ceiling.

Where soffits are used for the installation of sidewall sprinklers, they must not exceed 8" in width or projection from the wall; and where soffits exceed 8", additional sprinklers must be installed below the soffit.

4.2.3 Clearance

Sidewall sprinklers have their own obstruction criteria. See *Figure 19* and *Table 9*. They must be installed no closer than 4' from beams or similar obstructions. Beams or similar obstructions located more than 4' from the sprinkler must conform to NFPA standards.

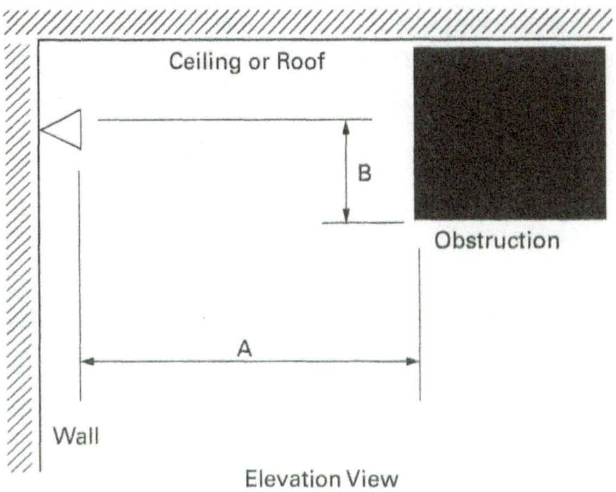

Figure 19 Positioning of sprinklers to avoid obstructions (standard sidewall spray sprinklers) (NFPA 13, *Figure 10.3.6.1.3*).

Table 9 Positioning of Sprinklers to Avoid Obstructions (Standard Sidewall Spray Sprinklers) (NFPA 13, *Table 10.3.6.1.3*)

Distance from Sprinklers to Side of Obstruction (A)	Maximum Allowable Distance of Deflector Above Bottom of Obstruction (in) (B)
Less than 4'	Not Allowed
4' to less than 5'	1
5' to less than 5'-6"	2
5'-6" to less than 6'	3
6' to less than 6'-6"	4
6'-6" to less than 7'	6
7' to less than 7'-6"	7
7'-6" to less than 8'	9
8' to less than 8'-6"	11
8'-6" or greater	14

Notes:
For A and B, refer to *Figure 19*.

4.2.4 Obstructions

There are other considerations that need to be addressed when locating sidewall sprinklers. The obstruction criteria shown in *Figure 20* and *Table 10* show the requirements for obstructions along a wall. *Figure 21* shows the requirements for horizontal and vertical obstructions. *Figure 22* and *Table 11* show the requirements for floor-mounted obstructions. Always follow the NFPA requirements, product Listings, manufacturer's instructions, and approved shop drawings to ensure that the spray pattern of the sprinkler allows the water to reach the hazard.

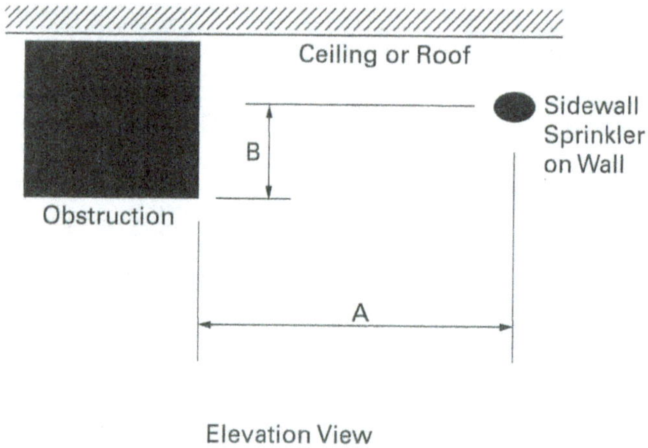

Figure 20 Positioning of sprinklers to avoid obstructions along a wall (standard sidewall spray sprinklers) (NFPA 13, *Figure 10.3.6.1.4*).

Table 10 Positioning of Sprinklers to Avoid Obstructions Along Wall (Standard Sidewall Spray Sprinklers) (NFPA 13, *Table 10.3.6.1.4*)

Distance from Sprinklers to Side of Obstruction (A)	Maximum Allowable Distance of Deflector Above Bottom of Obstruction (in) (B)
4" to less than 6"	1
6" to less than 1'	2
1' to less than 1'-6"	3
1'-6" to less than 2'	4½
2' to less than 2'-6"	5¾
2'-6" to less than 3'	7
3' to less than 3'-6"	8
3'-6" to less than 4'	9¼
4' to less than 4'-6"	10
4'-6" to less than 5'	11½
5' to less than 5'-6"	12¾
5'-6" to less than 6'	14
6' to less than 6'-6"	15
6'-6" to less than 7'	16¼
7' to less than 7'-6"	17½

Notes:
For A and B, refer to *Figure 20*.

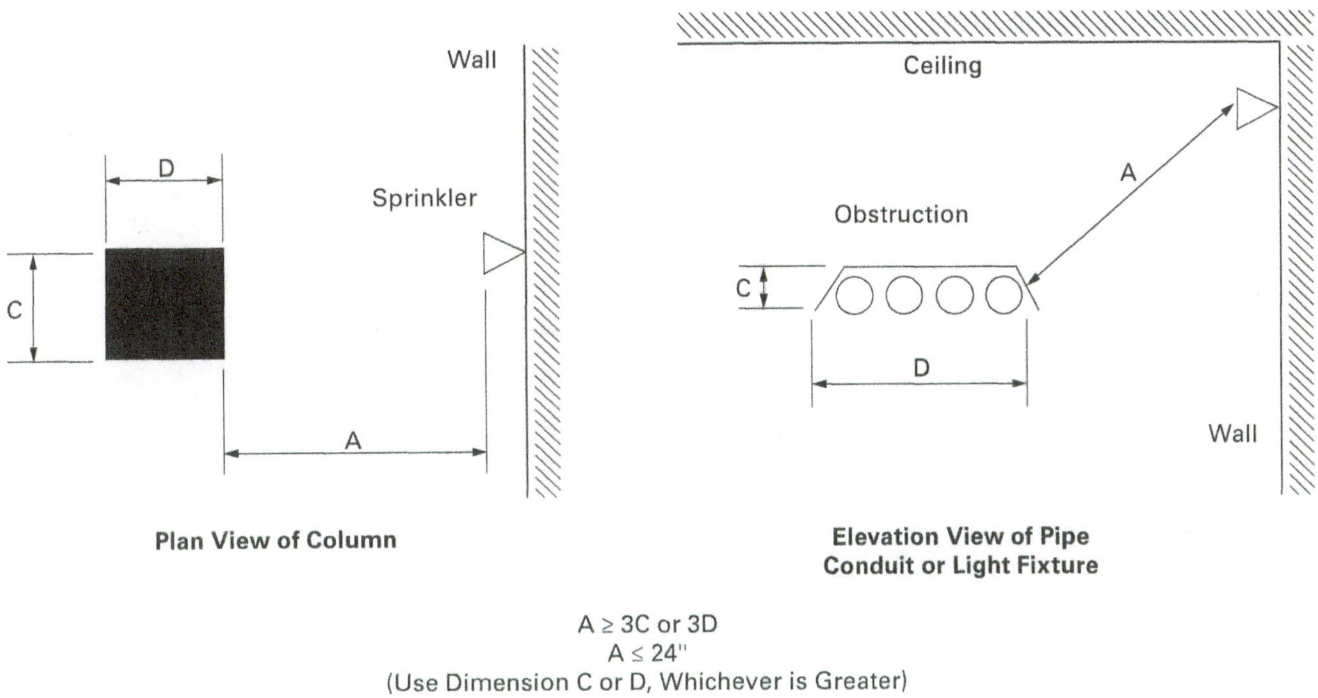

Figure 21 Positioning of sprinklers to avoid obstructions (standard sidewall spray sprinklers) [NFPA 13, *Figure 10.3.6.2.1.3(a) and (b)*].

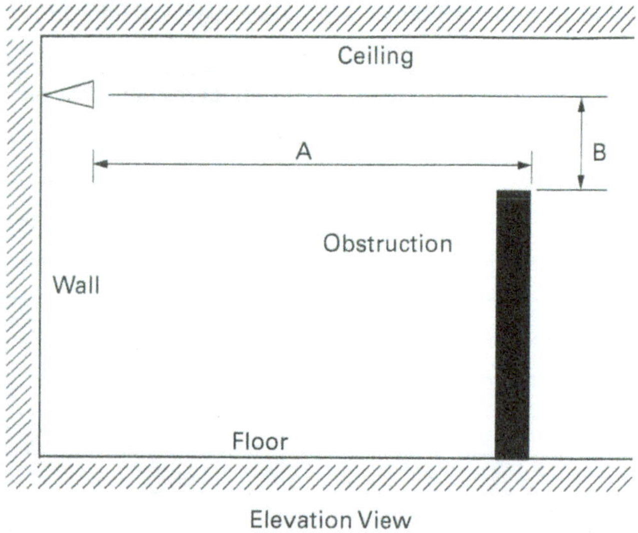

Figure 22 Suspended or floor-mounted obstructions (standard sidewall spray sprinklers) in light hazard occupancy only (NFPA 13, *Figure 10.3.6.2.2*).

Table 11 Suspended or Floor-Mounted Obstructions (Standard Sidewall Spray Sprinklers) in Light Hazard Occupancies Only (NFPA 13, *Table 10.3.6.2.2*)

Horizontal Distance (A)	Minimum Vertical Distance Below Deflector (in) (B)
6" or less	3
More than 6" up to 9"	4
More than 9" up to 12"	6
More than 12" up to 15"	8
More than 15" up to 18"	9½
More than 18" up to 24"	12½
More than 24" up to 30"	15½
More than 30"	18

Notes:
For A and B, refer to *Figure 22*.

4.0.0 Section Review

1. The *maximum* protection area of a hydraulically calculated system with noncombustible unobstructed construction is _____.
   a. 130 ft^2
   b. 168 ft^2
   c. 200 ft^2
   d. 225 ft^2

2. Sidewall spray sprinklers have their own obstruction criteria.
   a. True
   b. False

Review Questions

1. The *maximum* ambient ceiling temperature for an ordinary temperature-rated sprinkler is _____.
   a. 100°F
   b. 120°F
   c. 180°F
   d. 212°F

2. A sprinkler with a red glass bulb has a temperature classification of _____.
   a. ordinary
   b. high
   c. extra high
   d. very extra high

3. An example of a light hazard occupancy is a(n) _____.
   a. bakery
   b. cannery
   c. office
   d. laundry

4. An example of an ordinary hazard occupancy is a _____.
   a. bakery
   b. museum
   c. hospital
   d. kennel

5. To determine if a type of construction is obstructed or unobstructed, you must consider both the water distribution and _____.
   a. sprinkler K-factor
   b. sprinkler temperature rating
   c. combustability of the materials being used
   d. heat flow across the ceiling

6. Unobstructed construction has horizontal structural members that are not solid and have openings *at least* _____.
   a. 40 percent of their cross-sectional area
   b. 50 percent of their cross-sectional area
   c. 70 percent of their cross-sectional area
   d. 80 percent of their cross-sectional area

7. A type of obstructed construction includes _____.
   a. open-grid ceiling construction
   b. smooth ceiling construction
   c. bar joist construction
   d. wood joist construction

8. Sprinkler spacing guidelines state that the distance from sprinklers to walls must *not* exceed _____.
   a. ¼ the allowable distance between sprinklers
   b. ½ the allowable distance between sprinklers
   c. ⅔ the allowable distance between sprinklers
   d. ¾ the allowable distance between sprinklers

9. Baffles are required when standard spray sprinkler spacing is less than _____.
   a. 3'
   b. 6'
   c. 9'
   d. 12'

10. The *maximum* sprinkler spacing between branch lines for ordinary hazard occupancies is _____.
    a. 10'
    b. 12'
    c. 13'
    d. 15'

11. When determining end branch line sprinkler spacing using the NFPA 13 S × L method, the S distance that must be used is _____.
    a. twice the distance to the wall
    b. three times the distance to the wall
    c. four times the distance to the wall
    d. equal to the distance to the wall

12. The main reason for the small room rule in NFPA 13 is to _____.
    a. increase the number of sprinklers as required
    b. adjust for rooms under 500 ft^2
    c. ensure that the maximum distance off a wall is never exceeded
    d. allow for lights and diffusers in small rooms

13. Under obstructed construction, the sprinkler deflector must be located 1" to 6" below the structural member and a *maximum* of _____.
    a. 12" below the ceiling
    b. 22" below the ceiling
    c. 28" below the ceiling
    d. 30" below the ceiling

14. Horizontal and vertical sidewall sprinklers are permitted to be positioned with their deflectors less than _____.
    a. 4" from the wall
    b. 6" from the wall
    c. 8" from the wall
    d. 10" from the wall

15. A sidewall sprinkler installed under a sloped ceiling *must* _____.
    a. have its deflector parallel to the sloped ceiling
    b. be baffled
    c. be located at the high point and positioned to discharge upwards
    d. have a high K-factor

Trade Terms Introduced in This Module

Cold soldering: A condition that occurs when sprinklers are positioned in close proximity to one another. It results when one sprinkler distributes water on another sprinkler, preventing or delaying the adjacent sprinkler's activation by cooling it below the fusing temperature.

Combustible: Refers to materials that can burn.

Fusing (operating) temperature: The temperature at which a sprinkler activates.

Joist: A structural member that provides support.

Limited-combustible: A material with a structural base of noncombustible material; an example is gypsum wallboard.

Noncombustible: A material that, in the form in which it is used, and under the conditions anticipated, will not ignite, burn, support combustion, or release flammable vapors when subjected to fire or heat.

Obstructed construction: Beams, trusses, or other structural members that impede heat flow or water distribution and affect the ability of sprinklers to control or suppress a fire.

Pendent: Refers to a sprinkler installed so that the deflector is below the fusible element and designed to direct the water stream downward against the deflector.

Sidewall: Refers to a sprinkler installed so that the deflector discharges most of the water away from the nearby wall in a pattern resembling one quarter of a sphere, with a small portion of the discharge directed at the wall behind the sprinkler.

Small room: A compartment of light hazard occupancy classification having unobstructed construction and a floor area not exceeding 800 ft^2.

Unobstructed construction: Construction where beams, trusses, or other members do not impede heat flow or water distribution in a manner that materially affects the ability of sprinklers to control or suppress a fire.

Upright: Refers to a sprinkler installed so that water distribution is directed upward against a deflector.

Typical Data Sheets

Bulletin 067 April 2021

Reliable®

F1-42 Series
Standard-Response Sprinklers

K-factor 4.2 (60)

Features

- Standard coverage standard-response sprinklers
- Upright and pendent deflectors
- Low profile, compact design
- Available in a wide variety of finishes

Product Description

Reliable Model F1 series sprinklers are standard-response standard spray automatic fire sprinklers utilizing a 5.0 mm glass bulb thermal element.

Pendent and horizontal sidewall sprinklers may be installed exposed or surface mounted using escutcheons such as the Reliable Models B, C, or HB (reference Technical Bulletin 204). When installed recessed, the Model F1-42 series sprinklers are specifically listed with and may only be installed with listed Reliable escutcheons. Refer to the technical information on the following pages for specific listings for recessed installations and refer to Figure 3 for dimensional information.

When fitted with an approved water shield, these sprinklers may considered intermediate sprinklers for use in racks, below grated walkways, and other areas where intermediate level sprinklers are required.

Table A provides a summary of the approvals and availability of specific Model F1-42 series sprinkler configurations. Additional technical information for each sprinkler model is provided on the following pages.

Note: This bulletin may contain information on New and Legacy sprinklers that reflects a dimensional change only. Sprinkler Identification Number (SIN), application, performance, and listings/approval are not otherwise affected. Sprinklers with new frames can be identified by a rotoclip indent on one side of the wrench boss and in some cases an installed rotoclip. Sprinklers with new frames will include the suffix "N" in the order. For reference only, legacy version of this sprinkler can be found on Technical Bulletin 014.

Model F1-42 Pendent

Model F1-42 Upright

F1-42 Series Sprinklers Summary — **Table A**

Sprinkler Model	K-Factor gpm/psi$^{1/2}$ (lpm/bar$^{1/2}$)	Orientation	Listings & Approvals	Max. Working Pressure psi (bar)	Sprinkler Identification Number (SIN)
F1-42	4.2 (60)	Upright	cULus	175 (12)	RA1323
		Pendent		175 (12)	RA1313

Model F1-42 Upright Sprinkler	**SIN RA1323**

Technical Specifications
Style: Upright
Threads: 1/2" NPT or ISO 7-R1/2
Nominal K-Factor: 4.2 (60)
Max. Working Pressure: 175 psi (12 bar)

Material Specifications
Thermal Sensor: 5.0mm Glass Bulb
Sprinkler Frame: Brass Alloy
Cap: Bronze Alloy
Sealing Washer: Nickel with PTFE
Load Screw: Copper Alloy
Deflector: Brass Alloy

Sprinkler Finishes
(See Table B)

Sensitivity
Standard Response

Temperature Ratings
135°F (57°C)
155°F (68°C)
175°F (79°C)
200°F (93°C)
286°F (141°C)*
360°F (182°C)
500°F (260°C)

Guards & Shields (New Frames)
F-1 Guard Kit
F-3 Guard/Shield Kit

Guards & Shields (Legacy Frames)
D-1 Guard

Sprinkler Wrenches
Model W2
Model J (New frame with guard installed)
Model JD (Legacy frame with guard installed)

Listings and Approvals
cULus

Model F1-42 Upright Sprinkler Components and Dimensions **Figure 1**

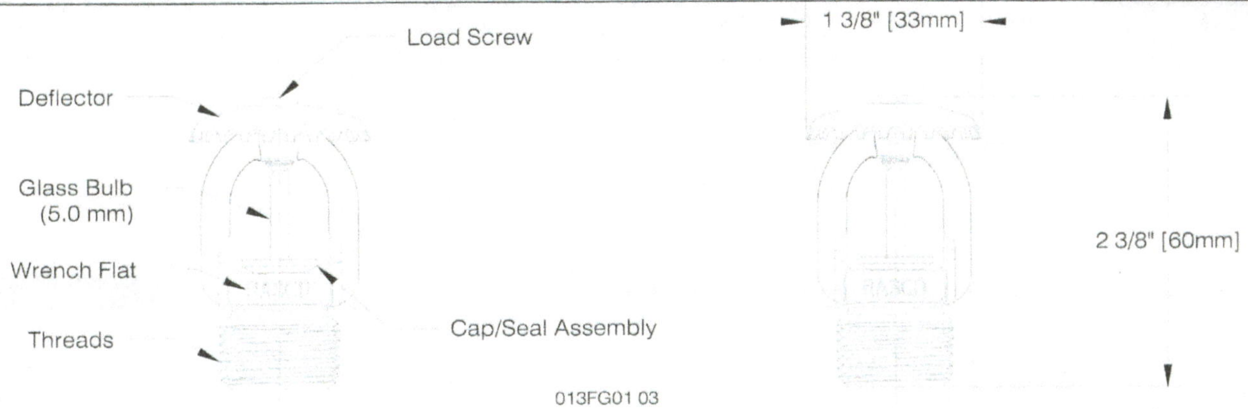

Model F1-42 Pendent Sprinkler		SIN RA1313
Technical Specifications **Style:** Pendent Recessed Pendent **Threads:** 3/4" NPT or ISO 7-R3/4 **Nominal K-Factor:** 4.2 (60) **Max. Working Pressure:** 175 psi (12 bar) **Material Specifications** **Thermal Sensor:** 5.0mm Glass Bulb **Sprinkler Frame:** Brass Alloy **Cap:** Bronze Alloy **Sealing Washer:** Nickel with PTFE **Load Screw:** Copper Alloy **Deflector:** Brass Alloy **Sprinkler Finishes** (See Table B) **Sensitivity** Standard Response	**Temperature Ratings**(1) 135°F (57°C) 155°F (68°C) 175°F (79°C) 200°F (93°C) 286°F (141°C) 360°F (182°C) 500°F (260°C) **Recessed Escutcheons** Model F1 Model F2 Model FP **Guards & Shields (New Frames)**(2) F-7 Guard Kit F-8 Guard/Shield Kit **Guards & Shields (Legacy Frames)**(2) D-1 Guard **Sprinkler Wrenches** Model W2 (pendent) Model W4 (recessed) Model J (New frame with guard installed) Model JD (Legacy frame with guard installed) **Listings and Approvals** cULus	

Notes:
1. 286°F (141°C) and higher temperature rated sprinklers not listed for recessed use.
2. Not suitable for recessed pendent installations.

Model F1-42 Pendent Sprinkler Components and Dimensions — **Figure 2**

Note: Please refer to Figure 3 for recessed installation.

Recessed Installation: Pendent Sprinklers

Figure 3

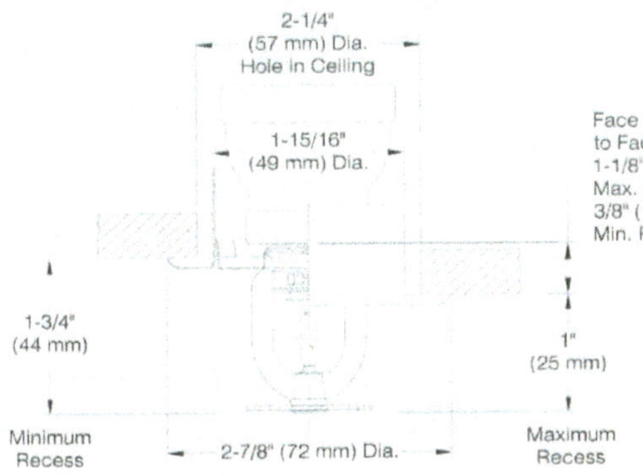

Model F1 Pendent Sprinkler
with Model F1 Recessed Escutcheon

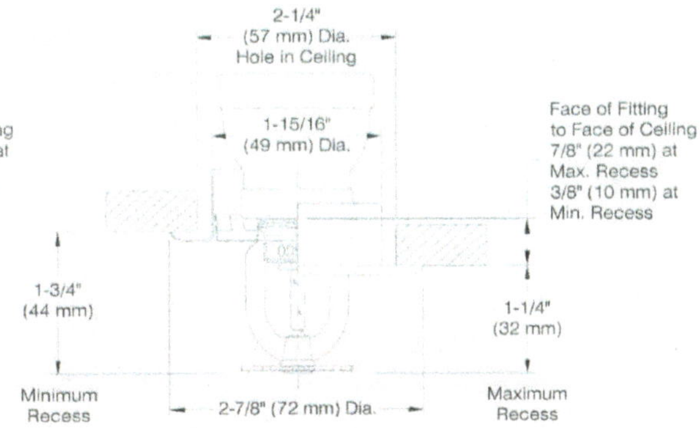

Model F1 Pendent Sprinkler
with Model F2 Recessed Escutcheon

013REC

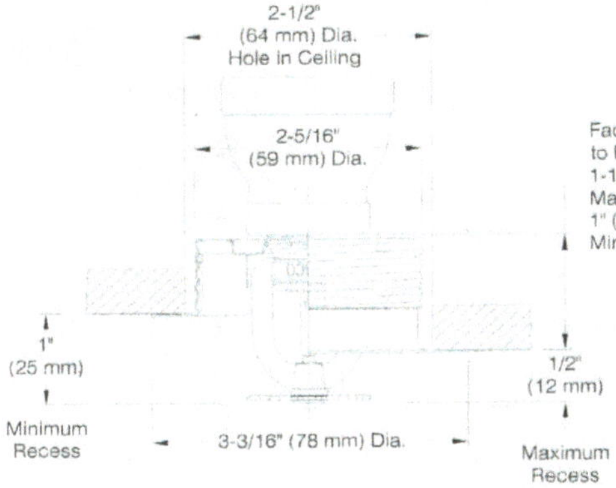

Model F1 Pendent Sprinkler
with Model FP Recessed Escutcheon

F1 Escutcheon
(F2 similar)

FP Escutcheon

Wrenches

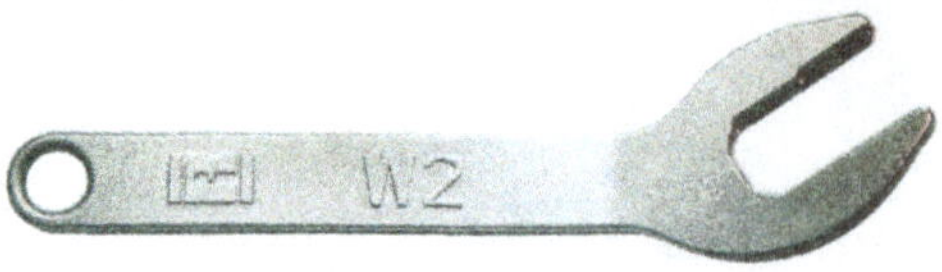

Model W2 (upright, pendent)

Model W4 (recessed pendent)

Model J (New frame with guard installed)

Model JD (Legacy frame with guard installed)

Finishes[1][2]

Table B

Standard Finishes		Special Application Finishes	
Sprinkler	**F1, F2 and FP**[3] **Escutcheons**	**Sprinkler**	**F1, F2 and FP**[3] **Escutcheons**
Bronze	Brass	Electroless Nickel PTFE[4]	Bright Brass
Chrome Plated	Chrome Plated	Bright Brass[5]	Black Plated
White Polyester[4]	White Polyester	Black Plated	Black Paint
		Black Paint	Off White
		Off White	Chrome Dull
		Chrome Dull	

Notes:

1. Other finishes and colors are available on special order. Consult your Reliable sales representative for details.
2. Paint or any other coating applied over the factory finish will void all approvals and warranties.
3. The Model FP assembly consists of a sprinkler mounted in a galvanized steel cup with a finished trim ring or cover plate.
4. cULus Listed as corrosion resistant.
5. For 200°F (93°C) maximum temperature rated sprinklers only.

Installation

Model F1 Series sprinklers must be installed with the Reliable sprinkler installation wrench identified in this Bulletin. Any other wrench may damage the sprinkler. A leak tight sprinkler joint can be obtained with a torque of 8 to 18 lb-ft (11 to 24 N-m). Do not tighten sprinklers over the maximum recommended installation torque. Exceeding the maximum recommended installation torque may cause leakage or impairment of the sprinkler.

Glass bulb sprinklers have orange bulb protectors or protective caps to minimize bulb damage during shipping, handling and installation. Reliable sprinkler installation wrenches are designed to install sprinklers with bulb protectors in place. Remove the bulb protector at the time when the sprinkler system is placed in service for fire protection. Removal of the bulb protector before this time may leave the bulb vulnerable to damage. Remove bulb protectors by undoing the clasp by hand. Do not use tools to remove bulb protectors.

Maintenance

Reliable Model F1 series sprinklers should be inspected and the sprinkler system maintained in accordance with NFPA 25, as well as the requirements of any Authorities Having Jurisdiction.

Prior to installation, sprinklers should remain in the original cartons and packaging until used. This will minimize the potential for damage to sprinklers that could cause improper operation or non-operation.

Do not clean sprinklers with soap and water, ammonia liquid or any other cleaning fluids. Remove dust by gentle vacuuming without touching the sprinkler.

Replace any sprinkler which has been painted (other than factory applied). A stock of spare sprinklers should be maintained to allow quick replacement of damaged or operated sprinklers. Failure to properly maintain sprinklers may result in inadvertent operation or non-operation during a fire event.

Additional Resources

This module presents thorough resources for task training. The following reference material is suggested for further study.

NFPA 13, *Standard for the Installation of Sprinkler Systems*, Latest Edition. Quincy, MA: National Fire Protection Association.

NFPA 13D, *Standard for the Installation of Sprinkler Systems in One- and Two-Family Dwellings and Manufactured Homes*, Latest Edition. Quincy, MA: National Fire Protection Association.

NFPA 13R, *Standard for the Installation of Sprinkler Systems in Low-Rise Residential Occupancies*, Latest Edition. Quincy, MA: National Fire Protection Association.

Figure Credits

Reliable Automatic Sprinkler Co., Inc., Module Opener, Figures 1, 4, A01–A05

Reproduced with permission of NFPA from NFPA 13, *Standard for the Installation of Sprinkler Systems*, 2019 edition. Copyright© 2018, National Fire Protection Association. For a full copy of NFPA 13, please go to www.nfpa.org., Figures 3, 12-22, Tables 1-11

Taylor Schumacher, Figure 5

Shutterstock.com/voyager, Figure 7

Fred Fuhrmeister / Alamy Stock Photo, Figure 8

Shutterstock.com/sockagphoto, Figure 9

Used with permission of Johnson Controls, Figures 10–11

Section Review Answer Key

Section 1.0.0

Answer	Section Reference	Objective
1. b	1.1.0; *Table 1*	1a
2. a	1.2.0; *Table 3*	1b
3. a	1.3.0	1c

Section 2.0.0

Answer	Section Reference	Objective
1. b	2.1.0	2a
2. a	2.2.0	2b

Section 3.0.0

Answer	Section Reference	Objective
1. c	3.1.1	3a
2. a	3.2.0	3b

Section 4.0.0

Answer	Section Reference	Objective
1. d	4.1.1; *Table 4*	4a
2. a	4.2.3	4b

User Update

NCCER makes every effort to keep its textbooks up-to-date and free of technical errors. We appreciate your help in this process. If you find an error, a typographical mistake, or an inaccuracy in NCCER's curricula, please submit a User Update form by visiting **https://www.nccer.org/olf**. You can also scan the QR code using the camera on your phone or mobile device to access the form.

Wet Pipe Sprinkler Systems

Overview

This module describes the operation and installation of wet pipe sprinkler systems. It covers various system controls and switches and explains how they operate. It also describes how to complete a Contractor's Material and Test Certificate for Aboveground Piping and provides an overview of basic system troubleshooting procedures.

Module 18206

18206

Wet Pipe Sprinkler Systems

Objectives

Successful completion of this module prepares you to do the following:

1. Identify the function and operation of wet pipe sprinkler system control valves.
   a. Explain the operation of an alarm check valve and related trim.
   b. Identify the components in a typical wet system riser assembly.
   c. Describe a typical Fire Department Connection (FDC).
2. Identify the types of switches used in wet pipe sprinkler systems.
   a. Describe how switches operate.
   b. Explain the purpose of paddle-type vane switches and describe how they are installed.
   c. Explain the purpose of pressure switches and describe how they are installed.
   d. Explain the purpose of tamper switches and describe how they are installed.
3. Describe how to test and troubleshoot wet pipe sprinkler systems.
   a. Explain the basic procedure for performing hydrostatic testing.
   b. Describe how to troubleshoot common wet pipe sprinkler system issues.

Performance Task

Under supervision, you should be able to do the following:

1. Complete a Contractor's Material and Test Certificate for Aboveground Piping for a wet pipe sprinkler system.

Trade Terms

Alarm check valve
Bypass
Clapper
Double-pole (DP) switch
Double-throw (DT) switch
False alarm
Hydrostatic test
Main drain
Pressure switch
Retard chamber
Riser
Single-pole (SP) switch
Single-throw (ST) switch
Turbine
Water motor gong

Industry Recognized Credentials

If you are training through an NCCER-accredited sponsor, you may be eligible for credentials from NCCER's Registry. The ID number for this module is 18206. Note that this module may have been used in other NCCER curricula and may apply to other level completions. Contact NCCER's Registry at 1.888.622.3720 or go to **www.nccer.org** for more information.

You can also show off your industry-recognized credentials online with NCCER's digital badges. Transform your knowledge, skills, and achievements into badges that you can share across social media platforms, send to your network, and add to your resume. For more information, visit **www.nccer.org**.

NOTE

Unless otherwise specified, references in parentheses following figure and table numbers refer to NFPA 13.

Contents

Figures and Tables

Section One

1.0.0 Wet Pipe Sprinkler System Control Valves

Objective

Identify the function and operation of wet pipe sprinkler system control valves.

a. Explain the operation of an alarm check valve and related trim.
b. Identify the components in a typical wet system riser assembly.
c. Describe a typical Fire Department Connection (FDC).

Trade Terms

Alarm check valve: A device installed in a wet sprinkler system that triggers an audible alarm when water flow is detected.

Bypass: Alarm check valve trim where pressures are equalized.

Clapper: A valve component that restricts water flow under normal pressure.

False alarm: A signal transmitted from a system that should not be in operation.

Main drain: The drain used to empty the system for servicing, to check the alarm check valve, and to perform main drain tests.

Pressure switch: A device used to detect changes in pressure within a system and to transmit a signal when fluctuations occur.

Retard chamber: A device in variable-pressure systems that is commonly used with pressure switches to prevent false alarms from occurring.

Riser: The vertical supply pipe in a sprinkler system.

Turbine: A component part of a water motor gong that produces the sound.

Water motor gong: An alarm that is used in wet pipe sprinkler systems.

Wet pipe sprinkler systems are networks of piping that are sized using hydraulic calculations or pipe schedules. They are installed in a building, structure, or area, generally overhead, with sprinklers attached. Wet pipe sprinkler systems use sprinklers on supply piping that is filled with water. The system is connected to a water supply so that water discharges immediately when a sprinkler is activated by a heat source.

Wet pipe sprinkler systems are very common because they are the least expensive type of sprinkler system to install, the easiest to maintain, and the most reliable. As with other sprinkler systems, a wet pipe sprinkler system must be designed and installed in accordance with NFPA 13.

Waterflow control in a wet pipe sprinkler system is provided by the system check valve. Two types of system check valves are alarm check valves and wet system riser assemblies.

1.1.0 Alarm Check Valves and Trim

An alarm check valve is one of the first devices water reaches in a wet pipe sprinkler system. Alarm check valves are waterflow control devices installed in the system supply piping. The valves are connected, using trim, to devices that actuate an alarm when a wet pipe sprinkler system is activated.

Trim is the piping that connects monitoring devices, gauges, alarm devices, and bypasses to alarm check valves. Trim is configured based on the sprinkler installation and the function of the alarm devices within the system.

Waterflow alarm check valves must be Listed and used only for the type of service for which they are designed. The valves must be installed so that the flow of water from a single sprinkler of the smallest K-factor installed in the system will result in an audible alarm within five minutes after the water flow begins. Listed alarm check valves or other waterflow-detecting alarm devices with horns, sirens, or bells are commonly used. Paddle-type waterflow alarm switches are restricted to wet pipe sprinkler systems only.

Alarm check valves are available in flanged-flanged, flanged-grooved, and grooved-grooved configurations. The alarm check valve, indicated in *Figure 1* as a solid black triangle, may also serve as a check valve when the Fire Department Connection (FDC) is in use. Indicating valves are located beneath the alarm check valve in the assembly and used to shut off the system for service and maintenance.

Some piping systems require multiple wet risers. These systems must be installed so that they can be isolated for maintenance, meaning a Listed indicating valve must be installed at each riser stack. This installation allows an individual valve or system to be shut down for maintenance or repair while other systems remain active. *Figure 2* shows the valve arrangement with multiple wet system riser assemblies.

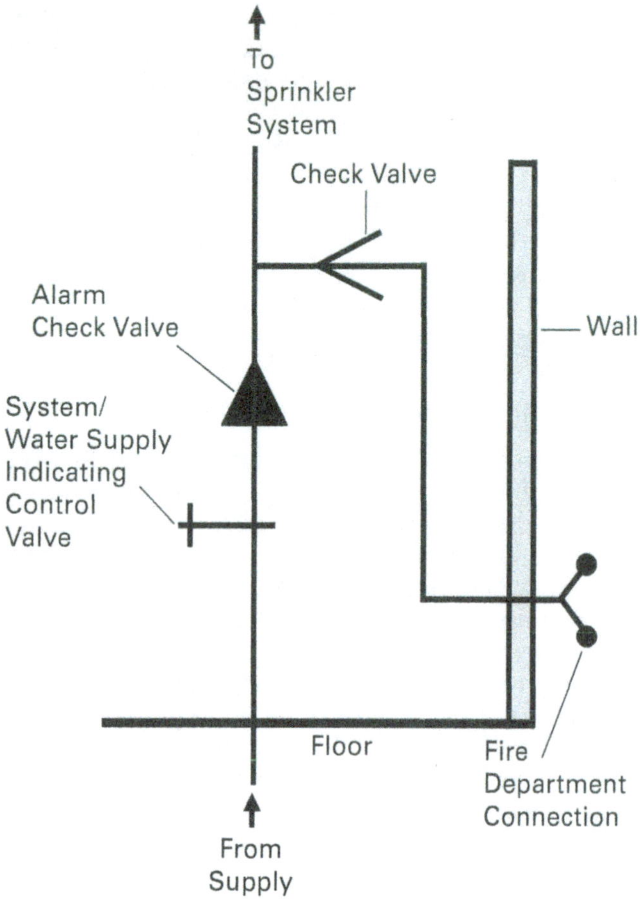

Figure 1 Wet pipe sprinkler riser as shown on an approved shop drawing.

1.1.1 Alarm Check Valve Operation

Like other check valves, alarm check valves contain a clapper that keeps the valve closed under normal conditions. The clapper rests on an intermediate chamber and forms a seal when the clapper is in the closed position. Some valves also contain a spring that provides positive closure of the valve.

When the pressure is equal above and below the clapper, the clapper rests on the seat and seals off the water pressure from the openings in the intermediate chamber. When the pressure above the clapper is higher than the water pressure below the clapper, the alarm check valve operates as a check valve and prevents the water above the clapper from reentering the potable water supply. *Figure 3* shows a cross section of an alarm check valve in the closed position.

When the pressure above the clapper drops below the pressure of the water supply, the pressure difference across the clapper forces it off its seat into the open position. This pressurizes the system above the clapper to more closely match the pressure below the clapper. The difference in pressure must be sufficient to overcome the weight of the clapper and the tension of the spring. When the clapper is open, water flows to both the system and the intermediate chamber. As water is transported, the alarm sounds, indicating water flow. *Figure 4* shows an alarm check valve in the open position.

1.1.2 Pressure Gauges

Wet pipe sprinkler systems require two approved pressure gauges in each system riser assembly. Pressure gauges are installed above and below the alarm check valve seat to measure pressure above and below the clapper. The gauges must have a maximum pressure limit that is at least twice the normal working pressure of the system in which the gauge is installed. Pressure gauges must be installed so that they can be removed without taking the system out of service.

Alarm check valves are used in systems that have constant and variable pressures and that require different trim configurations for each type of system pressure. Most sprinkler systems are installed where the water pressure is variable, and therefore require a retard chamber to absorb excess pressure surges.

1.1.3 Retard Chamber

A retard chamber is a device used to prevent a false alarm when there is variable pressure in the water supply. Retard chambers delay water from reaching the alarm provisions when minor pressure fluctuations occur. *Figure 5* shows the operation of a retard chamber.

Small fluctuations in water pressure cause water to partially fill the chamber, which drains through an orifice or causes the clapper valve to close. No water is sent to the alarm switch unless the chamber completely fills so that water flows through the top of the chamber. If the alarm check valve remains open (1), water flows through the chamber faster than it can drain. When the chamber fills, water flows through the alarm port (2) to the retard chamber (3), which then begins to fill. Once the retard chamber overflows, it activates the alarm devices (4).

A drain or drip cup is used to catch the overflow from the orifice in the retard chamber. The retard drain line must be piped to an area where the discharge causes no damage. One way is to pipe the retard drain line into the main drain. *Figure 6* shows a retard chamber with valve trim.

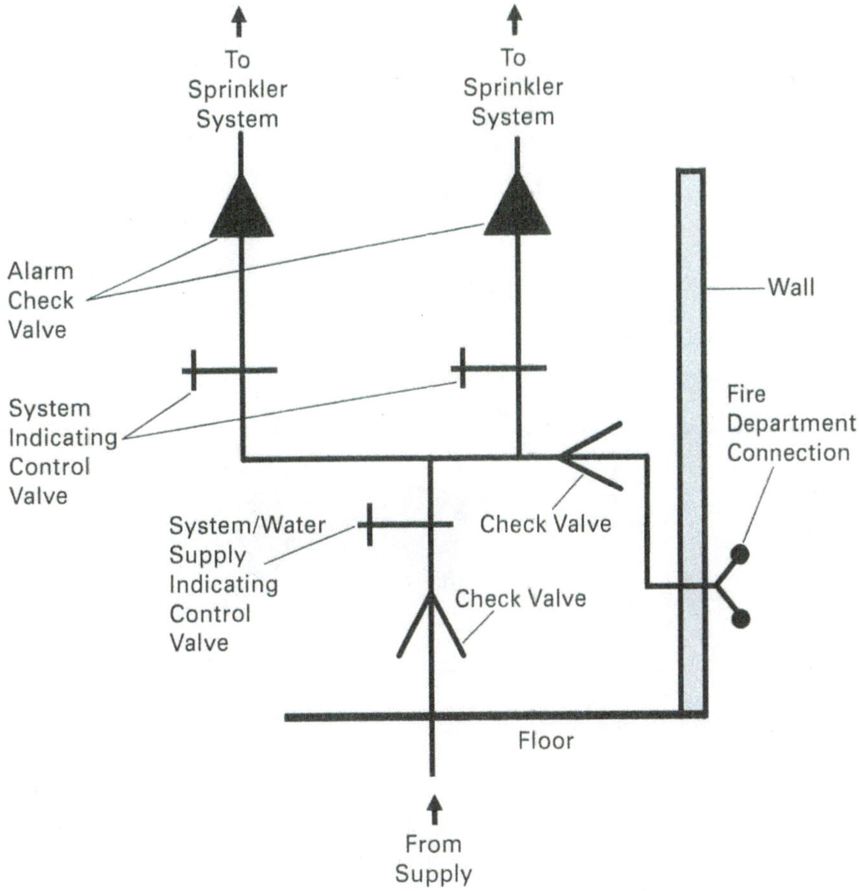

Figure 2 Valve arrangement with multiple wet sprinkler risers.

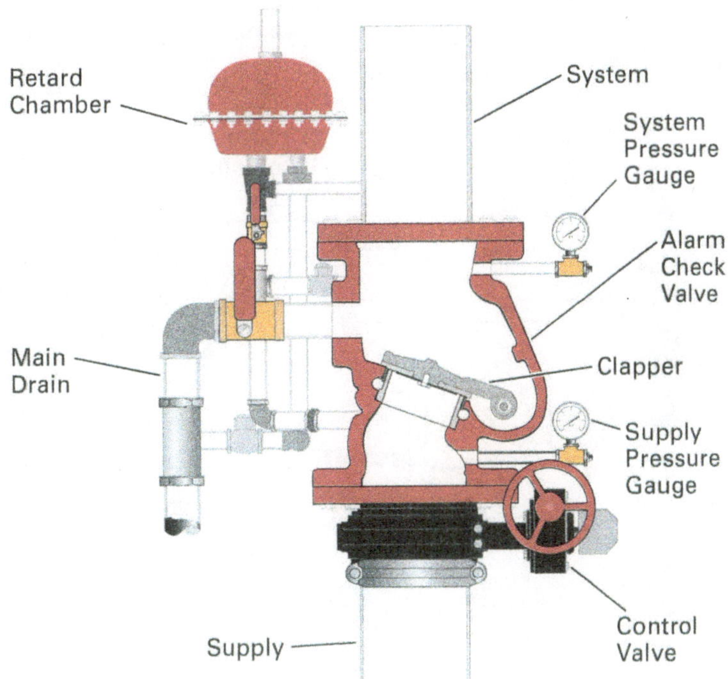

Figure 3 Wet sprinkler alarm check valve in the closed position.

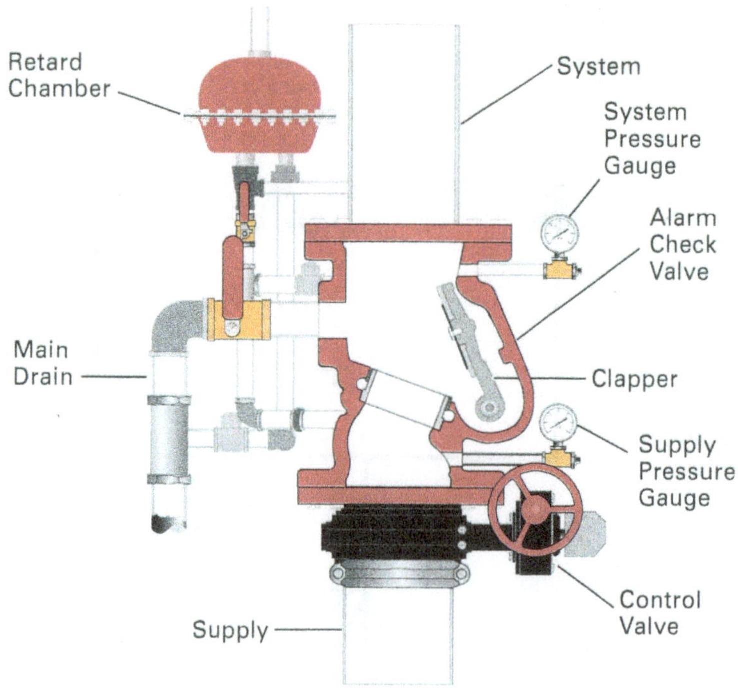

Figure 4 Wet sprinkler alarm check valve in the open position.

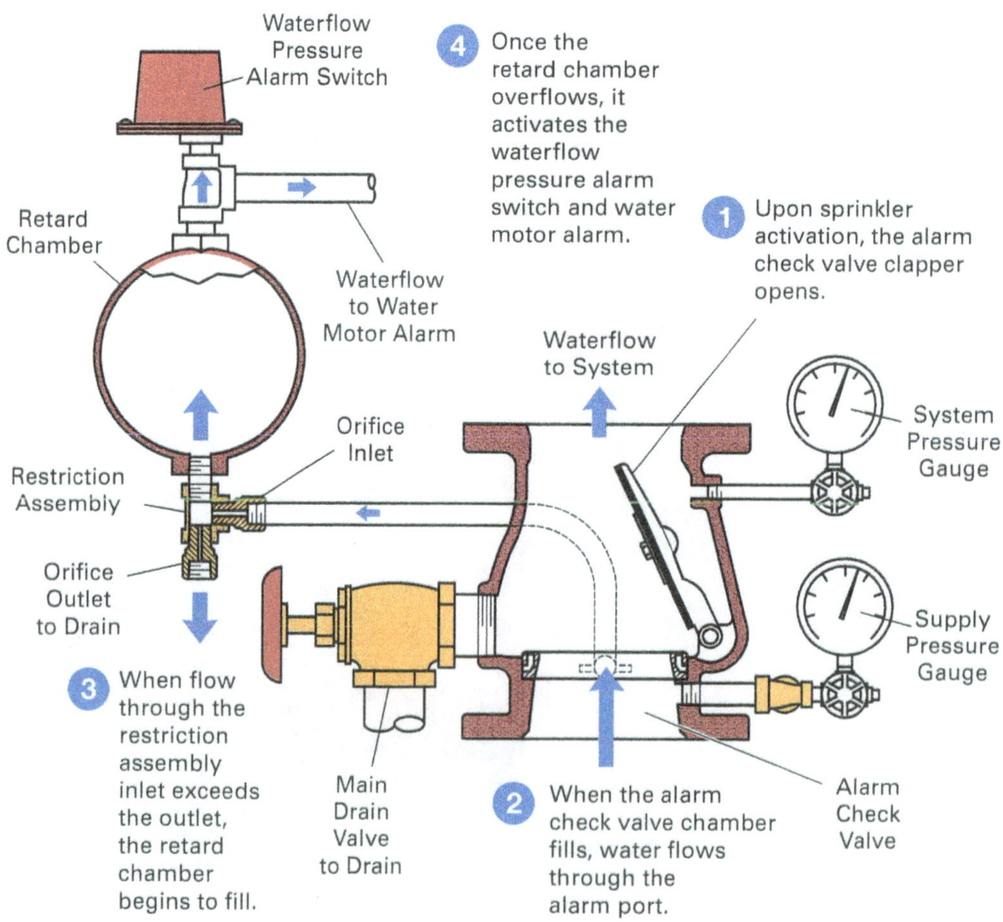

Figure 5 Retard chamber operation.

Figure 6 Retard chamber with valve trim.

Figure 7 Main drain on a fully trimmed alarm check valve.

Sometimes two retard chambers are used in series to accomplish an extended retarding action to prevent false alarms in complex installations. When multiple alarm check valves are used, they may share a common alarm line, retard chamber, and water motor gong. Consult the manufacturer's literature to determine when additional trim must be added.

1.1.4 Main Drain

The main drain on an alarm check valve serves a number of purposes. It is used when the alarm check valve is tested, when the system is drained for servicing, when performing a main drain test, and as a catch drain.

Drain connections for system supply risers and mains must be sized in accordance with NFPA 13 requirements. Drain connections of $\frac{3}{4}$" or larger must be used for risers or mains up to 2". Drain connections $1\frac{1}{4}$" or larger must be used for risers or mains between $2\frac{1}{2}$" and $3\frac{1}{2}$". For risers or mains 4" and larger, a 2" or larger drain connection is used.

Alarm check valves are typically manufactured in $2\frac{1}{2}$", 3", 4", 6", and 8" sizes and usually have a threaded opening for attaching the main drain into the body of the alarm check valve. *Figure 7* shows a main drain on a fully trimmed alarm check valve.

Proper arrangement of the main drain discharge is important because of the large amount of water discharging from the main drain. It can overflow a floor drain or mop sink very rapidly. The most common drain discharge location for alarm check valves is over a parking lot or other paved area. The pavement surface will not erode and is large enough for the water to disperse without causing damage. The drain outlets have a 45-degree elbow at the end of the drain connection to direct the water onto the pavement. This arrangement also provides an air gap between the drain line and any potential source of contamination.

If a parking lot cannot be used for drainage, a new location must be selected to avoid damaging the property and potential for contamination of the potable water supply. Always check local codes on drain discharge.

1.1.5 Alarm Lines

Alarm lines connect the intermediate chamber of the alarm check valve and the alarm. When the clapper lifts off its seat, water flows into the intermediate chamber and into the alarm line. With a constant water supply, the alarm check valve is connected directly to a water motor gong and/or pressure switch. With a variable water supply, a retard chamber is added between the devices to prevent false alarms.

Normally, an alarm control valve is located in the trim piping. This valve is normally in the open position. If a pressure switch is installed that is connected to a building fire alarm system or off-site monitoring company, then the alarm control valve must be supervised in the open condition or not installed.

The alarm line connection provides a bypass that is connected below the alarm check valve to the alarm line. When the valve on this line is opened, supply line pressure is provided to the alarm line to test the alarm. The bypass is installed with a small swing check valve to prevent or limit false alarms. Any pressure differential between the system side and the supply side of the alarm check valve will be equalized through the checked bypass. *Figure 8* shows an alarm line with bypass check valve.

1.1.6 Excess Pressure Pump

An excess pressure pump is recommended when water pressure fluctuates and a retard chamber is not sufficient to prevent false alarms. An excess pressure pump pressurizes the portion of the system above the clapper so that it is approximately 20–30 psi greater than the pressure below the clapper but not greater than the maximum operating pressure of the system. The pump constantly monitors and maintains the pressure difference to eliminate false alarms. In areas where the freezing of drain lines can be a problem, excess pressure pumps are sometimes used instead of a retard chamber. *Figure 9* shows an excess pressure pump.

Excess pressure pumps operate on electrical power. Retard chambers do not, and are therefore more reliable because they are not defeated by power outages or the location of wiring.

1.1.7 Water Motor Alarm

The most common mechanical alarm device is the water motor alarm, or *water motor gong*. This device is a small water-powered turbine that turns a striker inside a gong that is attached to the alarm line. When water fills the retard chamber, replacing the air in the chamber, water pressure enters the alarm line and turns the turbine to sound the alarm. This alarm alerts the Fire Department to the location of the sprinkler riser serving a building and can also alert passersby to the existence of a fire or water discharge emergency.

The discharge of water through the turbine creates another need for a piped drain. The drain line must be routed to prevent freezing and ensure the discharged water causes no property damage. An air gap is required on the line between the outlet and the discharge area to avoid contaminating the potable water supply.

A Listed ¾" strainer must be installed at the alarm outlet of the waterflow-detecting device unless a retard chamber is used. A strainer must also be installed at the outlet of the retard chamber unless one is already provided. Follow the manufacturer's instructions for exact installation requirements. *Figure 10* shows a typical water motor gong.

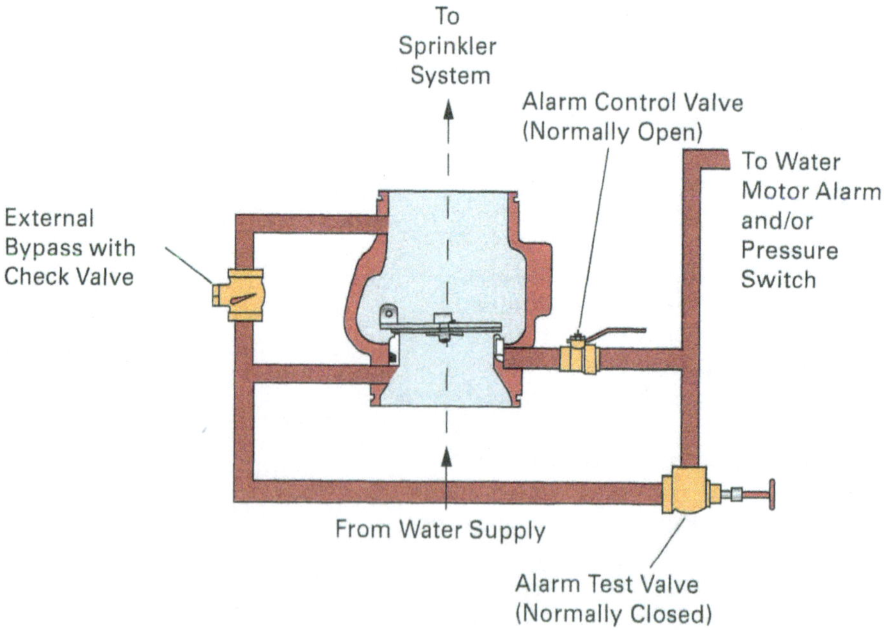

Figure 8 Alarm line with bypass check valve.

Figure 9 Excess pressure pump.

1.1.8 Pressure Switches

Pressure switches may be used in conjunction with an alarm check valve in a wet pipe sprinkler system. The switches are located on alarm lines and are used to electrically transmit an alarm to a remote and/or supervisory location. These switches open or close when the pressure in the alarm line increases or decreases to a specified value.

1.2.0 Wet System Riser Assemblies

A wet system riser assembly is used in most new installations in place of an alarm check valve. A system riser assembly uses a type of check valve that is drilled and pre-tapped for water gauges and the main drain valve, eliminating the need for fabricating threaded outlets when alarm check valves will not be used. A paddle-type or *vane-type* flow switch is used instead of a pressure switch or water motor gong. This configuration is generally less expensive than an alarm check valve and is acceptable to most Authorities Having Jurisdiction (AHJs). *Figure 11* shows a typical vertical system riser assembly. Horizontal or floor control valve assemblies are also available.

Both vertical system risers and floor control valve assemblies can function to provide zone control. Zone control allows part of the sprinkler system to be shut off to allow for service and maintenance without losing protection in other areas of the building. In addition, having a waterflow switch in each zone assists the Fire Department or other responders in identifying where a sprinkler has been activated.

In the past, system riser assemblies were not used on systems with fluctuating water supply pressures as they lacked a retard chamber to act as a hydraulic time delay to prevent false alarms.

Figure 10 Water motor gong.

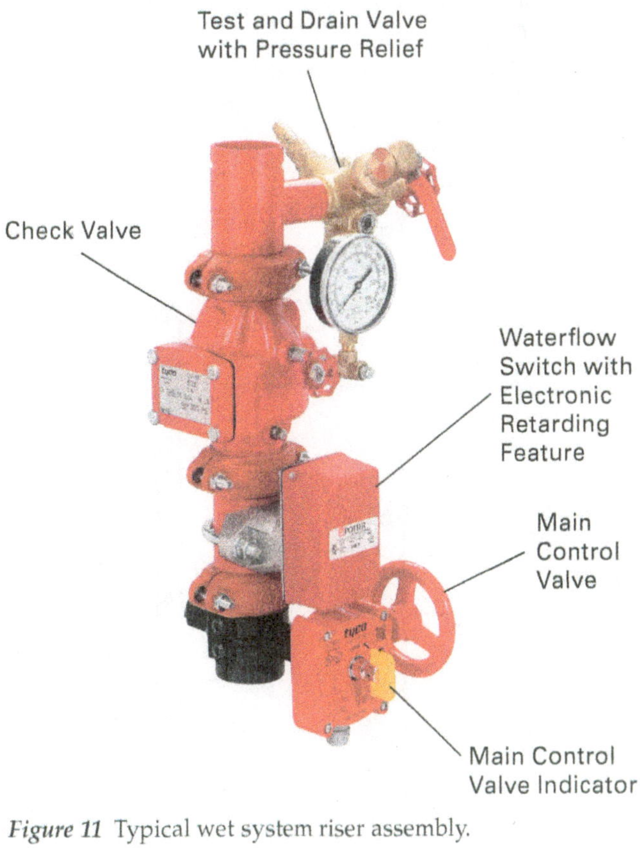

Figure 11 Typical wet system riser assembly.

Today, system riser assemblies can be ordered with waterflow switches that provide an electronic time delay to prevent false alarms in the event of a temporary surge.

1.3.0 Fire Department Connections (FDCs)

Fire Department Connections (FDCs) are used to supplement the fire suppression provided by a sprinkler system. FDCs must be prominently visible to firefighting personnel and located where hose lines can be connected immediately. Connections must be positioned with no interference from nearby buildings, fences, posts, landscaping, or obstructions of any kind. A clearance of 4' on either side of the FDC is a good rule of thumb.

An FDC should be located not less than 18" nor more than 4' above the adjacent grade level or access level. The local Fire Department may have its own, more restrictive specifications. Locations relative to streets and fire hydrants will be stipulated by the AHJ. A general requirement is that FDCs be adjacent to a fire lane and at the closest location to Fire Department access. Another requirement is that they be close enough to a fire hydrant to permit efficient pumper operations. The permissible separation from a fire hydrant varies by jurisdiction. Where hydrants are not available, other water supplies such as natural bodies of water and tanks or reservoirs should be utilized.

1.3.1 Components of FDCs

The main components of FDCs are the piping, connections, and valves (*Figure 12*). Different systems have specific requirements for the piping, check valves, and automatic drain (ball drip) valves used in the system. These requirements are based on the configuration of the system and type of connection.

FDCs are put in place to supplement pressure to a sprinkler system, not to provide all necessary pressure and flow. Therefore, the FDC piping is not based on system demand. For fire engine connections, pipe size is specified as a minimum of 4". For fireboat connections, use 6" or greater pipe. If the sprinkler system is hydraulically calculated and serving only one system riser assembly, the FDC must be no less than the pipe size of the riser. When piped to a $2^1/_2$" or smaller system riser assembly, a single-outlet FDC is acceptable.

FDC Configurations

Many types of FDCs are available, including single-inlet, double-inlet, and multiple-inlet connections in various configurations. The specific type required is determined by the type of system in use.

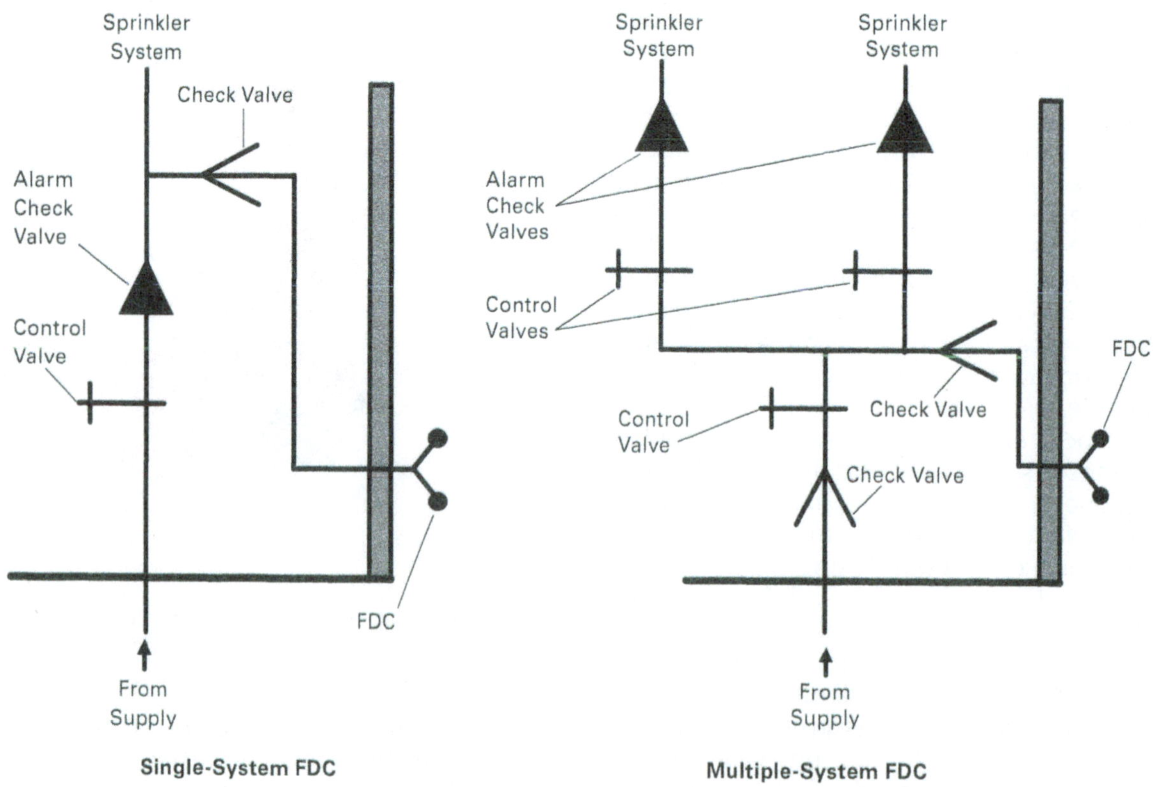

Figure 12 FDC underground piping configurations.

Thread sizes for FDCs are specified in NFPA 1963, *Standard for Fire Hose Connections*. The connections have an internally threaded swivel fitting with National Hose (NH) standard threads. Some FDCs do not conform to NFPA 1963. In these cases, thread size is determined by the AHJ.

- *Single-inlet connection* — A single-inlet connection may be used with wet pipe systems that have riser piping of $2^1/_2$" or less. Single-inlet connections are typically available with $2^1/_2$" hose nipples.
- *Double-inlet connection* — The double-inlet connection, sometimes referred to as a *Siamese connection*, is available in many configurations with various options. Double-inlet connections may be exposed, flush-mounted (*Figure 13*), or freestanding. In addition, they are available with a back outlet or a top/bottom outlet to facilitate installation. Double-inlet connections consist of two parts, the inlet itself and a cover plate or *escutcheon*. The escutcheon is usually separate, but some manufacturers cast the escutcheon with the FDC. One variation allows a choice of one or two internal clappers. These clappers restrict the water flow out of one inlet when water is pumped to the other inlet. Other assemblies are designed for a back outlet, and some are available with double clappers where the hose connections are exposed and angled down. Double-inlet FDCs are typically available with $2^1/_2$" hose nipples.

CAUTION

Exercise caution when installing FDCs to ensure they are not damaged. Some FDCs are designed as an architectural feature, with polished chrome or brass finishes that are easily susceptible to damage during installation.

Figure 13 Double-inlet flush connection.

- *Multiple-inlet connection* — There are many types of multiple-inlet connections, all of which come in exposed, flush, and freestanding configurations. All multiple-inlet configurations must contain a mechanism to prevent water flow through the other inlets when one is in use. Flush connections use $2^1/_2$" nipples for two-way and three-way configurations. In four-way and six-way configurations, some models use 3" nipples between the header or body and the inlets. The main problem in installing flush connections is cutting the nipples that go through the wall to fit both the inlet and outlet body. Always follow the manufacturer's instructions to ensure a leak-free installation.
- *Storz-type connection* — Large Storz-type FDCs (4" and over) allow for the quick connection of a single hose to move large amounts of water. *Figure 14* shows a Storz-type connection.

Figure 14 Storz-type connection.

Check Valves and Automatic Drain Valves

Listed check valves are required between the sprinkler system and the FDC to prevent water in the sprinkler system from leaking out of the connection. An automatic drain valve between the check valve and each FDC inlet is required to drain excess water. Note that an FDC *cannot* be equipped with a shutoff valve.

The piping between the FDC and the check valve must be equipped with an approved automatic drain valve. This portion of pipe is trapped and unless a drain valve is provided, the water will become stagnant and promote corrosion. An automatic drain valve must be piped to a drain using galvanized pipe, which may be through a wall. Treat it like any other drain.

1.3.2 Installation Requirements

NFPA 13 states that both single systems and multiple systems may be attached to underground piping. Always follow the approved shop drawings when making system connections.

Each FDC to sprinkler systems must have a sign with raised letters at least 1" in height, cast on a plate or a fitting, reading the service design: AUTO SPKR, OPEN SPKR, or STANDPIPE. For standpipe systems, a sign must also indicate the pressure required at the inlets to deliver the greatest system demand where the system demand pressure is greater than 150 psi. Where an FDC services only a portion of a building, a sign must be attached to indicate the portion of the building served.

Wall-Mounted FDCs

A wall-mounted FDC must be on the system side of all shutoff valves to ensure that the FDC is operational when the water supply is turned off. *Figure 15* shows a wall-mounted FDC.

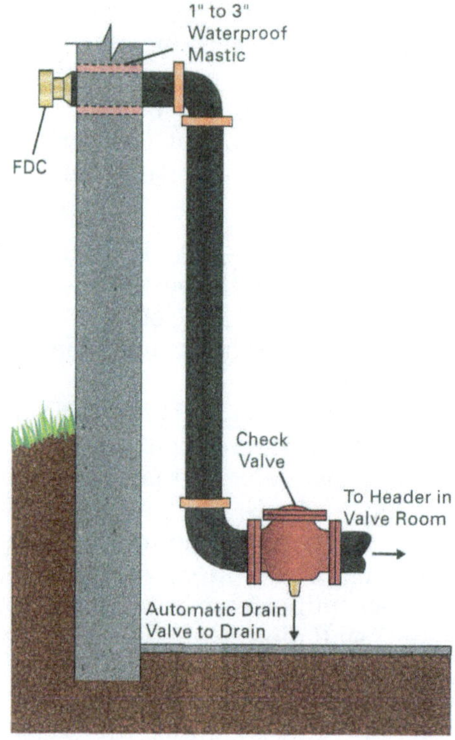

Figure 15 Wall-mounted FDC.

FDC in a Pit

Wet pipe sprinkler systems supplied by an FDC in the yard do not require the connection to be on the supply side of the shutoff riser valves. However, a check valve between the supply line and the FDC is required. The check valve must be located to maximize accessibility and minimize freezing potential. In yard installations, check valves in the water supply are located in a pit. Listed indicating valves are required around the check valve to allow servicing without disabling the FDC. *Figure 16* shows an FDC in a pit.

> **WARNING!**
>
> Never connect FDCs to the suction side of a fire pump. This can overpressurize the system and destroy the piping. In other words, a standard pit connection for an FDC cannot be used when a fire pump is used. However, a freestanding FDC can be used with a fire pump. The freestanding FDC must have its own line into the sprinkler system with a check valve between the fire pump and the FDC.

1.3.3 FDC Inspection and Maintenance

FDCs must be inspected quarterly in accordance with NFPA 25, *Standard for the Inspection, Testing, and Maintenance of Water-Based Fire Protection Systems*. Inspectors ensure that all caps, plugs, and thread protectors are in place. They also check inlets for blockage and ensure that automatic drains are functioning properly.

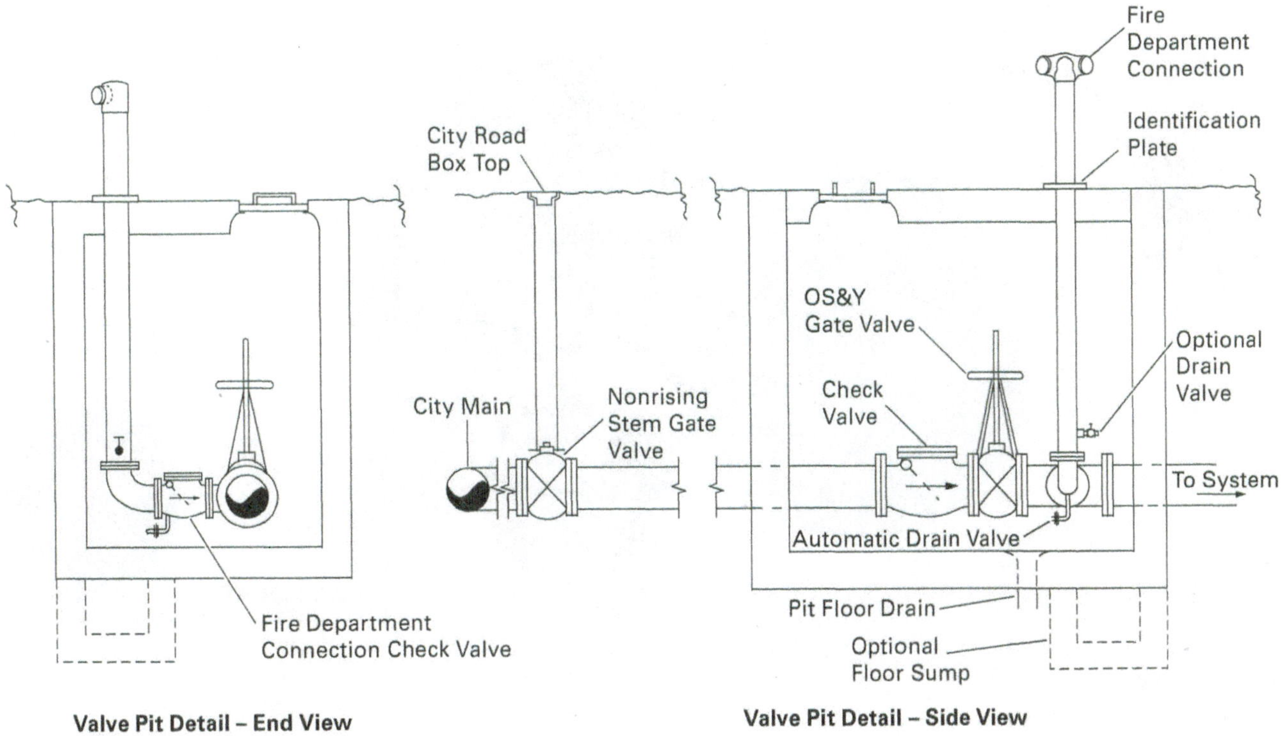

Figure 16 FDC in a pit.

Brass FDC plugs are prone to theft, which leaves the connection open to the accumulation of debris and other foreign material that could be pushed into the system when the connection is used. Inexpensive plastic covers deter theft and are often installed in place of brass plugs (*Figure 17*). Both threaded plugs and caps with break-away tabs are available.

Knox® locks (*Figure 18*) may also be installed by the local Fire Department to protect connections. These locks can only be opened using a Knox® FDC universal wrench.

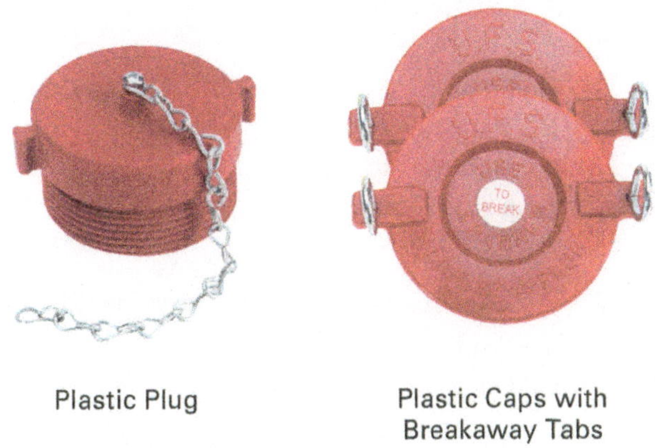

Figure 17 Plastic FDC covers.

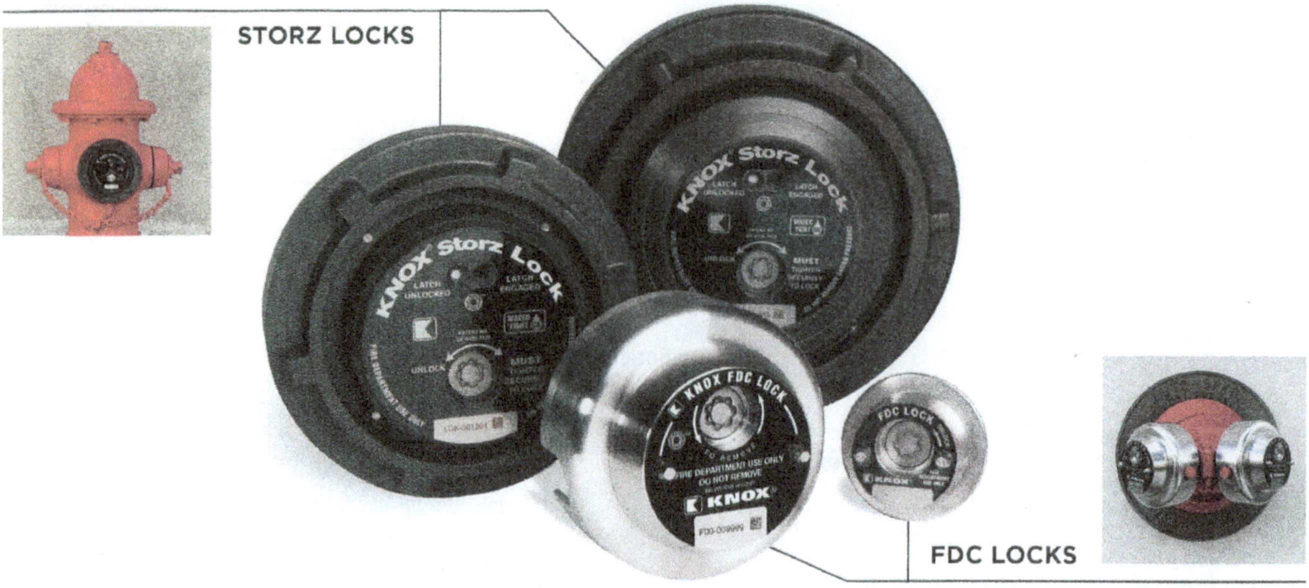

Figure 18 Knox® locks.

1.0.0 Section Review

1. On an approved shop drawing, an alarm check valve is *most* likely to be depicted as a _____.
   a. T shape
   b. V shape
   c. solid black triangle
   d. solid black circle

2. System riser assemblies can *never* be used on systems with fluctuating water supplies.
   a. True
   b. False

3. Per NFPA 25, FDCs must be inspected *at least* _____.
   a. daily
   b. weekly
   c. monthly
   d. quarterly

SECTION TWO

2.0.0 SWITCHES

Objective

Identify the types of switches used in wet pipe sprinkler systems.

a. Describe how switches operate.
b. Explain the purpose of paddle-type vane switches and describe how they are installed.
c. Explain the purpose of pressure switches and describe how they are installed.
d. Explain the purpose of tamper switches and describe how they are installed.

Trade Terms

Double-pole (DP) switch: A switch that can control two circuits.

Double-throw (DT) switch: A switch that transmits a signal in two directions.

Single-pole (SP) switch: A switch that can control only one circuit.

Single-throw (ST) switch: A switch that transmits a signal in one direction.

A major advantage of a sprinkler system is its automatic alarm capability. Flow switches, tamper switches, and pressure switches are used to transmit signals to alarms or to alert those responsible for supervising the system. There are many configurations for each type of switch.

For wet pipe sprinkler systems to function properly, all valves and their components must be in the correct on or off position. Switches are used to transmit an alarm and to monitor and supervise a system when a valve is placed in the wrong position, either on-site or remotely.

Some devices designed for wet pipe sprinkler systems have built-in flow switches and tamper switches. For example, butterfly valves and factory-assembled system riser assemblies contain such devices. The electrical connection of flow, tamper, and pressure switches is usually not the responsibility of sprinkler fitters, but the mechanical installation is often part of a fitter's work. Therefore, it is important to understand how these switches operate.

2.1.0 Switch Functions

When installing flow switches, tamper switches, and pressure switches, it is important to understand how a switch functions in the system for which it is being installed. To understand what a switch is and how it works, the following terminology must be understood:

- Poles
- Throws
- Switch positions
- Current and voltage ratings

2.1.1 Poles

Poles determine the number of circuits or electrical signal paths that can be controlled by a particular switch. A switch that can control only one circuit is called a single-pole (SP) switch. A switch that can control two circuits is called a double-pole (DP) switch. Where more than one pole is indicated, the circuits are normally designed to switch simultaneously. Sprinkler fitters mainly encounter SP and DP switches, but there are situations in which three-pole and four-pole switches are used. Three-pole switches can be controlled from two locations. Four-pole switches can be controlled from three locations.

2.1.2 Throws

Throws indicate the number of directions that an electrical signal can be sent by the movable arm within a switch. A single-throw (ST) switch transmits a signal in one direction and a double-throw (DT) switch transmits a signal in two directions. Common switches are the single-pole, single-throw (SPST); single-pole, double-throw (SPDT); and double-pole, double-throw (DPDT). *Figure 19* shows drawing symbols for SPST, SPDT, and DPDT switches.

2.1.3 Switch Positions

When a switch moves or throws from the open position to the closed position, a connection is made. Conversely, when a switch throws from the closed position to the open position, the connection is broken. These actions are referred to as *making* and *breaking* the circuit.

The normal position of a switch is its position under normal operating conditions. Switches in a circuit may be either normally open or normally closed.

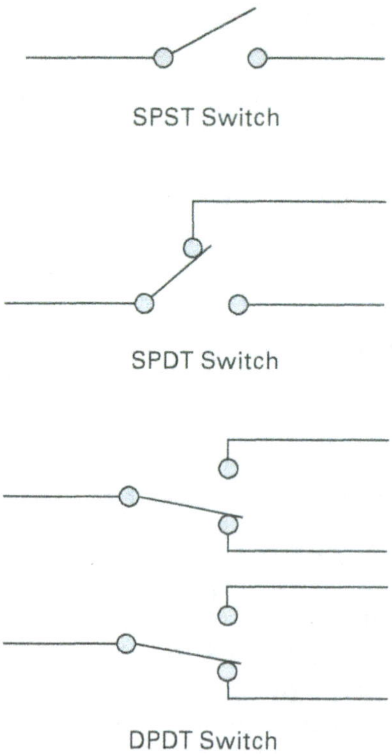

Figure 19 SPST, SPDT, and DPDT switch symbols.

- If a switch is normally open (N.O.), the contacts are open under normal operating conditions. When the switch is operated, the contacts close to make the circuit.
- If a switch is normally closed (N.C.), the contacts are closed under normal operating conditions. When the switch is operated, the contacts open to break the circuit.

2.1.4 Current and Voltage Ratings

Voltage is the difference in charge between two points in a circuit, while *current* is the flow rate of electric charge through the circuit. The unit of voltage is the volt (V). The unit of current is the ampere (A). A switch may have ratings for both alternating current (AC) and direct current (DC). AC is the power supplied by a utility and DC is the power supplied by a battery or DC generator.

Current and voltage have an inverse relationship. As voltage increases, current decreases. As voltage decreases, current increases. For example, a typical switch might have a rating of 10A at 125VAC and 2A at 30VDC. Every electrical device has both a current rating and a voltage rating printed on the device. You must be able to identify these ratings in order to select the proper switch for the application.

WARNING! Never install an electrical device unless it matches the voltage and current ratings indicated on the approved shop drawings. Installing a device with the wrong rating may result in tripping of associated circuit breakers as well as circuit overheating, failure, and fire.

2.2.0 Paddle-Type Flow Switches

A paddle-type flow switch is a device used to transmit a signal to an alarm when a sprinkler is activated and water moves through the system. Paddle-type flow switches are used only in wet pipe sprinkler systems. The force of moving water in dry pipe and preaction/deluge systems striking the paddle in a paddle-type flow switch would shear it off, destroying the switch and clogging the system.

Paddle-type flow switches are commonly used where alarm check valves are *not* used. They are used with swing check valves to produce what is called a *system riser assembly*. Paddle-type flow switches are also used in multistory buildings where each floor connects to the riser. The paddle-type flow switch follows the floor or zone control valve to provide an alarm for that floor or zone only.

2.2.1 Switch Mechanics

Paddle-type flow switches have a flexible vane that almost completely obstructs the pipe. When water flows through the pipe, the flexible vane is pushed forward, or deflected. When the flexible vane is deflected, it operates a switch to send a signal that water is flowing. Paddle-type flow switches are available for plastic and steel pipe and copper tubing from 1" through 8" in diameter.

Paddle-type flow switches under 2" are shipped with various paddles to fit different types of piping (steel, copper, or CPVC). The fitter must select the proper paddle for the piping in use. Failure to use the proper paddle will result in system damage and failure. If the paddle is too small for the pipe being monitored, it will delay switch operation. If the paddle is too large, it will cause interference in the waterway or become stuck in the flow position, sending a continuous alarm. The switch is mounted to the pipe using a tee (*Figure 20*).

Figure 20 Flow switch with thread mounting.

CAUTION

If you must install a paddle, make sure it is the right size and type. Paddle-type flow switches operate in only one direction; therefore, they must be installed so that the arrow points in the direction of the flow. Attach the paddle in accordance with the manufacturer's instructions and ensure that it is securely fastened. Failure to use the proper size and type of paddle will result in system damage and failure.

Paddle-type flow switches over 2" have the paddle preinstalled and must be ordered for the specific pipe size and material. These switches are installed by drilling a hole in the pipe to be monitored and then attaching the switch using a U-bolt (*Figure 21*). Consult the manufacturer's literature for proper location in relation to other system components. For example, flow switches must not be installed near changes in piping direction that could interfere with proper operation of the switch.

CAUTION

After drilling the hole, remove the disc or coupon. Do not allow it to fall into the pipe. Wire the coupon to the exterior of the pipe as evidence of removal. Failure to remove the coupon will clog pipes and may result in system failure.

Figure 21 Flow switch with hole mounting.

To install the switch, ensure that the arrow on the saddle points in the direction of the water flow. The flexible plastic paddle is rolled into a small diameter that will slip through the hole as the assembly is pushed into place. Clamp the assembly in place using the U-bolt provided with the switch. Ensure that the rubber gasket and hole are the correct size and the U-bolt is tightened properly. If the rubber gasket is too large, the gasket will not seal. If the hole is too small, the gasket will not fit. If the U-bolt is overtightened, the gasket may be cut and leak.

2.2.2 Retarding Devices

Retarding devices for wet pipe sprinkler systems are available in many configurations. Some retarding devices use a calibrated orifice for air discharge like the retard chamber discussed earlier. Another type of retarding device is a time delay built into a paddle-type flow switch. Since most pressure surges occur quickly, the time delay prevents false alarms. The time delay must be set in the field per the manufacturer's instructions and local AHJ requirements.

2.3.0 Pressure Switches

Standard pressure switches can detect changes in pressure within a system. *Figure 22* shows a standard pressure switch. Some specialized pressure switches allow adjustment to set the activation pressure. These switches are often used in conjunction with an alarm check valve to transmit a signal to an alarm when a fluctuation in pressure is detected. The switches are installed on top of the retard chamber in wet pipe sprinkler systems. The tamper-resistant housing may only be removed using a special key.

Switches may be enclosed in an explosionproof housing where required for use in hazardous atmospheres (*Figure 23*). When an explosionproof switch is indicated on the approved shop drawings, always check the nameplate on the switch before installing it to verify that it is an explosionproof device.

2.4.0 Tamper Switches

Tamper switches, also known as *supervisory switches*, are used with many types of valves. These switches are used to send signals indicating whether or not a specific valve is in the correct position for the sprinkler system to function properly.

2.4.1 OS&Y Valve Tamper Switches

Tamper switches for OS&Y gate valves fit into a groove in the unthreaded portion of the rising stem and are mounted using an adjustable clamp bar and carriage bolts (*Figure 24*). When the valve is operated, the actuator moves with the stem and trips the switch. The switch actuator length may be adjustable, or it may have to be cut in the field so that it sits in the groove properly.

Many new OS&Y valves come with a factory-installed groove, as shown in *Figure 25*. If the stem does not have a groove, use a file to cut a groove in the stem to the depth specified by the manufacturer. Take care to prevent valve damage when filing. After filing the groove, deburr the cut edges to allow smooth movement of the actuator. The groove must be near the threads, so the stem seal is not damaged when the valve is closed. It must also be of a sufficient depth, so the switch does not interfere with the normal operation of the valve.

Always follow the manufacturer's instructions for the installation and adjustment of tamper switches. *Figure 26* shows an OS&Y stem with an installed tamper switch. Another type of tamper switch uses an indentation in the valve stem rather than a groove and operates in a similar manner.

Figure 22 Standard pressure switch.

Figure 23 Explosionproof pressure switch.

Figure 24 Tamper switch for OS&Y gate valve.

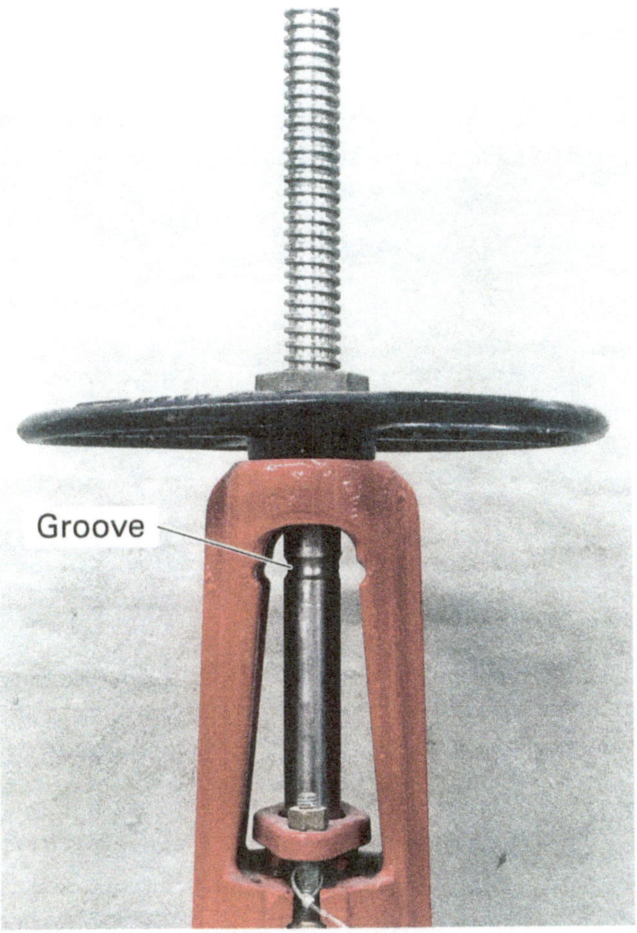

Figure 25 OS&Y stem with a factory-installed groove.

Figure 26 OS&Y valve with a tamper switch.

2.4.2 Post Indicator Valve Tamper Switches

Tamper switches are also available for post indicator valves. These switches have one or two sets of contacts. *Figure 27* shows a tamper switch for a post indicator valve. This device mounts in a tapped hole in the post indicator housing and is triggered by movement of the indicating assembly.

2.4.3 Plug-Type Tamper Switches

Plug-type supervisory (tamper) switches are used with certain types of control valves, such as nonrising stem valves and pit valves that are subject to flooding. These switches are only acceptable when no other supervisory method is available or practical. A plug-type supervisory switch has one set of SPST normally closed contacts. *Figure 28* shows a plug-type supervisory switch.

This switch is attached to the valve wheel so that the plug is pulled out of the switch when the wheel is turned. The plug cannot be reinserted without removing the cover using the special key provided.

Figure 27 Tamper switch for post indicator valve.

Figure 28 Plug-type supervisory switch.

> **NOTE**
> Some pressure and flow switches are also equipped with cover tamper switches. If the cover is removed, a signal is sent to a central station, proprietary station, or remote signaling unit to indicate that tampering has occurred.

2.0.0 Section Review

1. A type of switch that controls one circuit but can transmit a signal in two directions is a(n) _____.
   a. SPST switch
   b. SPDT switch
   c. DPST switch
   d. DPDT switch

2. Paddle-type flow switches are used *only* in wet pipe sprinkler systems.
   a. True
   b. False

3. The standard pressure switches used with alarm check risers are installed _____.
   a. on top of the retard chamber
   b. at the FDC
   c. at the system high point
   d. between two supervisory switches

4. Which type of switch may require that you file a groove in the valve stem in order to complete the installation?
   a. Paddle-type flow switch
   b. Explosionproof pressure switch
   c. Plug-type tamper switch
   d. OS&Y valve tamper switch

SECTION THREE

3.0.0 HYDROSTATIC TESTING AND SYSTEM TROUBLESHOOTING

Objective

Describe how to test and troubleshoot wet pipe sprinkler systems.

a. Explain the basic procedure for performing hydrostatic testing.
b. Describe how to troubleshoot common wet pipe sprinkler system issues.

Performance Task

1. Complete a Contractor's Material and Test Certificate for Aboveground Piping for a wet pipe sprinkler system.

Trade Term

Hydrostatic test: A wet pipe sprinkler system test that uses a test pump to increase the pressure of the system to 200 pounds per square inch (psi) or 50 psi above static pressure in excess of 150 psi for two hours to determine whether there are any leaks. If the system holds the pressure for the specified time period, it passes the test.

After a wet pipe sprinkler system is installed and before it is put into service, it must be tested. A **hydrostatic test** is performed by filling the sprinkler system with water and then increasing the pressure to check for leaks and other defects. The major tasks involved in performing hydrostatic tests include performing the pretest requirements, preparing the pump, sealing the system to be tested, and performing the test.

Test pumps are used to increase the system pressure to ensure that the system will hold up under the normal working pressure. The pumps may be gasoline, electric, or manual types.

3.1.0 Hydrostatic Testing

The sprinkler contractor may be required to make formal pressure tests in the presence of the AHJ or the owner's representative. Before these tests are made for official acceptance, preliminary tests should be made privately to make sure the installation will pass when the formal test takes place.

A pressure test is required for the entire system. This includes testing the FDC for working pressures that Fire Departments apply when supporting the sprinkler system (200 psi minimum). Any failures that occur must be corrected before the final demonstration and project completion.

After the preliminary test is complete, the AHJ observes a pressure test demonstration. Several people typically observe and scrutinize the system in operation. If there is a visible leak and drop in pressure, repairs must be made, and the test must be performed again. Hydrostatic tests must be made at a minimum of 200 psi for two hours or 50 psi above static pressure in excess of 150 psi for two hours. All aboveground piping leaks must be stopped.

An interim air test is used when weather does not permit a hydrostatic test. An interim pneumatic test is established at 40 psi air pressure and then the drop in pressure is measured. It must always be followed by a hydrostatic test when weather permits. The drop cannot exceed $1\frac{1}{2}$ psi in 24 hours.

All sprinkler systems also require a main drain test, during which the main drain valve is opened and remains open until the system pressure stabilizes. The static and residual pressures are recorded on the Contractor's Material and Test Certificate for Aboveground Piping and the sprinkler system general information sign. Note that every system must have a fully completed general information sign as well as a hydraulic information sign.

The contractor is responsible for completing all required acceptance tests and recording the results on the Contractor's Material and Test Certificate for Aboveground Piping (*Figure 29A*, *Figure 29B*, and *Figure 29C*). Witnesses for both the contractor and the property owner must sign the completed test certificate.

3.1.1 Purpose of Testing

The purpose of hydrostatic testing is to test for defects in components and the integrity of the system. While Listed components undergo rigorous testing under challenging conditions, manufacturing defects and shipping or handling damage can sometimes occur.

Contractor's Material and Test Certificate for Aboveground Piping

PROCEDURE

Upon completion of work, inspection and tests shall be made by the contractor's representative and witnessed by the property owner or their authorized agent. All defects shall be corrected and system left in service before contractor's personnel finally leave the job.

A certificate shall be filled out and signed by both representatives. Copies shall be prepared for approving authorities, owners, and contractor. It is understood the owner's representative's signature in no way prejudices any claim against contractor for faulty material, poor workmanship, or failure to comply with approving authority's requirements or local ordinances.

AFSA
American Fire Sprinkler Association

Property name			Date
Property address			
New Installation?	☐ Yes	☐ No	
Modification? If yes, complete applicable portions of the form.	☐ Yes	☐ No	
Provide a description of the scope of work on page 3.			

Plans	Accepted by approving authorities (names)		
	Address		
	Installation conforms to accepted plans	☐ Yes	☐ No
	Equipment used is approved If no, explain deviations	☐ Yes	☐ No
Instructions	Has person in charge of fire equipment been instructed as to location of control valves and care and maintenance of this new equipment? If no, explain	☐ Yes	☐ No
	Have copies of the following been left on the premises?	☐ Yes	☐ No
	1. System components instructions	☐ Yes	☐ No
	2. Care and maintenance instructions	☐ Yes	☐ No
	3. NFPA 25	☐ Yes	☐ No
Location of system	Supplies buildings		

Sprinklers

Make	Model	Year of manufacture	Orifice size	Quantity	Temperature rating

Pipe and fittings

Type of pipe

Type of fittings

Alarm valve or flow indicator

Alarm device			Maximum time to operate through test connection	
Type	Make	Model	Minutes	Seconds

Dry pipe operating test

Dry valve			Q. O. D.		
Make	Model	Serial no.	Make	Model	Serial no.

	Time to trip through test connection[a,b]		Water pressure	Air pressure	Trip point air pressure	Time water reached test outlet[a,b]		Alarm operated properly	
	Minutes	Seconds	psi	psi	psi	Minutes	Seconds	Yes	No
Without Q.O.D.									
With Q.O.D.									

If no, explain

[a] Measured from time inspector's test connection is opened

[b] *NFPA 13* only requires the 60-second limitation in specific sections

Figure 29A Example Contractor's Material and Test Certificate for Aboveground Piping (1 of 3).

Deluge and preaction valves

Operation	☐ Pneumatic	☐ Electric	☐ Hydraulics

Piping supervised ☐ Yes ☐ No | Detecting media supervised ☐ Yes ☐ No

Does valve operate from the manual trip, remote, or both control stations? ☐ Yes ☐ No

Is there an accessible facility in each circuit for testing? ☐ Yes ☐ No | If no, explain

Make	Model	Does each circuit operate supervision loss alarm?		Does each circuit operate valve release?		Maximum time to operate release	
		Yes	No	Yes	No	Minutes	Seconds

Pressure-reducing valve test

Location and floor	Make and model	Setting	Static pressure		Residual pressure (flowing)		Flow rate
			Inlet (psi)	Outlet (psi)	Inlet (psi)	Outlet (psi)	Flow (gpm)

Backflow device forward flow test

Indicate means used for forward flow test of backflow device: ____________

When means to test device was opened, was system flow demand created? ☐ Yes ☐ No ☐ N/A

Test description

Hydrostatic: Hydrostatic tests shall be made at not less than 200 psi for 2 hours or 50 psi above static pressure in excess of 150 psi for 2 hours. Differential dry-pipe valve clappers shall be left open during the test to prevent damage. All aboveground piping leakage shall be stopped.

Pneumatic: Establish 40 psi air pressure and measure drop, which shall not exceed 1½ psi in 24 hours. Test pressure tanks at normal water level and air pressure and measure air pressure drop, which shall not exceed 1½ psi in 24 hours.

Tests

All piping hydrostatically tested at ______ psi for ____ hours

Dry piping pneumatically tested ☐ Yes ☐ No

Equipment operates properly ☐ Yes ☐ No

If no, state reason

Do you certify as the sprinkler contractor that additives and corrosive chemicals, sodium silicate or derivatives of sodium silicate, brine, or other corrosive chemicals were not used for testing systems or stopping leaks?

☐ Yes ☐ No

Drain test | Reading of gauge located near water supply test connection: ______ psi | Residual pressure with valve in test connection open wide: ______ psi

Underground mains and lead-in connections to system risers flushed before connection made to sprinkler piping

Verified by copy of the Contractor's Material and Test Certificate for Underground Piping. ☐ Yes ☐ No

Flushed by installer of underground sprinkler piping ☐ Yes ☐ No

Other | Explain

If powder-driven fasteners are used in concrete, has representative sample testing been satisfactorily completed? ☐ Yes ☐ No | If no, explain

Blank testing gaskets

Number used	Locations	Number removed

Welding

Welding piping ☐ Yes ☐ No

If yes . . .

Do you certify as the sprinkler contractor that welding procedures used complied with the minimum requirements of *AWS B2.1, ASME Section IX Welding and Brazing Qualifications,* or other applicable qualification standard as required by the AHJ? ☐ Yes ☐ No

Do you certify that all welding was performed by welders or welding operators qualified in accordance with the minimum requirements of *AWS B2.1, ASME Section IX Welding and Brazing Qualifications,* or other applicable qualification standard as required by the AHJ? ☐ Yes ☐ No

Do you certify that the welding was conducted in compliance with a documented quality control procedure to ensure that (1) all discs are retrieved; (2) that openings in piping are smooth, that slag and other welding residue are removed; (3) the internal diameters of piping are not penetrated; (4) completed welds are free from cracks, incomplete fusion, surface porosity greater than 1/16 in. diameter, undercut deeper than the lesser of 25% of the wall thickness or 1/16 in.; and (5) completed circumferential butt weld reinforcement does not exceed 3/32 in.? ☐ Yes ☐ No

Figure 29B Example Contractor's Material and Test Certificate for Aboveground Piping (2 of 3).

Cutouts (discs)	Do you certify that you have a control feature to ensure that all cutouts (discs) are retrieved? ☐ Yes ☐ No	
Hydraulic data nameplate	Nameplate provided ☐ Yes ☐ No	If no, explain
Sprinkler contractor removed all caps and straps? ☐ Yes ☐ No		
Remarks	Date left in service with all control valves open	
Signatures	Name of sprinkler contractor	
	Tests witnessed by	
	The property owner or their authorized agent (signed)	Title Date
	For sprinkler contractor (signed)	Title Date
Additional explanations and notes		

Figure 29C Example Contractor's Material and Test Certificate for Aboveground Piping (3 of 3).

Most of the components in a wet pipe sprinkler system have a minimum pressure rating of 175 psi. This means that those materials in the system in the nominal 6" category or smaller have been type-tested at 875 psi before being Listed for a working pressure of 175 psi. For Underwriters Laboratories, this is five times the working pressure. For 8" and larger components, the type test is at 700 psi—four times the working pressure—before being Listed for a working pressure of 175 psi.

3.1.2 System Inspection

A system inspection is required by an AHJ before the system is tested and approved for operation. Several areas and components of wet pipe sprinkler systems must be inspected, including the base material, alignment joints, valves, instruments and supports. Every joint, component, and pipe attachment is scrutinized. If any pretest checks do not pass a visual inspection, tag the part, and write down why the part failed the inspection. Make adjustments and repairs before testing the system. A careful inspection is the first step in preventing leaks.

Electronic Records

System testing is often conducted using a jobsite tablet, such as the one shown here.

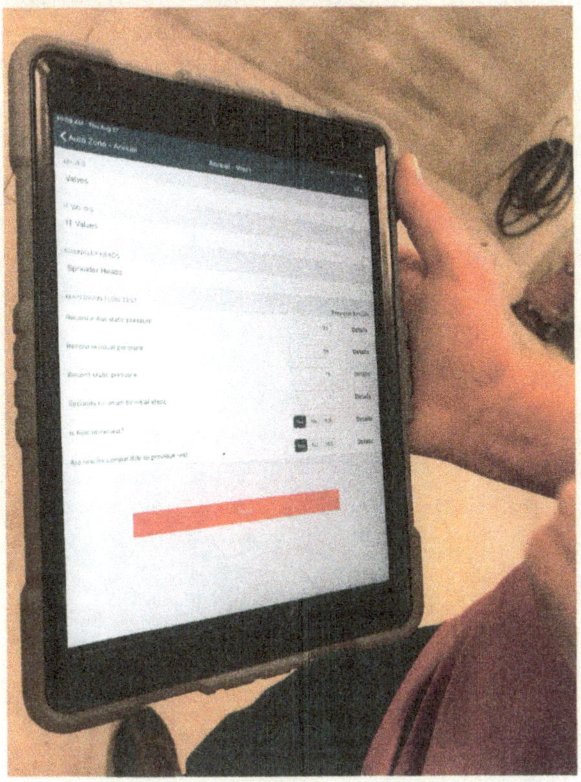

WARNING!

Be sure to wear ear protectors, safety glasses, and gloves. Make sure others in the area also wear personal protective equipment.

3.1.3 Testing Hazards

Hydrostatic testing at high pressures is dangerous. If a portion of the system fails during hydrostatic testing, serious injury and even death can result. Therefore, the test must be performed only by trained personnel. If you are not involved in the test, stay out of the area where the test is being conducted. If you are required to be in the area, pay close attention to the instructions given by your supervisor. Do not attempt to make repairs to a system while it is under pressure. If repairs are necessary, stop the test, release the pressure, and drain the system.

The two greatest testing hazards are air and water pressure. Air is compressible, and water is not. Therefore, when systems are pressurized with water and a break or leak occurs, the release of as little as a tablespoon to a cup of water relieves pressure from the system. Air is a different matter. Many cubic feet of air must be released before the pressure drops to safe limits.

With compressed air, a break causes an explosion. Air must never be used in place of hydrostatic testing, except when allowed by NFPA 13. One such exception is the interim test that allows a 24-hour, 40 psi air test when freezing conditions occur. The interim test must be followed by a full hydrostatic test when the weather permits. Even with this exception, CPVC manufacturers will not warrant their piping when it is tested under freezing conditions.

WARNING!

Always follow the manufacturer's instructions when testing CPVC systems. Some manufactures allow nominal air pressure testing (e.g., no more than 25 psi) using an oil-free compressor. Failure to follow the manufacturer's instructions will violate the product warranty and may result in system failure.

CAUTION

When doing an interim pressure test on a freezer that is already below freezing, the air supply must be drawn directly from the freezer because warm air holds more moisture, which will result in system expansion and possible failure. If freezer air cannot be used, nitrogen or dry air must be used.

Quality Control

Quality control inspections should be performed throughout an installation. This is especially true when it comes to pipe threads and pipe grooves. The threads must have a uniform appearance without scratches, tears, gouges, burrs, rust, dirt, foreign material, or corrosion. Pipe ends must be round, not egg-shaped. If the installation appears different from what you have experienced in the past, find out why. Never be afraid to ask questions.

Water damage is another hazard that is faced when performing testing. Whenever possible, perform testing early in the construction process to minimize any damage that may occur. On new construction, the first test should be conducted before the interior finish begins.

3.1.4 Test Pumps

Test pumps are precision devices used in performing pressure tests. The pumps are used to fill the system with the testing water and apply pressure to the system. The pumps depend on the closure of valves, clean water, and the operation of finely machined cylinders to function properly. To ensure proper function, some pumps have inlet strainers to filter sand, dirt, grit, and organic materials, such as grass and leaves. The strainers must be cleaned periodically to keep the water clean and clear.

Some situations require draft-testing of water from a barrel. Draft tests also require the water to be clean for the test pump to function properly. When drafting from a container, follow instructions carefully to keep air from entering the pump, which will result in pump failure.

Many types of pumps are available for testing wet pipe sprinkler systems. The test pump is normally connected to the system by a threaded connection protruding from the discharge side of the pump. The pressure ratings of all the connecting devices must be higher than the rated maximum output of the pump. The following types of pumps are used in pressure tests:

- Hand test pumps
- Gasoline-driven pumps
- Electric test pumps

Figure 30 shows typical test pumps.

WARNING! When operating these pumps, follow standard safety procedures such as staying clear of hot exhausts, not operating gasoline engines without adequate ventilation, and not operating pumps in hazardous locations.

Hand Test Pumps

Hand test pumps are simple pressure-testing devices fabricated with two check valves as one assembly. Some models have a platform used to mount the pump to a structure. The pumps are manufactured with various working pressure ratings. Some pumps measure a wider range of pressures than others. Pressure on the system must be maintained for a specified amount of time and then released. Follow the manufacturer's instructions for the pump in use.

Follow these steps to operate a manual-force test pump:

Step 1 Connect the test pump to the line to be tested.

Step 2 Close the manually operated external valve on the pump.

Hand Pump | Electric Pump | Gasoline-Driven Pump

Figure 30 Typical test pumps.

Step 3 Fill the system to be tested with water and ensure that the pump is not leaking.

> **NOTE**
> The intake and outlet poppet valves will open automatically.

Step 4 Pull the pump handle back to let the cylinder fill with water.

Step 5 Open the external valve and push the pump handle forward with smooth, even strokes to exhaust the air, maintaining constant pressure on the pump intake to prevent air from entering the system.

> **CAUTION**
> Failure to remove the majority of the air in the system can result in a false test. The air will leak out slowly and may never completely exhaust, producing a water leak. Pushing the pump handle harshly and using abrupt strokes may blow out the poppet valves.

Step 6 Release pressure by slowly opening the external valve to bleed all air from the system.

Step 7 Actuate the pump until the desired test pressure is attained.

Step 8 Close the external manual valve and maintain pressure only long enough to test the system.

Step 9 Open the external valve to relieve pressure in the system upon completion of the test.

Step 10 Drain the water from the pump and the gauge to prevent it from freezing.

Step 11 Cover the open ports of the pump to keep out dirt.

Gasoline-Driven Test Pumps

Gasoline-driven pumps are also manufactured with various pressure ratings and flow rates. They are often used to test new installations where electricity has not been activated. Gasoline-driven pumps require less labor and are more efficient than hand test pumps.

One disadvantage of gasoline-driven pumps is that they require maintenance. They are also a greater danger on the jobsite since they are powered by flammable liquids. Another disadvantage is that they may run out of gas before testing is completed. It is essential that these pumps be maintained and inspected before each use.

> **CAUTION**
> Ensure that the crankcase oil-fill chamber has the required amount of oil and the gas tank is full before starting the motor.

> **WARNING!**
> Gasoline is flammable and any spills must be cleaned up immediately. Gas must be kept in approved containers and stored properly. Do not leave a fuel container at the pump. Gas pumps must only be used in well-ventilated areas.

Electric Test Pumps

Electric test pumps operate on the same principles as the manual-force and gasoline-driven pumps. They are commonly used to test existing systems and new installations where electrical lines have already been run. These pumps are manufactured with various volume capabilities depending on the pressure capabilities. Some electrical models are equipped with an adjustable pressure regulator that feeds the bypass water back to the inlet connection.

3.1.5 Testing Procedures

Inspecting wet pipe sprinkler systems for defects is critical during testing. Testing procedures should be followed systematically to ensure that the systems are functioning properly. A general inspection procedure is as follows:

Step 1 Ensure that all pretest requirements have been completed and that the proper documentation has been filled out.

Step 2 Barricade the area around the system to be tested.

Step 3 Open the highest inspector's test valve when there are no zone control valves, or when testing a multilevel system to bleed the air out of the system.

> **WARNING!**
> Be sure to keep your face away from the valve when opening it.

Step 4 Open the main supply control valve to fill the system with water or the required testing medium, allowing the air to escape through the inspector's test connection.

Step 5 Monitor the water and air coming from the connection, and leave the valve open until water flows smoothly from the valve.

Step 6 Close the inspector's test valve.

Step 7 Attach the test pump to the system according to project specifications.

Step 8 Allow the pressure to build up slowly to 50 percent of the required test pressure.

> **NOTE**
> Monitor the pressure of the system at the pressure gauge. The person operating the pump must be able to see the pressure gauge at all times.

> **WARNING!**
> Never leave an operating hydrostatic pump unattended.

Step 9 Increase the test pressure in increments of 10 percent of the test pressure until the final pressure is reached.

Step 10 Close the valve between the test pump and the piping system.

Step 11 Disconnect the test pump from the system.

Step 12 Allow the system to remain at full pressure for two hours.

Step 13 Inspect the entire system for leaks.

> **NOTE**
> If a joint leaks, the system must be drained, repaired, and tested again. If there are no leaks, the system can be drained after the proper personnel have approved the test.

Step 14 Crack the inspector's test valve to relieve pressure from the system.

> **WARNING!**
> Stand to the rear side of the valve when releasing pressure from the system.

Step 15 Open the inspector's test valve completely to allow air back into the system.

Step 16 Open the drain valve to allow the system to drain.

> **NOTE**
> Antifreeze systems are wet pipe sprinkler systems that contain an antifreeze solution and are connected to a water supply. Antifreeze systems are used when a portion of a system is installed in an area that cannot be reliably maintained at or above 40°F. Antifreeze, along with water, is discharged immediately after sprinklers are opened by heat from a fire. Refer to the latest NFPA 13 and NFPA 25 rulings and recommendations for any antifreeze application. These can be found at **www.nfpa.org/antifreeze**. Always follow the manufacturer's instructions as well as NFPA requirements. New systems must use Listed antifreeze only.

3.2.0 Troubleshooting Wet Pipe Sprinkler Systems

Problems in wet pipe sprinkler systems occur when established maintenance and operating procedures are not followed. Sprinkler fitters or licensed contractors must inspect wet pipe sprinkler systems annually to ensure that the system is functioning properly and identify any problems. Since wet pipe sprinkler systems are continuous systems, problems in one section can affect the whole system. Problems can be compounded, or new problems can appear in other areas when incorrect adjustments are made. Troubleshooting a wet pipe sprinkler system requires a systematic approach that begins by identifying the problem.

3.2.1 Identify the Problem

Effective troubleshooting begins by determining how a particular system operates under normal conditions and then identifying the areas or components in which the system is no longer operating as expected. The customer's view of a problem can be very different from your view. "It doesn't work right," can mean anything from a closed valve to lack of knowledge of how the system works. Ask questions of the customer and be specific. Listen to the description of the problem and document all problem areas.

When searching for the problem, proceed from the simplest types of problems to the more complex problems. Eliminate the simple failures first. Determine the root cause of failures in order to prevent repeat failures.

3.2.2 Collect Data

After determining the problem, data must be collected and evaluated so possible causes may be determined. The data may be collected from parts manufacturers, maintenance records, and any personnel with operating knowledge of the system. Review the manufacturer's data for product specifications.

Another good way to collect data is by placing the system into operation. If a system is unfamiliar, installed by another company, or not maintained regularly, inspect the whole system first to become familiar with it. Use a standard inspection procedure or checklist to document the system. Familiarize yourself with what is present before seeking to solve a specific problem. There may be other problems that have to be solved before the original problem can be solved. Find out how different components in the system relate to one another during operation. Also, consider any alternatives that might be the cause of problems such as faulty materials, misapplication of materials, or lack of system maintenance. Note any obvious design or installation problems by inspecting for maladjustments or any other visual clues. Document all equipment readings.

3.2.3 Analyze Data

When analyzing data, it is important to consider all possibilities. Refer to manufacturer's troubleshooting charts to help identify the root cause of the problem.

As a system ages, defects in materials become a more frequent cause of problems. However, defects in workmanship can also go undetected for a long time. There is a known instance where a sprinkler system had no water connected to it, and this was not detected for 10 years. Regular system maintenance would have prevented this problem. Analyze the data carefully and exhaust all possible causes before deciding on the necessary action required to solve the problem.

3.2.4 Take Action

After all data has been collected and analyzed, the action required to solve a problem is determined. System components may need to be cleaned, modified, repaired, or replaced.

If system maintenance or repair will disable the sprinkler system, follow the proper procedures to inform building occupants and the supervisory service. A fire watch may be necessary to avoid a potential catastrophe. It is surprising how many fires occur when sprinkler systems are out of service. Protect yourself and your company from liability by obtaining a signed release before taking any action. If piping must be disconnected or sprinklers removed, use proper draining procedures to avoid flooding and damage. Where there are sectional control valves, disable and drain only the section involved, leaving the rest of the system in service. Otherwise, draining procedures start with opening the main drain. Opening the inspector's test valve will expedite draining. At low points, use a five-gallon bucket. For large volumes of trapped pipe, a garden hose may be used.

After maintenance or repairs have been performed, fill the system and retest according to system requirements. Document the actions that were taken. Perform a visual inspection and ensure pressure readings are accurate. Ensure that the actions performed solved the problem. If the initial problem is not solved by the actions performed, troubleshoot the system again, analyzing the data for any possible causes that were overlooked. *Table 1A*, *Table 1B*, *Table 1C*, and *Table 1D* show a sample troubleshooting chart for wet pipe sprinkler systems.

Table 1A Troubleshooting Chart for Wet Pipe Sprinkler Systems (1 of 4)

Machine Component or Problem Area	Possible Cause	Corrective Action
False alarms from normal pressure fluctuations	Drain port on retard chamber plugged	Clean retard chamber, drain piping, and inspect strainer to be certain it is in place, intact, and clean.
	External bypass check valve defective (blocked, defective seat, or defective clapper)	Replace external bypass check valve.
	External alarm bypass valve installed backwards	Install bypass check valve with arrow pointing in direction of water flow.
	Pressure changes greater than can be effectively handled by one retard chamber	Install two retard chambers in series or install an excess pressure pump.
	Bypass check valve too small	Increase bypass valve and trim from ½" to ¾" and remove tuberculation (buildup of organic matter) from line.
	Defective alarm valve seat	Drain system and replace valve seat. Hydrostatically test system. If pressure gauges equalize during testing, a warped clapper is indicated.
	Warped valve clapper	Close supply valve, manually pressurizing above alarm valve, and observe pressure readings of two pressure gauges. If pressure readings equalize, bypass check valve is holding, and valve seat is in good condition. Replace clapper.
	Foreign material in valve seat	Fully open main drain valve to raise clapper and flush out foreign material. If unsuccessful, shut down and drain system, remove valve inspection plate, and clean manually.
	Normal pressure changes excessive	Have water purveyor install a graphical recorder to determine time and magnitude of pressure variations. Install pressure-reducing valve. Adjust on/off pressure ranges if fire pump used.

Table 1B Troubleshooting Chart for Wet Pipe Sprinkler Systems (2 of 4)

Machine Component or Problem Area	Possible Cause	Corrective Action
Water motor alarm not operating	Strainer in water motor alarm line plugged	Clean or replace strainer.
	Impeller orifice plugged	Clean impeller orifice or replace.
	Water motor drain line plugged	Remove insect nests from drain line (check 45-degree elbow at lower end of line).
	Alarm valve port to auxiliary chamber clogged	Follow procedures to shut off and drain system. Remove alarm valve faceplate. Follow manufacturer's instructions to clean valve seat. Make sure valve ports (small holes in valve seat) are open. Clogged valve ports restrict water flow and prevent water motor gong from operating.
	Water motor alarm impeller blocked	Check impeller to ensure freedom of movement. If blocked, disassemble impeller to determine cause and correct. Replace if necessary.
	Impeller disconnected from striker	Check for rigid attachment of striker to impeller. Replace missing nut or bolt or if striker is sheared, replace striker.
	Striker in water motor alarm blocked	Remove alarm and clean out insect/bird nests. If gong screen is missing, install a new one.
Continuous automatic drain leak at FDC	Leaking valve seat on FDC check valve	Shut down and drain system. Remove any obstructions that prevent proper closure of FDC check valve. Rust or other foreign objects may prevent valve from seating. Inspect clapper and seat for deterioration or damage.
Excessive pressure in sprinkler system	Water supply exceeds maximum operating pressure	Reduce water supply pressure by adjusting pump supply pressure or installing one or more pressure reducing valves.
	Temperature increases causing increased system pressure in a system with no trapped air	Install a pressure relief valve above alarm valve internal and external checks.

Table 1C Troubleshooting Chart for Wet Pipe Sprinkler Systems (3 of 4)

Machine Component or Problem Area	Possible Cause	Corrective Action
Frequent intermittent alarms after system testing	Retard chamber not draining properly	Clean retard chamber drain piping and inspect strainer to ensure it is in place, intact, and clean.
	Flow switch retard mechanism defective	Replace flow switch or its retard mechanism.
Leaks in water motor alarm line or retard drain, or continuous water drainage from alarm line	Alarm valve gasket defective	Follow procedures to shut off and drain system. Remove alarm valve faceplate. Replace alarm valve gasket per manufacturer's instructions.
	Foreign material on valve seat preventing valve closure	Follow procedures to shut off and drain system. Remove alarm valve faceplate. Wipe gasket(s) and seat using a clean, dry cloth. Do not use solvents or sharp instruments. Avoid clogging drain ports. Examine upper chamber for foreign material that might drop back onto seat. Evaluate foreign material to see if additional action is required to clear the system piping.
	Valve seat damaged	If possible, replace valve seat. Otherwise, replace valve.
Trouble alarms from undisturbed tamper switch	Loose electrical connection on switch	Check and tighten connections. Do not overtighten.
	Tamper switch not installed securely on valve	Mount switch properly per manufacturer's instructions.
	Tamper switch adjusted improperly	Adjust switch properly per manufacturer's instructions.
Paddle-type flow switch stays in alarm once activated	Incorrect paddle type, causing it to drag on pipe and lock in flow position	Remove and replace with correct paddle. There should be enough clearance that small changes in pressure will not activate switch.
	Foreign material caught on paddle, blocking it open (may result from inadequate flushing following initial installation or work on underground supply after system in service)	Remove switch and clear any foreign material. Reverse flushing of system may be necessary to remove material that has passed switch and presents potential system blockage.
	Buildup of organic matter on pipe walls causing paddle to drag	Remove and replace affected pipe section. Evaluate system for widespread issue.

Table 1D Troubleshooting Chart for Wet Pipe Sprinkler Systems (4 of 4)

Machine Component or Problem Area	Possible Cause	Corrective Action
Alarm connected to a pressure switch does not operate when alarm valve opens	Pressure line to switch plugged	Clear pressure line obstruction.
	Switch mechanical failure	Replace switch.
	Switch loose or broken electrical connection	Call an electrician to repair switch.
	Wiring or signal system defective beyond switch	Call an electrician to repair wiring or signal system.
No flow when inspector's test valve open	Water supply turned off at floor control valve or main water supply	Check all supply valves to ensure water supply to test valve is on.
	Inspector's test line blocked with gravel, sand, or rust	Shut down system and remove test valve and piping as necessary to clear obstructions. Consider flushing entire system.
Pinhole leaks in piping	Corrosion of piping system causing pinholes	Shut down and drain system, then remove and replace affected piping. Remove trapped air to prevent further corrosion. Note that any corrosion inhibitors added to system must be Listed for use in fire protection systems.

3.0.0 Section Review

1. A hydrostatic test on a system with a static pressure of 175 psi must be tested at a *minimum* pressure of _____.
   a. 150 psi
   b. 175 psi
   c. 200 psi
   d. 225 psi

2. A continuous automatic drain leak at an FDC is *most* likely to be caused by a(n) _____.
   a. leaking valve seat
   b. excessive pressure fluctuation
   c. incorrect paddle
   d. defective retard mechanism

Review Questions

1. When a wet pipe sprinkler system is activated, an alarm sounds due to the functioning of the _____.
   a. FDC
   b. alarm check valve
   c. water supply control valve
   d. excess pressure pump

2. Which of the following is required in system riser assemblies of wet pipe sprinkler systems?
   a. Excess pressure pump
   b. OS&Y valve
   c. Retard chamber
   d. Two approved pressure gauges

3. One type of device used to minimize pressure surges (and thus false alarms) in the alarm line is a(n) _____.
   a. alarm check valve
   b. retard chamber
   c. pressure switch
   d. paddle-type flow switch

4. Which of the following is used in place of a retard chamber in areas where freezing drain lines can be a problem?
   a. Alarm check valve
   b. Excess pressure pump
   c. Paddle-type flow switch
   d. Heated check valve

5. A switch that can control two circuits and send current in two directions is called a(n) _____.
   a. DPST switch
   b. DPDT switch
   c. SPDT switch
   d. SPST switch

6. Which of the following is used where alarm check valves are *not* used?
   a. Paddle-type flow switch
   b. Plug-type tamper switch
   c. DPDT switch
   d. SPDT switch

7. Hydrostatic tests shall be made at not less than 200 psi or 50 psi above static pressure in excess of 150 for _____.
   a. one hour
   b. two hours
   c. three hours
   d. four hours

8. When using a test pump to hydrostatically test a system, a failed test can result from _____.
   a. using too many pump strokes
   b. using a pump rated over 500 psi
   c. failure to remove air in the system
   d. failure to drain the system

9. If the strainer in the water motor alarm line is plugged, which of the following is *most* likely to result?
   a. The water motor alarm will not operate.
   b. Frequent intermittent alarms will occur after system testing.
   c. Continuous leakage of the automatic drain at the FDC.
   d. No flow occurs when the inspector's test valve is opened.

10. If the alarm connected to a pressure switch does not operate when the alarm valve opens, which of the following is *least* likely to be the problem?
    a. Plugged pressure line to the switch
    b. Water pressure too high
    c. Mechanical failure of the switch
    d. Loose or broken electrical connection at the switch

Trade Terms Introduced in This Module

Alarm check valve: A device installed in a wet sprinkler system that triggers an audible alarm when water flow is detected.

Bypass: Alarm check valve trim where pressures are equalized.

Clapper: A valve component that restricts water flow under normal pressure.

Double-pole (DP) switch: A switch that can control two circuits.

Double-throw (DT) switch: A switch that transmits a signal in two directions.

False alarm: A signal transmitted from a system that should not be in operation.

Hydrostatic test: A wet pipe sprinkler system test that uses a test pump to increase the pressure of the system to 200 pounds per square inch (psi) or 50 psi above static pressure in excess of 150 psi for two hours to determine whether there are any leaks. If the system holds the pressure for the specified time period, it passes the test.

Main drain: The drain used to empty the system for servicing, to check the alarm check valve, and to perform main drain tests.

Pressure switch: A device used to detect changes in pressure within a system and to transmit a signal when fluctuations occur.

Retard chamber: A device in variable-pressure systems that is commonly used with pressure switches to prevent false alarms from occurring.

Riser: The vertical supply pipe in a sprinkler system.

Single-pole (SP) switch: A switch that can control only one circuit.

Single-throw (ST) switch: A switch that transmits a signal in one direction.

Turbine: A component part of a water motor gong that produces the sound.

Water motor gong: An alarm that is used in wet pipe sprinkler systems.

Additional Resources

This module presents thorough resources for task training. The following reference material is suggested for further study.

NFPA 13, *Standard for the Installation of Sprinkler Systems*, Latest Edition. Quincy, MA: National Fire Protection Association.

NFPA 13D, *Standard for the Installation of Sprinkler Systems in One- and Two-Family Dwellings and Manufactured Homes*, Latest Edition. Quincy, MA: National Fire Protection Association.

NFPA 13R, *Standard for the Installation of Sprinkler Systems in Low-Rise Residential Occupancies*, Latest Edition. Quincy, MA: National Fire Protection Association.

NFPA 25, *Standard for the Inspection, Testing, and Maintenance of Water-Based Fire Protection Systems*, Latest Edition. Quincy, MA: National Fire Protection Association.

NFPA 1963, *Standard for Fire Hose Connections*, Latest Edition. Quincy, MA: National Fire Protection Association.

Figure Credits

Taylor Schumacher, Module Opener

Used with permission of Johnson Controls, Figures 3–6

Courtesy of Viking Group Inc, Figure 7

LynCar Canada Inc., Figure 9

Reliable Automatic Sprinkler Co., Inc., Figure 10

Courtesy of Victaulic Company, Figure 11

Courtesy of Dixon Valve & Coupling Company., Figure 13

Courtesy of FireResQ, Inc., Figure 17

Courtesy of Knox Company, Figure 18

Courtesy of Potter Electric Signal, Figures 20–24, 27–28

Courtesy of Reed Manufacturing, Figure 30 (first two images)

Wheeler-Rex, Figure 30 (third image)

Section Review Answer Key

Section 1.0.0

Answer	Section Reference	Objective
1. c	1.1.0	1a
2. b	1.2.0	1b
3. d	1.3.3	1c

Section 2.0.0

Answer	Section Reference	Objective
1. b	2.1.2	2a
2. a	2.2.0	2b
3. a	2.3.0	2c
4. d	2.4.1	2d

Section 3.0.0

Answer	Section Reference	Objective
1. d	3.1.0	3a
2. a	3.2.4; *Table 1B*	3b

User Update

NCCER makes every effort to keep its textbooks up-to-date and free of technical errors. We appreciate your help in this process. If you find an error, a typographical mistake, or an inaccuracy in NCCER's curricula, please submit a User Update form by visiting **https://www.nccer.org/olf**. You can also scan the QR code using the camera on your phone or mobile device to access the form.

Dry Pipe Sprinkler Systems

Overview

This module describes the operation and installation of dry pipe sprinkler systems. It covers differential and mechanical dry pipe valves and trim, including switches, gauges, air sources, accelerators, and drains. It also provides an overview of basic system troubleshooting and acceptance testing procedures, including the completion of a Contractor's Material and Test Certificate for Aboveground Piping for a dry pipe sprinkler system.

Module 18207

18207
Dry Pipe Sprinkler Systems

Objectives

Successful completion of this module prepares you to do the following:

1. Identify the function and operation of dry pipe sprinkler system valves.
   a. Explain the operation of differential dry pipe valves.
   b. Explain the operation of mechanical dry pipe valves.
   c. Identify the installation considerations for valve trim and accessory devices.
2. Make calculations for installing piping at the correct pitch to drains.
   a. Calculate pitch.
   b. Identify the installation considerations for various drains.
3. Describe the air sources and air maintenance devices used in dry pipe sprinkler systems.
   a. Identify the air sources used to supply pressure in dry pipe sprinkler systems.
   b. Identify the air maintenance devices used to regulate airflow in dry pipe sprinkler systems.
4. Describe how to test and troubleshoot dry pipe sprinkler systems.
   a. Explain the basic procedures for testing dry pipe sprinkler systems.
   b. Describe how to troubleshoot common dry pipe sprinkler system issues.

Performance Task

Under supervision, you should be able to do the following:

1. Complete a Contractor's Material and Test Certificate for Aboveground Piping for a dry pipe sprinkler system.

Trade Terms

Accelerator
Air maintenance device (AMD)
Air receiver
Alarm test valve
Differential
Differential dry pipe valve
Drip check
Dry pipe sprinkler system
Intermediate chamber
Mechanical dry pipe valve
Nitrogen generator
Pitch
Pneumatic test
Quick-opening device (QOD)
Restricted orifice
Trip pressure

Industry Recognized Credentials

If you are training through an NCCER-accredited sponsor, you may be eligible for credentials from NCCER's Registry. The ID number for this module is 18207. Note that this module may have been used in other NCCER curricula and may apply to other level completions. Contact NCCER's Registry at 1.888.622.3720 or go to **www.nccer.org** for more information.

You can also show off your industry-recognized credentials online with NCCER's digital badges. Transform your knowledge, skills, and achievements into badges that you can share across social media platforms, send to your network, and add to your resume. For more information, visit **www.nccer.org**.

NOTE

Unless otherwise specified, references in parentheses following figure and table numbers refer to NFPA 13.

Contents

Figures and Tables

Section One

1.0.0 Dry Pipe Sprinkler System Valves

Objective

Identify the function and operation of dry pipe sprinkler system valves.

a. Explain the operation of differential dry pipe valves.
b. Explain the operation of mechanical dry pipe valves.
c. Identify the installation considerations for valve trim and accessory devices.

Trade Terms

Accelerator: A type of quick-opening device (QOD) used to speed up the trip time of a dry pipe valve.

Air maintenance device (AMD): A Listed assembly of valves, nipples, fittings, and actuators used to automatically control the air pressure in the piping of dry pipe sprinkler systems.

Alarm test valve: A shutoff valve used to allow the testing of system alarms without activating the dry pipe valve.

Differential: The ratio between the surface area on the top of the clapper, where the supply air applies pressure, and the bottom of the clapper, where the supply water applies pressure.

Differential dry pipe valve: A valve that is held in the closed position by the system air/nitrogen pressure exposed to the larger surface area on the air/nitrogen side of the clapper where it is at least five times that of the surface area on the water supply side.

Drip check: A spring-loaded check valve that is primarily used to evaluate the seating of the dry pipe valve clapper, as well as to retain atmospheric pressure in the intermediate chamber of a differential dry pipe valve.

Dry pipe sprinkler system: A sprinkler system employing automatic sprinklers that are attached to a piping system containing air or nitrogen under pressure, the release of which (as from the opening of a sprinkler) permits the water pressure to open a valve known as a *dry pipe valve*, and the water then flows into the piping system and out the opened sprinklers.

Intermediate chamber: A chamber in a differential dry pipe valve that water flows into when the clapper is unseated. A port in the intermediate chamber allows water to flow into the alarm line, where it activates the waterflow alarm pressure switch. In addition, the intermediate chamber is used to introduce pressure below the clapper when an accelerator is used.

Mechanical dry pipe valve: A valve that uses a series of mechanical devices such as levers, springs, diaphragms, and latches to hold the valve in the closed position using air/nitrogen pressure and without using the clapper surface areas to provide a differential between the air/nitrogen and water pressures.

Pitch: The ratio of rise over run.

Quick-opening device (QOD): A device used in dry pipe sprinkler systems to reduce the trip time of the dry pipe valve.

Restricted orifice: The orifice through which most automatic sources of air are supplied to a dry pipe valve. It prevents system air from being replenished faster than the air that is being discharged through an activated sprinkler or test connection.

Trip pressure: The air pressure at which a dry pipe valve operates.

A dry pipe sprinkler system consists of a network of sprinkler piping containing pressurized air. When a sprinkler activates, the pressurized air is released from the piping system. The drop in pressure trips the dry pipe valve and water then flows through the system and discharges from the sprinklers.

The most common type of piping used in dry pipe sprinkler systems is black steel pipe with Listed grooved couplings and fittings. Piping must be installed at the correct pitch and use the hanger spacing indicated in NFPA 13. Cut grooving is preferred to roll grooving. Cut-grooved pipe is smooth internally and water will easily drain back to the riser, while roll-grooved pipe has an internal distortion that can trap water and increase the rate of corrosion. Flush seal gaskets are recommended for grooved couplings used in dry pipe sprinkler systems. These gaskets differ from standard gaskets in that they close off the gap between the two pipe ends, preventing moisture from collecting in the gap and reducing the risk of corrosion. Some dry pipe sprinkler systems may use heavier piping (e.g., Schedule 40 and up) to increase the amount of material that the corrosion would have to eat through, while others may use corrosion-resistant piping, such as copper tubing with brazed joints or Listed connectors.

Dry pipe sprinkler systems are installed less often than wet pipe sprinkler systems due to their increased complexity, higher rates of corrosion, and time delay in applying water to a fire. This delay is produced by two factors. First, the air pressure must be reduced to the trip pressure of the dry pipe valve. Second, once tripped, the water must travel through the piping system to the activated sprinklers. For these reasons, dry pipe sprinkler systems should only be installed in locations where the temperature cannot be reliably maintained above 40°F and the area is subject to freezing. They are typically used in unheated warehouses, exterior sheds and canopies, freezers, loading docks, and similar locations.

The primary function of a dry pipe valve is to prevent water from entering the piping until a sprinkler is activated. The secondary function is to provide an alarm signal, which results from a waterflow alarm pressure switch that senses pressure changes in the piping. When the waterflow alarm pressure switch is activated, it sends an alarm signal to a supervisory system and may also be connected to a water motor gong and/or local alarm bell. There are two main types of dry pipe valves: differential dry pipe valves and mechanical dry pipe valves.

1.1.0 Differential Dry Pipe Valves

A differential dry pipe valve uses a pressure difference between the system air and water supply to hold the valve closed. The term differential refers to the ratio between the surface area on the top of the clapper, where the *supply air* applies pressure, and the bottom of the clapper, where the *supply water* applies pressure (*Figure 1*). This must be at a minimum ratio of 5:1. When a sprinkler activates, the air pressure in the system decreases until the valve trips. This is known as the *trip pressure*. For example, the trip pressure in a 5:1 dry pipe valve at 60 psi of water pressure is 12 psi (60 ÷ 5 = 12). When the air pressure dips below 12 psi, the valve trips, allowing water to enter the sprinkler piping and eventually exit through the open sprinkler. A safety factor is added to help prevent false trips. An air compressor or other means is used to automatically maintain adequate air pressure.

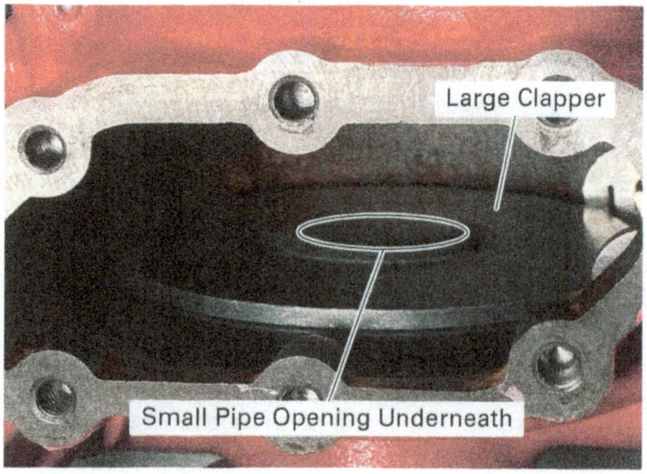

Figure 1 Differential dry pipe valve clapper and pipe opening comparison.

1.1.1 Differential Dry Pipe Valve Operation

Figure 2 shows a typical differential dry pipe valve and its components. A typical sequence of operation is as follows:

Step 1 When a sprinkler activates, the fusible linkage melts or the bulb shatters. The linkage or bulb is blown out of the way, releasing the compressed system air through the open sprinkler.

Step 2 As air escapes the riser, air pressure is reduced in the valve's upper chamber. The compressed air escapes from the piping more rapidly than the air supply can be replenished.

Step 3 If the dry pipe sprinkler system is equipped with an accelerator, the accelerator senses the drop in air pressure and sends compressed system air to the valve's intermediate chamber to speed up operation of the valve actuator by forcing air pressure under the clapper.

Step 4 A pressure decrease of 10 psi activates the high/low air pressure supervisory switch and alarm.

Step 5 At the trip pressure of the valve, the compressed air on top of the clapper becomes insufficient to hold back the water. The higher water pressure then forces the clapper off its seat and pivots it to the back of the valve where it latches in the open position.

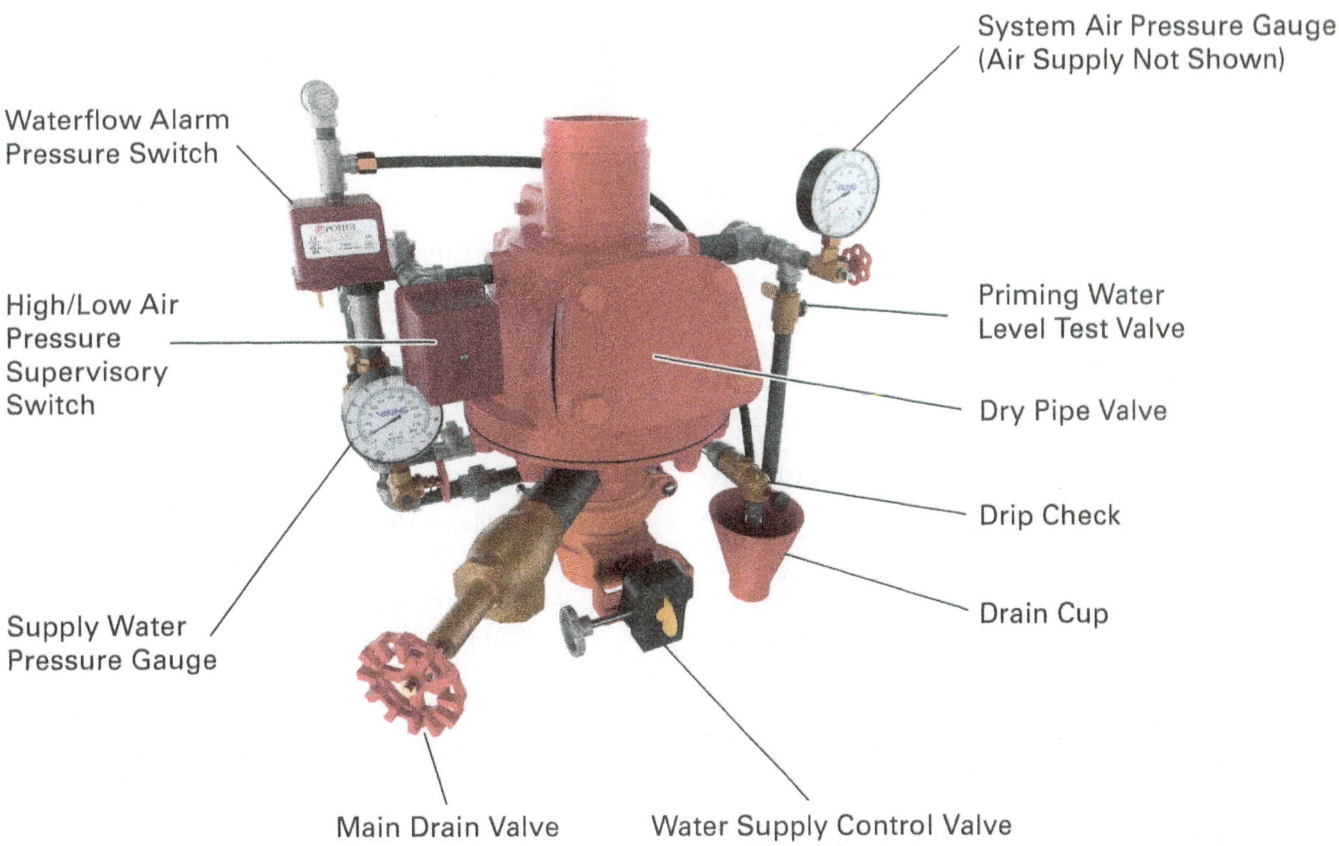

Figure 2 Differential dry pipe valve and trim.

Step 6 When the clapper is unseated, it allows the water to flow into the intermediate chamber of the dry pipe valve. A port in the intermediate chamber (*Figure 3*) allows water to flow into the alarm line, where it activates the waterflow alarm pressure switch. The waterflow alarm pressure switch provides an alarm signal to a supervisory system and may also be connected to a water motor gong and/or local alarm bell.

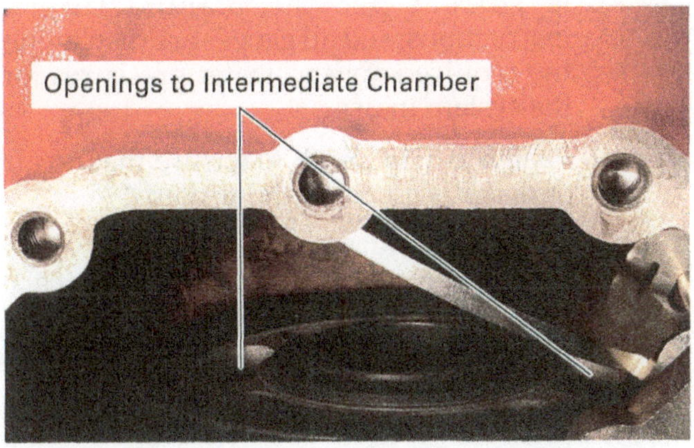

Figure 3 Differential dry pipe valve intermediate chamber.

Step 7 With the valve in the open position, water enters the system piping and is discharged through any activated sprinklers.

Watch How It Works: Differential Dry Pipe Valve Operation

Scan this QR code to observe the sequence of operation for a differential dry pipe valve.

Watch Now

The use of a differential in a dry pipe valve is a practical measure because obtaining leak-free air piping is more difficult than obtaining leak-free water piping. As air pressure is increased in a piping system, the integrity of the piping is more difficult to achieve. The use of the differential allows a lower air pressure to be used to hold back the water. Typical differential ratios are 5:1 and 6:1.

Extra air pressure is applied to hold the clapper closed against normal pressure variations and surges. For example, if the city water pressure on the water side of a dry pipe valve clapper is 120 psi and the clapper area differential is 6:1, an air pressure of 20 psi should keep the clapper on its seat and prevent water from entering the system. In practice, however, a margin of safety is necessary to accommodate for surges in water pressure. NFPA 13 requires that the system air pressure be maintained in accordance with the manufacturer's instructions or at least 20 psi above the calculated trip pressure of the dry pipe valve, based on the highest normal water pressure of the system supply. In the example above, this would require a minimum air pressure of 40 psi.

1.1.2 Resetting a Differential Dry Pipe Valve

Differential dry pipe valves may be reset by removing the faceplate and moving the clapper back down onto its seat. Some valves require the use of a special reset tool, while others are reset by hand or using an external reset feature. Always follow the manufacturer's reset instructions for the valve in use.

A general reset procedure is as follows:

Step 1 Shut off the water supply by closing the water supply control valve.

Step 2 Silence the alarm by closing the alarm shutoff valve.

Step 3 Shut off the air supply by closing the air supply control valve.

Step 4 Inspect and replace any portion of the system subjected to fire conditions.

Step 5 Drain the sprinkler riser by opening the main drain.

Step 6 Remove the valve faceplate and clean the air and water seats using a clean, dry cloth. Any debris must be removed to ensure a proper seal.

CAUTION

Never use grease or oily substances on the rubber seals. Oil deteriorates rubber and will destroy the seals, allowing the valve to flood the piping and resulting in extensive damage to piping, fittings, and devices.

Step 7 Reset the valve clapper per the manufacturer's instructions.

CAUTION

Never insert your hand inside the valve after it has been reset as accidental clapper release can cause injury.

Step 8 Replace the faceplate gasket cover and bolts. Do not overtighten.

Step 9 Verify that the accelerator air pressure is at 0 psi (if used). Bleed the accelerator if necessary.

Step 10 Open the bypass valve to raise the system pressure to within 5 psi of the set pressure, then shut the valve.

Step 11 Open the air supply valve. The air pressure and accelerator pressure gauges should come up to pressure. (Note that the accelerator pressure builds more slowly than the air pressure.)

Step 12 Slightly open the water supply control valve to allow water to slowly enter the valve. Introducing water too quickly may trip the valve.

Step 13 Slowly close the main drain valve.

Step 14 Continue opening the water supply control valve until it is fully open.

Step 15 Verify that the static water pressure is correct per the manufacturer's instructions, and all gauges are displaying the required air and water pressures for the system in use.

Step 16 Push the drip check to ensure that no air or water is escaping the system.

Step 17 Open the alarm valve. The dry pipe sprinkler system is now back in service.

Watch How It Works: Differential Dry Pipe Valve Reset

Scan this QR code to observe the procedure for resetting a typical differential dry pipe valve.

Watch Now

1.2.0 Mechanical Dry Pipe Valves

A mechanical dry pipe valve uses a series of devices such as levers, springs, diaphragms, and latches to hold the valve in the closed position using air pressure without using the clapper surface areas to provide a differential. While these valves are slightly more complicated than differential dry pipe valves, they are more forgiving of fluctuations in supply pressure. They also operate at a lower air pressure than conventional differential dry pipe valves. This means that less expensive air sources can be used, and the reduced pressure may allow for faster water delivery, eliminating the need for accelerators in some cases. *Figure 4* shows a typical mechanical dry pipe valve and its components.

1.2.1 Mechanical Dry Pipe Valve Operation

A mechanical valve uses a dry pilot actuator (*Figure 5*) to operate a pushrod that opens the valve. In the closed position, air pressure on the actuator prevents the release of water pressure

Figure 4 Mechanical dry pipe valve and trim.

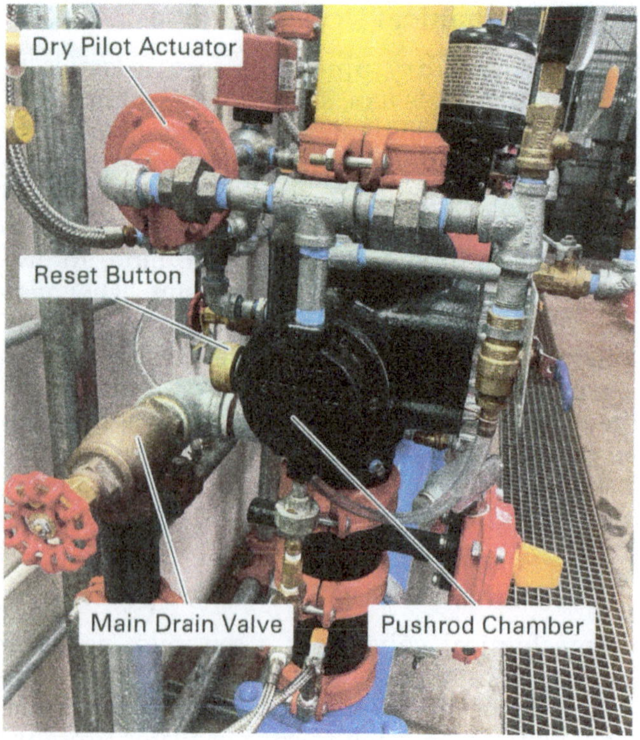

Figure 5 Mechanical dry pipe valve (side view).

Watch How It Works: Mechanical Dry Pipe Valve Reset

Scan this QR code to observe the reset of a mechanical dry pipe valve.

Watch Now

from the pushrod chamber. The supply water pressure acts simultaneously on the underside of the clapper and on the pushrod through the pushrod chamber restricted inlet. The resultant force on the pushrod is multiplied by the mechanical advantage of the lever and holds the clapper closed against normal pressure surges in the water supply.

A typical sequence of operation is as follows:

Step 1 When a sprinkler activates, the fusible linkage melts or the bulb shatters. The linkage or bulb is blown out of the way, allowing the air to discharge through the open sprinkler.

Step 2 The loss of air pressure in the sprinkler system causes the diaphragm and seal in the dry pilot actuator to move away from the water seat, allowing the release of water from the pushrod chamber.

Step 3 Since water cannot be replenished through the inlet restriction as rapidly as it is vented, the pushrod chamber pressure falls instantaneously. When the pushrod chamber pressure approaches approximately one-third of the supply pressure, the upward force of the water pressure acting beneath the clapper overcomes the force applied to the lever, opening the clapper.

Step 4 Water then flows through the dry pipe valve into the system piping and into the alarm outlet, activating the alarm device(s). Once the clapper has opened, the lever acts as a latch preventing the clapper from returning to the closed position.

Step 5 As the water fills the main riser, bulk mains, and cross mains, the original air is pushed out of the activated sprinkler(s) and onto the fire.

Step 6 If the dry pipe sprinkler system is equipped with an accelerator, the accelerator senses the rapid drop in air pressure caused by the activation of a sprinkler and opens the dry pipe valve faster by forcing system air pressure under the clapper.

NOTE

In areas where high temperatures are expected, a pressure relief kit can be installed in the pushrod chamber release line to limit the pressure to 250 psi.

1.2.2 Resetting a Mechanical Dry Pipe Valve

The following is a general reset procedure for a mechanical dry pipe valve. Always read and follow the manufacturer's instructions for the valve in use.

Vacuum Systems

Vacuum systems are also available. These systems prevent corrosion by using a vacuum pump to remove air and moisture from the system piping. They also provide a very fast response time as the release of the vacuum upon sprinkler activation acts to pull water through the system. Special sprinklers are required for use on vacuum systems.

Figure Credit: Courtesy of Viking Group, Inc.

Step 1 Close the main water supply control valve, pushrod chamber supply valve, and valve(s) controlling the air supply to the system.

Step 2 Open the main drain valve and allow the system to fully drain. Also open all drain valves and vents at low points throughout the system, closing them when the flow of water has stopped.

Step 3 Inspect and replace any portion of the system subjected to fire conditions.

Step 4 Open the manual emergency release valve to relieve pressure in the pushrod chamber of the valve and leave open.

Step 5 With the alarm line valve open, push in the plunger of the drip check valve and drain the alarm line, then close the alarm line valve.

Step 6 Push in and rotate the external reset knob counterclockwise until the clapper releases and lowers into position. Note that the reset knob cannot be rotated with the pushrod chamber pressurized.

Step 7 Close the main drain valve.

Step 8 Open the pushrod chamber supply valve and allow water to flow through the pushrod chamber. Purge all air from the actuation piping, then close the emergency manual release valve.

Step 9 Close the dry pilot actuator by opening the air supply quick fill valve. Allow the pressure to build to the manufacturer-specified level, then close the quick fill valve and set the air supply to automatic operation.

Step 10 If there is an accelerator installed, reset it per the manufacturer's instructions.

Step 11 Open the alarm line valve. Slightly open the main drain valve. Slowly open the main water supply control valve until water flow is heard flowing from the main drain, then close the main drain valve. Observe if water leaks through the drip check valve. If no leak occurs, the clapper is sealed.

Step 12 Slowly open the main water supply control valve until it is fully opened and verify that it is properly monitored.

Step 13 Verify that the pushrod chamber supply valve and alarm line valve are open. The pushrod chamber supply valve must remain open to maintain hydraulic pressure in the pushrod chamber after the dry pipe valve has been reset.

Step 14 Secure the manual emergency release valve handle in the closed position.

Step 15 Notify all concerned parties that the system has been placed into service.

1.3.0 Valve Trim and Accessory Devices

A dry pipe valve has both airside components and waterside components, each of which fulfills a specific purpose and must be installed properly to ensure the system operates as expected.

The following sections provide general guidance on the installation of these components. Always follow the manufacturer's instructions for the equipment in use.

1.3.1 Airside Connections

The piping connections on the air side of the valve clapper include a system air pressure gauge, a relief valve, and an air supply check valve. The air pressure gauge indicates the air pressure applied to the top of the dry pipe valve clapper in a differential valve or the air side of the dry pilot actuator in a mechanical valve. The air supply check valve allows air into the system and prevents back pressurization of the air compressor or other air supply piping. In addition, an approved relief valve must be installed between the air supply and the shutoff valve and set to relieve pressure at no less than 10 psi above the system air pressure. The check and relief valves may be installed as a component of an air maintenance device (AMD) (*Figure 6*).

> **NOTE**
> Gauges are not required to be Listed. When selecting gauges, choose those with larger numbers that can be easily read at a distance.

Exercise caution when installing pressure gauges as they are precision instruments and easily damaged by rough handling. Apply the wrench to the square wrenching boss, never to the body. Take care not to clog the small gauge inlet with pipe joint compound. The use of Teflon® tape is recommended. Note that air pressure gauges look similar to water pressure gauges, but they are not always interchangeable. An air pressure gauge designed for fire protection use typically reads from 0 to 80 psi with a retard setting to 250 psi, while a water pressure gauge typically reads from 0 to 300 psi (*Figure 7*). Always check the gauge type before installing it.

A high/low air pressure supervisory switch is used to monitor the air pressure in the system and must be adjusted per the manufacturer's instructions. This alarm provides a warning if there is a loss of air pressure, as when a sprinkler activates.

Pressurized air may be supplied from several sources, including the owner's plant air (shop air), a dedicated air compressor provided by the sprinkler contractor, compressed nitrogen, or other approved gases. The air source may be automatically operated by a pressure-sensing switch or an AMD.

All automatic sources of air that utilize an AMD are supplied to the dry pipe valve through a restricted orifice. This orifice must be much smaller than the K-factor of the smallest sprinkler in the system to prevent the air supply from providing more air than would be discharged by an activated sprinkler. Otherwise, the loss of pressure would not be sufficient to operate the valve.

Maintaining the proper air pressure is critical to the operation of a dry pipe valve. The system air pressure must be reduced to the trip level of the dry pipe valve as rapidly as possible in case of a fire. Excessive air pressure will delay the discharge of water onto the fire and potentially damage the valve as well as other system components. The air pressure is adjusted using a valve on the AMD and a regulator on systems that use nitrogen. Follow the dry pipe valve manufacturer's instructions when making air pressure adjustments.

Another area where caution should be exercised is in the selection and use of pipe thread compound. Avoid using liquid pipe thread compound, as it is likely to be blown out of the joint when air pressure is applied. The use of Teflon® paste or Teflon® tape is preferred for threaded joints. Extra care during the installation of the system will save rework when leak testing.

1.3.2 Waterside Connections

The waterside connections consist of the main drain valve, drip check, drain cup, alarm test valve, alarm shutoff valve, and supply water pressure gauge. The main drain valve is the primary water drain and is piped to the exterior of the building. Its function is to drain water from all system piping that is pitched toward the riser. All other trapped water is drained through the auxiliary drains and drum drips at the system piping low points. *Figure 8* shows a drum drip.

The drip check is a spring-loaded check valve used to determine whether or not the dry pipe valve clapper is seated properly, and to retain atmospheric pressure in the intermediate chamber of a differential dry pipe valve. Dirt or other foreign material under the clapper will allow supply water to seep into the drip check. If the drip check fails to discharge any water when its spring-loaded plunger is depressed, the clapper is seated properly.

The drain cup serves as a funnel to catch the water from the drip check and direct it into the drain piping. Carefully align and permanently affix the drain cup and the discharge from the drip check. Misalignment will discharge water onto the floor.

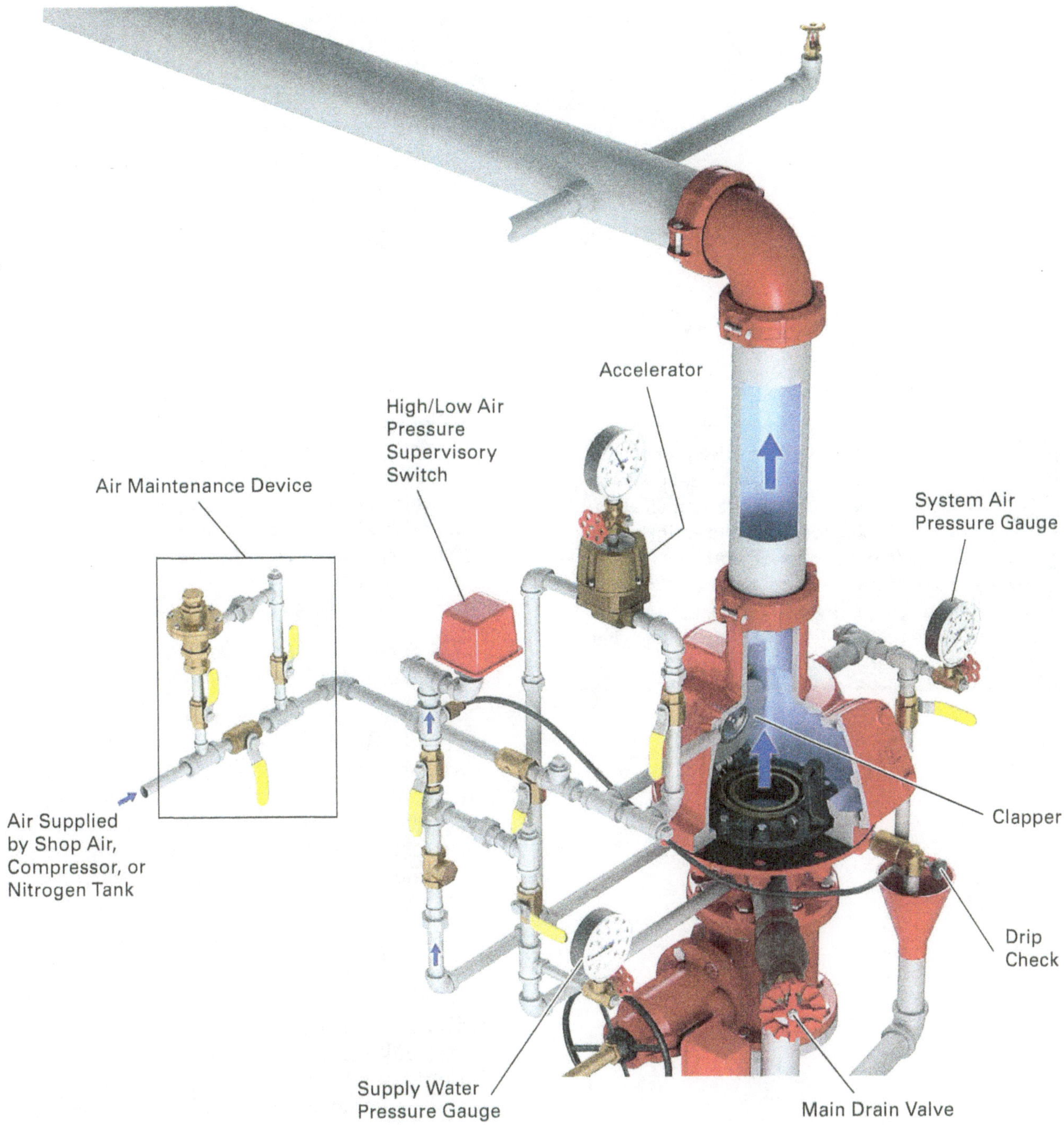

Figure 6 Dry pipe sprinkler system trim (differential valve shown).

The alarm shutoff valve allows the alarm to be silenced after operation. The alarm test valve allows the alarm to be tested without tripping the dry pipe valve.

The supply water pressure gauge indicates the amount of water pressure on the bottom of the dry pipe valve clapper. As with air pressure gauges, the inlet to the water pressure gauge is small and care must be taken to prevent pipe joint compound away from entering it. The use of Teflon® tape is recommended. The pressure gauge is also easily damaged by torque applied to its body. Always apply wrenches to the square wrenching boss, never to the gauge body.

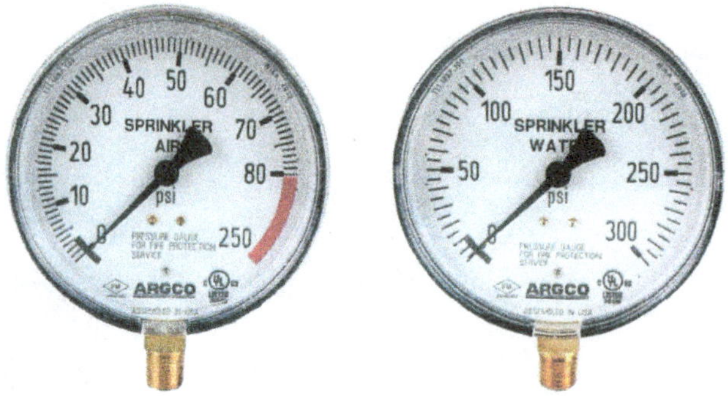

Air Pressure Gauge   Water Pressure Gauge

Figure 7 Pressure gauges.

> **NOTE**
>
> Differential dry pipe valves have pressurized water on the supply side of the clapper. Older dry pipe valves may also have a small amount of water (priming water) on the sprinkler system side of the clapper. Priming water is used to ensure the clapper is sealed tightly. When placing a dry pipe valve into service, the valve must be set according to the valve manufacturer's instructions.

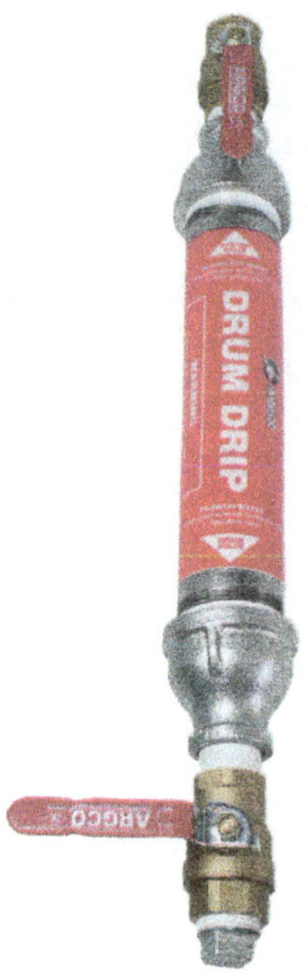

Figure 8 Drum drip.

> **Did You Know?**
>
> Pressure gauges below the system valve clapper are referred to as *supply pressure gauges*. Gauges above the clapper are called *system pressure gauges*.

1.3.3 System Volume and the Use of Quick-Opening Devices (QODs)

One concern with using a dry pipe sprinkler system is the delayed delivery of water compared to a wet pipe sprinkler system. Water delivery times are measured with the system starting at the normal air pressure and fully opening the inspector's test connection, then measuring the time it takes the water to reach the test connection. With some exceptions, the requirements are as follows:

- Dry pipe sprinkler systems with a capacity of less than 500 gallons have no time requirement.
- Dry pipe sprinkler systems with a capacity of 500 gallons or more have a maximum water delivery time of 60 seconds. NFPA 13 requires the installation of a Listed quick-opening device (QOD) on any system with a capacity of 500 to 750 gallons that cannot meet the required water delivery time.
- Systems with a capacity of 750 gallons or more must meet the required delivery time of 60 seconds, using a QOD if needed.

> **NOTE**
>
> Dry pipe system air pressure, volume, and piping system complexities can all delay water delivery times.

CAUTION

Dry pipe systems protecting dwelling unit portions of any occupancy have a maximum water delivery time of 15 seconds. Residential sprinklers must be Listed for dry pipe applications.

Calculating System Volume

The pipe capacity chart in *Table 1* can be used to calculate the capacity of 1' of pipe of different sizes and types. To find the capacity of an entire system, add the linear feet of each pipe size and multiply the footage by the factors given in the chart. Pipe lengths must be calculated using center-to-center dimensions. This calculation is typically made by the layout technician and indicated on the approved shop drawings. Design software can also be used to determine dry pipe system water capacities.

NOTE

NFPA 25, *Standard for the Inspection, Testing, and Maintenance of Water-Based Fire Protection Systems*, requires that water flow trip tests of dry pipe sprinkler systems be done regularly, and that the water delivery time to the inspector's test connection be documented. Failure to meet the 60-second limit is very expensive to correct after installation.

Types of QODs

Early QODs were known as *exhausters*. An exhauster was used to speed up the operation of a dry pipe valve by providing a large opening for the air to escape. However, pipe scale and other contaminants quickly fouled the exhauster and required extensive maintenance. For that reason, exhausters are no longer manufactured. The most common type of QOD in use today is known as an *accelerator*. An accelerator speeds up the trip time of a dry pipe valve and is typically used on larger systems. *Figure 9* shows an accelerator on a dry pipe sprinkler system.

Table 1 Capacity of 1 Foot of Pipe (Based on Actual Internal Pipe Diameter)

Nominal Pipe Diameter (Inches)	Schedule 40
1	0.045
1¼	0.078
1½	0.106
2	0.174
2½	0.248
3	0.383
4	0.660

The operation of an accelerator depends on the type of valve installed. The basic series of operation for differential and mechanical dry pipe valves is as follows:

- On a differential dry pipe valve, when the accelerator senses a rapid drop in system air pressure, it opens an internal exhaust valve that supplies air pressure from the upper chamber of the dry pipe valve to the intermediate chamber. This overcomes the differential characteristic of the dry pipe valve, allowing the air pressure to push the clapper off its seat, followed by water, which locks the clapper open.
- On a mechanical dry pipe valve, the accelerator outlet port is vented to the atmosphere, and speeds up purging of the air side of the dry pilot actuator, which in turn vents the pushrod chamber pressure of the dry pipe valve, causing its clapper to open and fill the system piping with water.

A mechanical accelerator (*Figure 10*) operates on the principle of unbalanced pressures. In the set position, air pressure is the same in all chambers. When a sprinkler or release device operates, air pressure in the middle and lower chambers is reduced at the same rate as the system pressure. An orifice in the cover diaphragm restricts airflow from the upper chamber, temporarily causing a higher air pressure in the upper chamber. The higher air pressure forces the cover diaphragm and actuator rod down, releasing

Dry Pipe Water Delivery/QOD Requirements

The system capacity determines the requirement for a 60-second delivery time and/or the use of QODs.

- ≤500 gallons: None
- >500 to 750 gallons: 60-second delivery *or* QOD required
- >750 gallons: Must meet 60-second delivery using QOD if needed

Figure 9 Accelerator on a dry pipe sprinkler system.

pressure from the lower chamber to the outlet vent and allowing residual pressure in the inlet chamber to force the clapper diaphragm open. After the clapper diaphragm opens, any pressure remaining in the inlet chamber and associated piping is vented to the atmosphere.

Electronic accelerators are also available for use with certain differential dry pipe valves. An electronic accelerator (*Figure 11*) monitors air pressure continuously (two times per second). The accelerator automatically adjusts to small or slow changes in system pressure, but trips when there is a steady drop in pressure, such as when a sprinkler activates. When a sustained drop in air pressure is sensed, the release switch signals the releasing panel, which in turn energizes the solenoid valve and sends system air into the differential dry pipe valve intermediate chamber.

Proper operation of an accelerator depends on proper installation. Always follow the manufacturer's instructions. General installation guidelines are as follows:

- Install accelerators as close to the dry pipe valve as possible to decrease the response time.
- Install a soft disc globe valve, angle valve, or ball valve between the system and the accelerator to facilitate repairs and maintenance of the device.

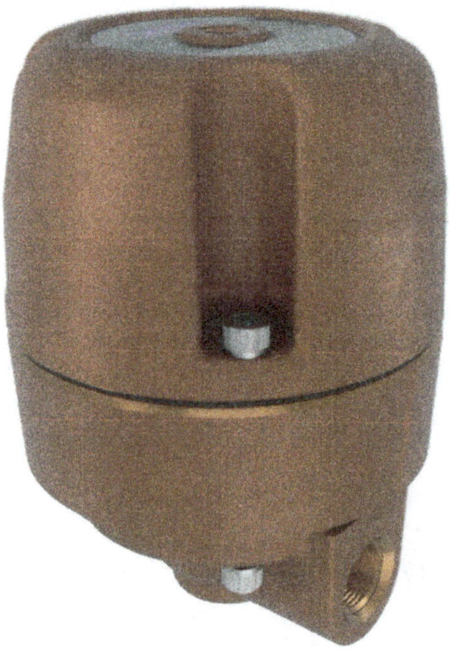

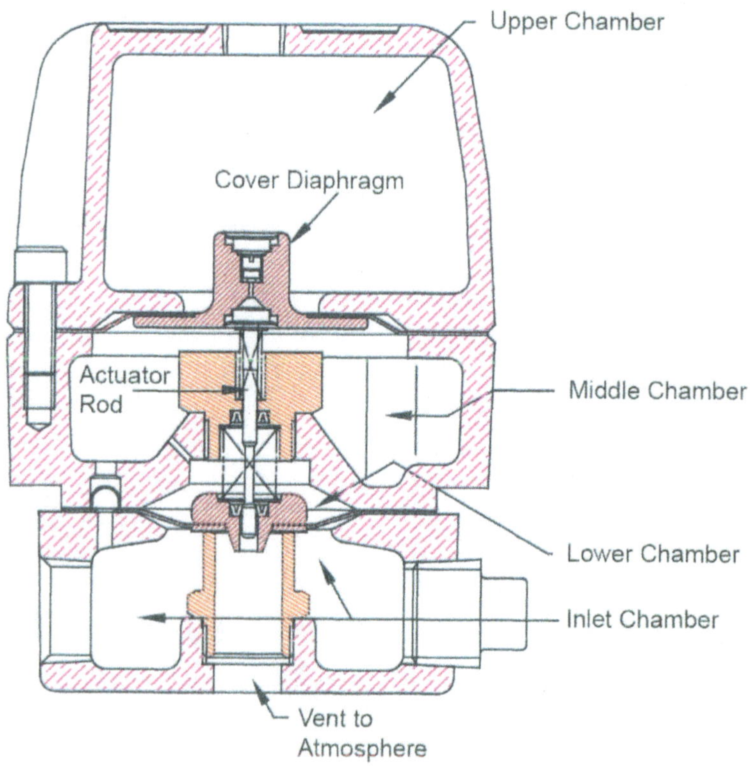

Figure 10 Mechanical accelerator.

- Install a check valve between the intermediate chamber of a differential dry pipe valve and the accelerator to prevent flooding of the accelerator when the intermediate chamber fills with water.
- A Listed anti-flooding device should be installed between the system piping and the accelerator device unless the accelerator has its own anti-flooding capabilities.
- A strainer should be installed in the piping between the system connection and the accelerator unless the accelerator has its own built-in strainer.
- Many accelerators are Listed for use with one specific manufacturer's dry pipe valve. Check the manufacturer's instructions for compatibility.
- Follow the manufacturer's instructions for inspection and scheduled maintenance. When servicing the accelerator, it is a good practice to isolate dry pipe sprinkler systems.

Figure 11 Electronic accelerator and valve trim.

1.3.4 Fire Department Connections (FDCs)

Fire Department Connections (FDCs) for both wet pipe and dry pipe sprinkler systems are almost identical, except for the point where they enter the system. As shown in *Figure 12*, the FDC on a single-zone dry pipe sprinkler system is located on the supply side of the dry pipe valve rather than the discharge side of the alarm check valve, as in a wet pipe sprinkler system. Multizone systems are designed in the same way as wet pipe sprinkler systems.

NOTE

NFPA 13 does not allow hose stations on dry pipe sprinkler systems.

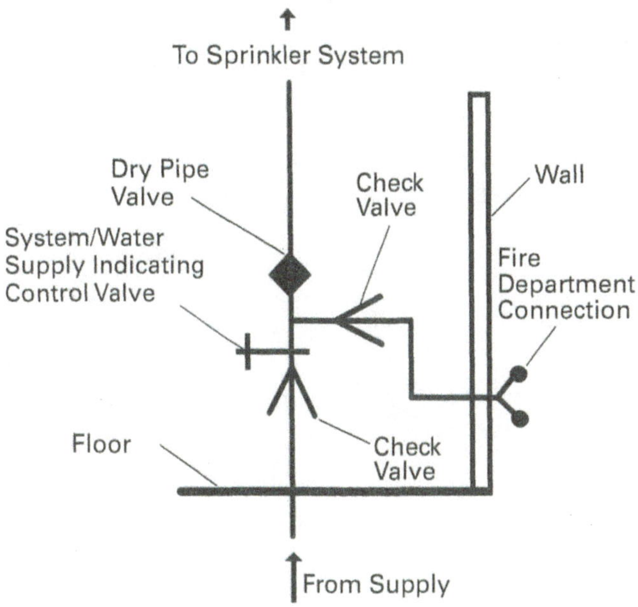

Figure 12 Dry pipe sprinkler system FDC and valve arrangement.

1.0.0 Section Review

1. In dry pipe valves, the term *differential* refers to the ratio of _____.

   a. piping water temperature to supply water temperature
   b. total air volume in the system to total water volume in the system
   c. surface area of the clapper air side to the clapper water side
   d. inlet port size to outlet port size

2. The operation of a mechanical dry pipe valve is initiated by a loss of air pressure detected by the _____.

   a. valve clapper
   b. dry pilot actuator
   c. air maintenance device
   d. alarm line

3. Fire Department Connections (FDCs) for wet pipe and dry pipe sprinkler systems both enter the system in the same way.

   a. True
   b. False

SECTION TWO

2.0.0 PITCHING PIPING AND INSTALLING DRAINS

Objective

Make calculations for installing piping at the correct pitch to drains.

a. Calculate pitch.
b. Identify the installation considerations for various drains.

In a wet pipe sprinkler system, the piping is run in a conditioned space and can generally be installed level as the decreasing pipe size away from the riser is sufficient to achieve drainage. This is not the case in dry pipe sprinkler systems, which are typically installed in unconditioned spaces where there may be a large temperature difference between the air inside and outside the pipe. Compressed air introduces a considerable amount of moisture into the piping system that condenses on the inside walls of the pipe and will accumulate unless adequate drainage is provided. This may result in freezing, burst pipes, and flooding. Another common issue with dry pipe sprinkler systems is corrosion. Both issues can be controlled by installing the piping at the correct slope or pitch.

2.1.0 Calculating Pitch

The pitch of pipe is the ratio of rise over run. The rise and run must be measured in the same measurement units to have a dimensionless pitch number. For instance, if a roof pitches 1" in 1', convert the foot into inches (by dividing by 12) so the quantity is defined as pitch in inches. In this case, the pitch would be defined as 1" ÷ 12" per foot = 0.0833'. *Figure 13* shows the concept of pitch.

NFPA 13 requires that all piping in a dry pipe sprinkler system be pitched in order to drain properly. The pitch varies by location, as follows:

- In nonrefrigerated applications, branch lines must be installed with a minimum pitch of ½" (0.5") per 10' of run, while main piping must be pitched at least ¼" (0.25") per 10' of run.
- In refrigerated areas, both branch lines and main piping must be pitched at ½" (0.5") per 10' of run.

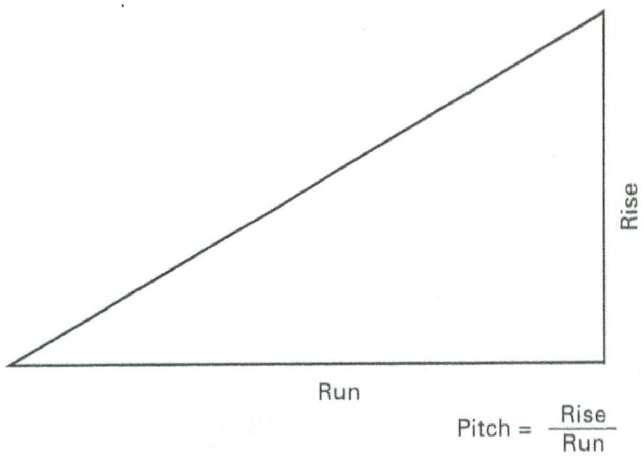

Figure 13 Pitch relationships.

To calculate the amount of rise required in a linear run of piping, divide the pipe length by 10' to find the number of sections, then multiply the result by 0.25" or 0.5", as applicable. Add this distance to the starting elevation to determine the end elevation.

Example 1: A branch line on a dry pipe sprinkler system in a nonrefrigerated area is 20' long. Find the end elevation if the starting elevation is 14' above the finished floor (AFF).

Step 1 Divide the length of the run by 10 to determine the number of 10' sections:

20' ÷ 10'/section = 2 sections

Step 2 Multiply the number of sections by the minimum pitch. In this case, 0.5" is used because the example refers to a branch line:

2 sections × 0.5" per section = 1" change in elevation

Step 3 Add the change in elevation to the starting elevation to find the end elevation:

14' starting elevation + 1" = 14'-1" end elevation

Example 2: A sprinkler main on a dry pipe sprinkler system in a nonrefrigerated area is 80' long with a starting elevation of 19' AFF. Find the end elevation.

Step 1 Divide the length of the run by 10 to determine the number of 10' sections:

80' ÷ 10'/section = 8 sections

Step 2 Multiply the number of sections by the minimum pitch expressed as a decimal. In this case, 0.25" is used because the example refers to a sprinkler main in a nonrefrigerated area:

8 sections × 0.25" per section = 2" change in elevation

Step 3 Add the change in elevation to the starting elevation to find the end elevation:

19' starting elevation + 2" = 19'-2" end elevation

2.2.0 Drains

Ideally, all piping on a dry pipe system should be pitched toward the system riser so the main drain can serve the entire system. In practice, however, varying structural elevations, roof pitches, mechanical equipment, and other obstructions prevent pitching all piping toward the riser. This can create areas of trapped piping where water can collect and result in corrosion. When sections of piping are trapped, auxiliary drains must be used and provided with signage that reads "AUXILIARY DRAIN." In addition, systems with low point drains must have a sign at the dry pipe valve indicating the number of low point drains and the location of each drain.

2.2.1 Auxiliary Drains

The size and arrangement of auxiliary drains depends on the volume of the trapped piping. There are two categories: small trapped sections (under five gallons) and large trapped sections (over five gallons). Each type has specific requirements.

Small Pitch = Big Elevation Change

The impact of pitch is significant. Note the change in elevation with increasing distance on a branch line:

- 0.5" per 10'
- 1" per 20'
- 5" per 100'
- 1' per 240'

Small Trapped Sections

If the capacity of the trapped piping section is five gallons or less, the auxiliary drain may consist of a valve of at least ½" in size and a plug. This is different than in a wet pipe sprinkler system, where only a plug or a nipple and cap are required. In a dry pipe sprinkler system, rapid air pressure changes might cause the dry pipe valve to trip. A globe-type throttling valve can be used to slowly drain trapped areas without activating the system. *Figure 14* shows an auxiliary drain configuration for a small trapped section.

Where a single sprinkler is trapped on a drop nipple using a return bend, and both the sprinkler and return bend are in an area not subject to freezing, the drain requirement can be satisfied by removing the sprinkler to drain that portion of the system. (Note that any time a sprinkler is removed, it must be replaced with a new sprinkler of the same size and type.) In a heated space, there is no requirement to drain one sprinkler. The normal method of installation, however, is to use a dry pendent sprinkler so that no moisture is trapped by the sprinkler drop.

Large Trapped Sections

If the capacity of the trapped piping section exceeds five gallons, NFPA 13 requires a special type of auxiliary drain known as a *drum drip*. Standard drum drips consist of two 1" valves and a 2" by 12" nipple installed in a vertical configuration (*Figure 15*). An equivalent assembly may also be used.

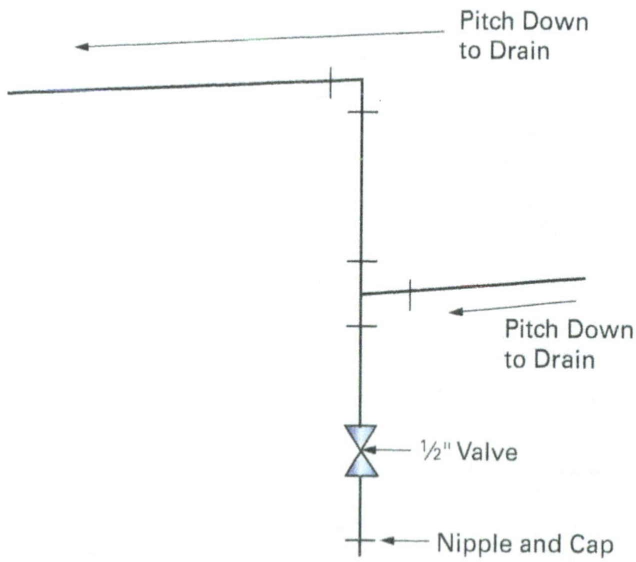

Figure 14 Auxiliary drain configuration, capacity of five gallons or less.

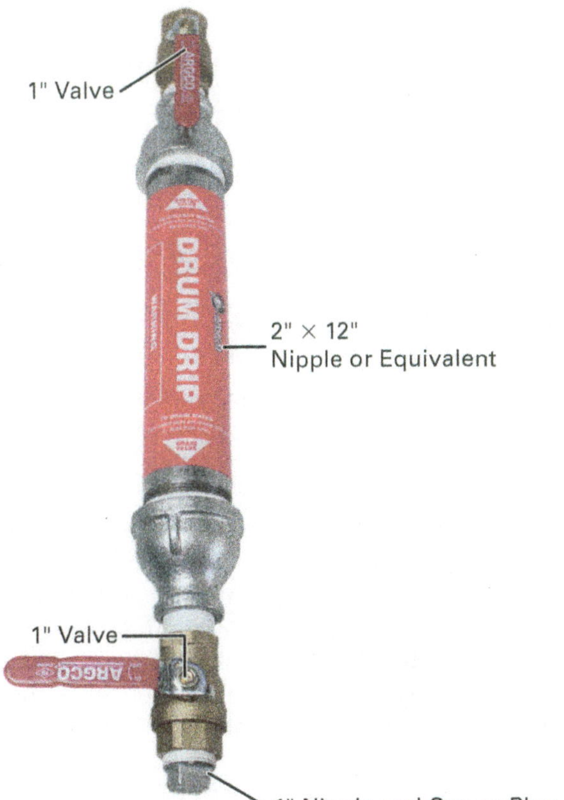

Figure 15 Dry pipe sprinkler system drum drip.

> **CAUTION**
> A dry pipe sprinkler system with a QOD may trip while servicing auxiliary drains. To prevent this and speed up drainage, remove the QOD and service the drains per the manufacturer's instructions.

A drum drip is basically a small storage container. Water in the piping drips in through the top, where it collects until it is drained through the bottom. Drum drips must be drained following system installation, before the onset of cold weather, and periodically as part of an ongoing system maintenance program. To drain a drum drip, proceed as follows:

Step 1 Close the top valve to prevent the loss of air pressure in the system.

Step 2 Remove the plug or nipple and cap from the bottom valve. Prepare for the release of water by positioning a bucket below the nipple.

Step 3 Open the bottom valve to drain the water, then close the valve.

Step 4 Open the top valve slowly to prevent the dry pipe valve from tripping.

Step 5 Replace the plug or nipple and cap to prevent air leakage from the system.

Drum drips should be protected against freezing and installed in a heated area if possible. Preventing a drum drip from bursting during cold weather requires frequent inspection and draining. Drum drips that fill quickly may require service daily, while those that fill more slowly may be serviced less frequently. A frozen drum drip will burst like any other piece of pipe, causing the dry pipe valve to trip and flood the area. A flood eliminator can be installed above a drum drip to prevent flooding due to a burst auxiliary drain (*Figure 16*). This device is designed to shut off the flow of water that would result from a failed drum drip, eliminating flooding and damage.

> **CAUTION**
> If heat tape is used on piping or drum drips, it must be monitored. NFPA 13 does not allow heat tape to be used on dry pipe sprinkler system valves. Heated valve enclosures are used for this purpose.

2.2.2 Tie-In Drains

In many cases, branch piping cannot be pitched to drain toward the cross main. For example, this may occur in a rigid-frame steel building, where the roof peaks in the middle. If the roof pitch is sufficiently long, a single cross main may be installed and located parallel to the peak and near the center of the pitching roof. This lends itself to a center-fed piping arrangement. All the branch piping on the high side drains to the cross main, but the piping on the low side is trapped. Instead of installing an auxiliary drain at the low end of each branch line, a tie-in drain can be installed to drain every branch line to a single location (*Figure 17*).

Tie-in piping must be pitched at ½" per 10' of run. Threaded unions can be used to connect the tie-in drain piping to the branch lines but are prone to leakage. Grooved couplings Listed for dry pipe sprinkler systems provide a tight gasketed connection and are preferred. A major disadvantage of tie-in drains is that they delay the air evacuation / water delivery in an activated dry pipe sprinkler system. For this reason, NFPA 13 limits the pipe size for tie-in drains to 1".

Figure 16 Drum drip with flood eliminator.

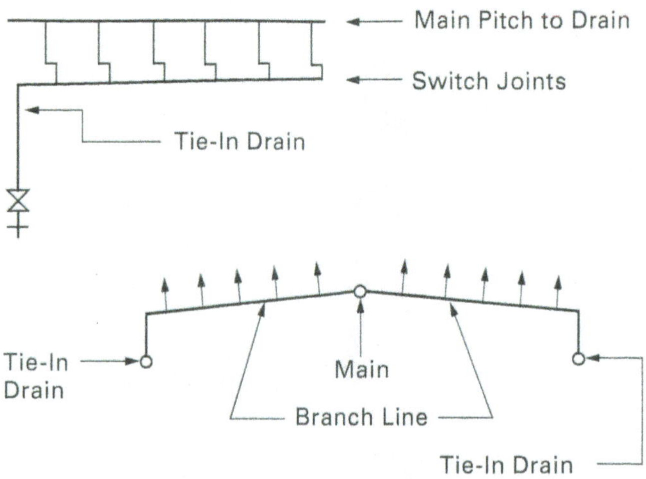

Figure 17 Typical tie-in drain arrangement.

2.2.3 Inspector's Test Connections

Inspector's test connections for dry pipe sprinkler systems are similar to those in wet pipe sprinkler systems. One difference is the location of the inspector's test connection. In a wet pipe sprinkler system, the inspector's test connection can be located anywhere downstream of the waterflow switch and is typically mounted on the system riser. In a dry pipe sprinkler system, the inspector's test connection must be located at the furthest outlet in order to measure the time required to deliver water to the most remote sprinkler in the event of a fire. Another difference is that a dry pipe sprinkler system has a vertical nipple installed at the connection to the branch line to minimize the amount of water that collects in the vertical run of pipe (*Figure 18*).

During a hydrostatic test, the first piece of drain piping below the valve is removed and the valve discharge outlet is plugged. The primary purpose of the plug is to block air leakage in the event of a valve seat or gasket leak. The plug also prevents cold outside air from cooling the piping above the valve. If the piping loses heat, the water could freeze and accumulate in the down pipe. The disadvantage of this arrangement is that the valve must be unplugged and the filler piece reinstalled to operate the inspector's test connection.

NFPA 13 requires that the trip test connection or manifold be equipped with an accessible shutoff valve and a plug not less than 1", of which at least one shall be brass. This is because corrosion of steel plugs can cause them to adhere to a steel fitting, making removal difficult or impossible. If either the valve or plug is made of brass, this will not occur. In lieu of a plug, a nipple and cap may be used. The inspector's test connection should terminate with a 45-degree elbow turned toward the ground and include a smooth-bore, corrosion-resistant orifice that provides a flow equivalent to one sprinkler of a type installed on the system. This orifice is often simply a sprinkler with the frame sawed off.

The location of the inspector's test connection and drains are usually determined by the layout technician and should consider potential discharge water damage. For example, planting beds are not good choices because the water is discharged at a rate that can damage plants and bedding material. A concrete pad should be used to prevent erosion. Discharge onto sidewalks and parking lots can produce an icing hazard to personnel if the discharge occurs during the winter. The discharged water may be black or rusty, depending on the age of the system. Either condition may stain concrete.

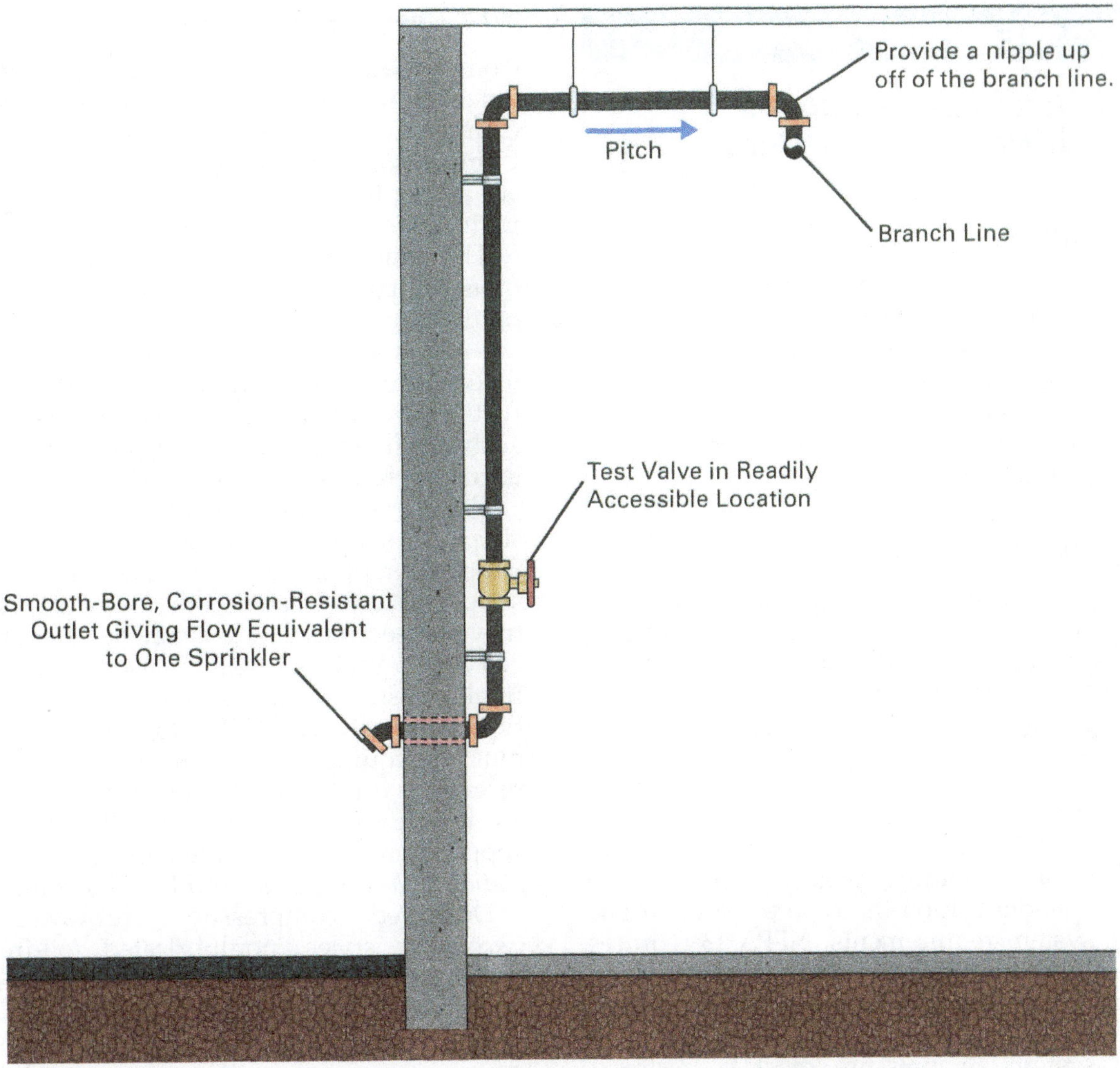

Figure 18 Acceptable arrangement of inspector's test connection on a dry pipe sprinkler system.

2.0.0 Section Review

1. In nonrefrigerated applications, main piping *must* be pitched *at least* _____.
   a. ⅛" per 10' of run
   b. ¼" per 10' of run
   c. ⅜" per 10' of run
   d. ½" per 10' of run

2. Which of the following is *true* with regard to trapped sections of five gallons or less in a dry pipe sprinkler system?
   a. A drain is not required.
   b. A minimum ½" drain valve and plug are required.
   c. A plug may be used by itself to meet the requirement.
   d. A nipple and cap are required.

SECTION THREE

3.0.0 AIR SOURCES AND AIR MAINTENANCE DEVICES

Objective

Describe the air sources and air maintenance devices used in dry pipe sprinkler systems.

a. Identify the air sources used to supply pressure in dry pipe sprinkler systems.
b. Identify the air maintenance devices used to regulate airflow in dry pipe sprinkler systems.

Trade Terms

Air receiver: A chamber, compatible with an air compressor, that can store air under pressure that is higher in pressure than that in the dry pipe or preaction system piping.

Nitrogen generator: A device used to extract nitrogen from compressed air.

Dry pipe sprinkler systems require a variety of valve holding pressures, which are dependent upon the type of valve in use and the system requirements. NFPA 13 requires that a system be brought up to the required air pressure in 30 minutes, except in refrigerated spaces that are maintained below 5°F, where normal system air pressure must be restored within 60 minutes.

Regardless of the pressure source, the layout technician should verify the following parameters to select the correct compressor or package to match the system:

- System design volume (gallons)
- Required pump up time (minutes)
- System design pressures (minimum, holding, and maximum)
- Power source available (single- or three-phase, amperage rating, and the voltage source)
- Location of the equipment
- Type of facility being sprinklered

3.1.0 Air Sources

Pressurized air may be supplied from several sources, include the owner's plant air (shop air), a dedicated air compressor provided by the sprinkler contractor, or bottles/tanks of compressed nitrogen or other approved gas.

3.1.1 Compressed Air Sources

Compressed air is derived by taking atmospheric air, drawing it into a compressor, and then compressing it to the desired (end-use) pressure. Compressors offer a simple, reliable method of supplying the correct system operating pressure in dry pipe sprinkler systems.

The atmosphere contains varying amounts of moisture, particulates, and other gases depending on the exact position from which it is drawn into the compressor. In most cases, the air is ready to use after being compressed. If the air needs further processing due to the end-use location, such as a freezer room, drying equipment is used to further reduce the moisture levels.

Compressed air sources fall into two categories: shop air supplies and dedicated compressors. Shop air supplies use air drawn from existing supply systems located at the facility. This supply may be used only if it is available 24 hours a day, 365 days a year. An AMD is used to regulate the final pressure to any system connected to this type of supply. Per NFPA 13, an air compressor must be equipped with an **air receiver**. An air receiver is a tank capable of holding compressor air. Where an air compressor is the dedicated air supply, it must be installed in accordance with *National Electrical Code®* (NFPA 70®) requirements.

Dedicated compressors are available in two basic styles: nonlubricated (oil-free) and lubricated. Either style can be base-mounted or tank-mounted.

Oil-Free Compressors

Oil-free compressors are direct drive, which means that the compression and pump sections are driven directly off the shaft of the motor. These compressors can be mounted horizontally or vertically because there is no concern about oil pressure requirements. They are often mounted directly onto the riser. *Figure 19* shows a direct-drive oil-free compressor designed for riser mounting. A tank can also be installed on the riser along with this compressor.

A pressure switch controls compressor start and stop. Oil-free units have built-in head unloading and do not require an unloader on the pressure switch.

Oil-free compressors are also available in tank-mounted versions (*Figure 20*). Tank-mounted compressors should be used when supplying multiple dry pipe systems.

Figure 19 Riser-mounted direct-drive oil-free compressor.

Lubricated Compressors

Lubricated compressors are belt-driven units that require oil to lubricate the cylinders during operation. Belt-driven means that a belt is used to turn the shaft. These compressors are mounted horizontally to allow the oil to remain in a position where it can deliver lubrication to the cylinders. Lubricated compressors are typically less expensive than oil-free compressors but require more maintenance. They are typically tank-mounted (*Figure 21*).

3.1.2 Nitrogen Sources

Industrial-grade nitrogen can be used in dry pipe sprinkler systems instead of compressed air to prevent moisture from accumulating in pipes and to remove the oxygen required to produce rust. Nitrogen is available in high-pressure tanks/bottles and can also be supplied using a nitrogen generator. Since nitrogen is a gas that must be replenished, leaks on these systems are more costly than in systems using air. A low-pressure alarm must be provided on gas storage containers to signal the need for refilling.

A typical 5½' nitrogen tank holds approximately 1,900–2,300 gallons of nitrogen gas. This is enough gas to fill a 750-gpm sprinkler system several times. When the tank is full, the pressure is 2,000–2,250 psi.

A regulator is used to control the output pressure to the dry pipe sprinkler system. There are two gauges on the regulator. The gauge on the output of the tank is likely to have a reading range of 0–4,000 psi. The gauge on the output of the regulator will typically read up to 200 psi. The minimum pressure for a tank to remain in service can be no lower than approximately 50 psi, but the tank should be replaced when the residual pressure reaches approximately 100 psi. It is important for maintenance personnel to monitor pressure readings to determine when a new tank is needed. *Figure 22* shows a regulator used to control the nitrogen pressure in a sprinkler system. The main disadvantages of using nitrogen are safety concerns and the need to replace tanks periodically. Two or more tanks are often manifolded (connected) so that a full tank can immediately be placed in service when one is depleted.

Figure 20 Tank-mounted direct-drive oil-free compressor.

Figure 21 Belt-drive, lubricated, tank-mounted compressor.

Quiet Compressors

Air compressors are very noisy and may disturb occupants, particularly when they are installed in or near occupied spaces rather than a dedicated mechanical room. Quiet compressors are available for use in occupied spaces.

WARNING!

Nitrogen is a high-pressure gas. At full cylinder pressure (about 2,000 psi), nitrogen can rupture piping. Always use a pressure regulator on the nitrogen cylinder to limit the pressure. Also, make sure to use an overpressure relief valve between the cylinder pressure regulator and the system.

Nitrogen generators extract nitrogen from compressed gas and offer the advantage of a continuous supply (*Figure 23*). The nitrogen is then stored in a tank to supply the system air maintenance device.

3.2.0 Air Maintenance Devices (AMDs)

Listed air maintenance devices (AMDs) are used to regulate airflow and limit the incoming air pressure to a level that will not affect the response time of the system. AMDs minimize false trips and are required for all systems except very small systems. When more than one system is connected to a common air supply, each system requires an individual AMD. The AMD in use must be capable of controlling the required air pressure on, and maximum airflow to, the dry pipe system.

NOTE

Where the air compressor supplying the dry pipe sprinkler system has a capacity less than 5.5 ft^3/min at 10 psi, an air receiver or AMD is not required.

An AMD is factory set to a specified nominal system pressure that can be adjusted by loosening the locknut at the top of regulator and turning the adjustment screw clockwise to increase pressure and counterclockwise to decrease pressure (*Figure 24*). Refer to the approved shop drawings and dry pipe valve manufacturer's instructions for specific values. The resulting pressure can be read on the sprinkler system air pressure gauge. The locknut must be tightened when the adjustment is complete.

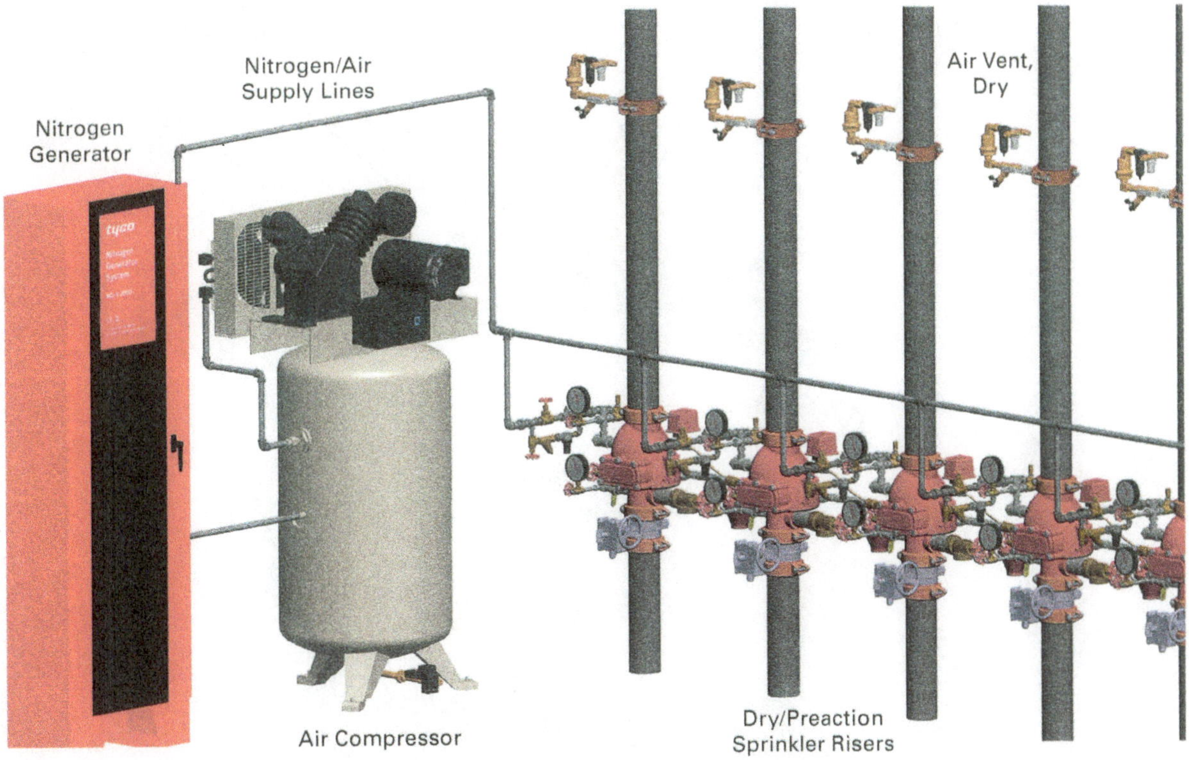

Figure 22 Nitrogen regulator.

Figure 23 Nitrogen generator system.

Did You Know?

Prepackaged cabinet units are available that include a dry pipe valve, nitrogen generator, and trim in a factory-assembled cabinet.

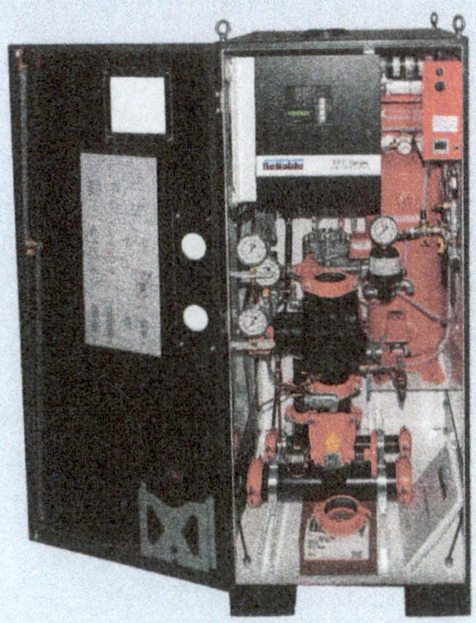

Figure Credit: Reliable Automatic Sprinkler Co., Inc.

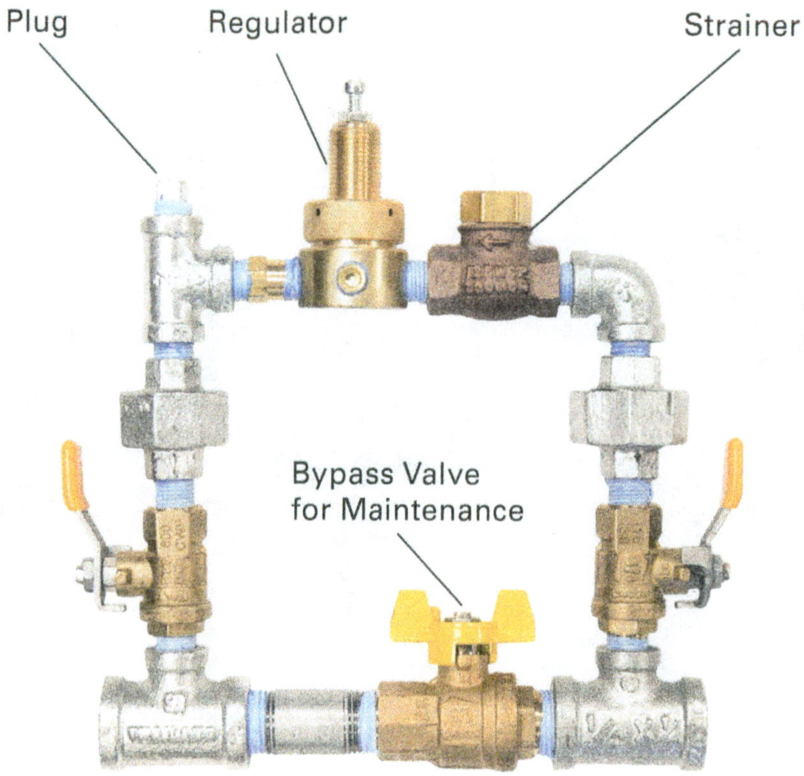

Figure 24 Regulator-style air maintenance device.

3.0.0 Section Review

1. A device used to extract nitrogen from compressed air is known as a nitrogen _____.
   a. compressor
   b. regulator
   c. distiller
   d. generator

2. Air maintenance devices are required in each system of a multi-system fire sprinkler installation.
   a. True
   b. False

SECTION FOUR

4.0.0 TESTING AND TROUBLESHOOTING DRY PIPE SPRINKLER SYSTEMS

Objective

Describe how to test and troubleshoot dry pipe sprinkler systems.

a. Explain the basic procedures for testing dry pipe sprinkler systems.
b. Describe how to troubleshoot common dry pipe sprinkler system issues.

Performance Task

1. Complete a Contractor's Material and Test Certificate for Aboveground Piping for a dry pipe sprinkler system.

Trade Term

Pneumatic test: A test that introduces air pressure into the system at no less than 40 psi for 24 hours. The pressure drop cannot exceed $1^1/_2$ psi in the 24-hour period.

After a sprinkler system is installed, various tests are conducted before it is put into service to ensure that the system is leak-free and will perform as expected in the event of a fire. These are known as *acceptance tests.*

4.1.0 Testing Dry Pipe Systems

Like wet pipe sprinkler systems, dry pipe sprinkler systems require acceptance tests. Hydrostatic and pneumatic tests are common to both types of systems. However, in wet pipe sprinkler systems, a pneumatic test is only performed as an interim test when conditions are not favorable for a hydrostatic test. In dry pipe sprinkler systems, a pneumatic test is required for acceptance.

- *Hydrostatic testing* — Hydrostatic testing is performed by filling the sprinkler system with water and then applying pressure to the water to check for pressure loss and leaks. Hydrostatic tests must be made at a minimum of 200 psi for two hours or 50 psi above static pressure in excess of 150 psi for two hours. Follow the valve manufacturer's instructions and safety precautions for hydrostatic testing. Some valves require that the clapper be removed during testing, while other valves do not. A hydrostatic test is normally performed only after the initial installation or after remodeling and additions involving 20 or more sprinklers.
- *Pneumatic testing* — On a dry pipe sprinkler system, an air pressure leakage or pneumatic test must be conducted for 24 hours at 40 psi. The pressure is held for 24 hours so the system is exposed to the full range of temperature variations throughout the day and night. Any loss of pressure in excess of $1^1/_2$ psi for the 24-hour period must be corrected.

In addition to the hydrostatic and pneumatic tests that also apply to wet pipe sprinkler systems, two types of trip tests are required on dry pipe sprinkler systems. These include a full flow trip test and a partial flow trip test.

- *Full flow trip test* — A full flow trip test is conducted with the water supply main control valve fully open. This test should be conducted in warm weather. A full flow trip test must be conducted at acceptance and then at least once every three years or as required by the Authority Having Jurisdiction (AHJ). It must be performed in accordance with NFPA standards and the valve manufacturer's instructions. To conduct a full flow trip test, the dry pipe valve is operated by opening the inspector's test connection to simulate the activation of a sprinkler in fire conditions. The sprinkler piping will be flooded with water and must be fully drained before the system is put back into service. Record the time required for the water to discharge from the inspector's test connection. A delay longer than 60 seconds is acceptable if the system capacity is 500 gallons or less, or 750 gallons or less if the system has a quick-opening device. If the system capacity is over 750 gallons and the delay is longer than 60 seconds, corrective action must be taken to meet the standard. Improper system design, excessive corrosion, or partially closed or damaged valves, are all possible causes of excessive time delay. After performing this test, drain the system and reset the dry pipe valve.
- *Partial flow trip test* — A partial flow trip test is conducted with the water supply main control valve partially closed to minimize the amount of water entering the system during the test. This test is used to exercise the valve and as soon as the valve trips, it is shut off again. (It does not simulate system operation in fire conditions.) A partial flow trip test should be performed in warm weather and at least annually unless a full flow trip test is conducted.

All sprinkler systems also require a main drain test, during which the main drain valve is opened and remains open until the system pressure stabilizes. The static and residual pressures are recorded on the Contractor's Material and Test Certificate for Aboveground Piping and the sprinkler system general information sign. Note that every system must have a a fully completed general information sign as well as a hydraulic information sign.

The contractor is responsible for completing all required acceptance tests and recording results on the Contractor's Material and Test Certificate for Aboveground Piping (*Figure 25A*, *Figure 25B*, and *Figure 25C*). It is important to fill out the form completely, including the dry pipe valve and AMD make, model, and serial numbers, time to trip through test connection, water/air/trip pressures, time water reached test outlet, and so on. Witnesses for both the contractor and the property owner must sign the completed test certificate.

4.2.0 Troubleshooting Dry Pipe Systems

As with wet pipe sprinkler systems, problems in dry pipe sprinkler systems occur due to poor installation or when established operating and maintenance procedures are not followed. The most common problem in dry pipe sprinkler systems is corrosion due to a failure to pitch the pipe properly when it is installed. Another common problem is false trips.

Dry pipe valves may trip unexpectedly due to burst pipes in cold weather where drain maintenance is not conducted regularly. Dry pipe sprinkler systems may also experience false trips due to temperature fluctuations, particularly when a dry pipe sprinkler system is used as an auxiliary system to a wet pipe sprinkler system. Examples include a covered garage attached to an office building, or the loading dock of a warehouse or big box store. When the weather is warm, any air in the wet pipe sprinkler system will expand and increase the system pressure, pressurizing the system up to the bottom of the dry pipe valve and eventually causing it to trip.

In addition, false trips may occur following an inspection, particularly if the system riser is in a small room that heats quickly in warm weather. When an inspector completes a main drain test, cold water from the underground supply piping enters the system. In a hot riser room, the cold water in the pipe warms up quickly and expands, causing pressure spikes. If the pressure increase is large enough, it can overcome the air pressure holding the dry pipe valve closed and cause the valve to trip.

The troubleshooting process for a dry pipe sprinkler system follows the same approach discussed for wet pipe sprinkler systems:

Step 1 Identify the problem. Determine how the system operates under normal conditions and then identify areas or components in which the system is no longer operating as expected.

Step 2 Collect data from parts manufacturers, maintenance records, and any personnel with operating knowledge of the system. If a system is unfamiliar, installed by another company, or not maintained regularly, inspect the whole system first to become familiar with it. Use a standard inspection procedure or checklist to document the system.

Step 3 Analyze the data. Refer to the manufacturer's troubleshooting charts to help identify the root cause of the problem.

Step 4 Take action. After all data has been collected and analyzed, the action required to solve a problem is determined. System components may need to be cleaned, modified, repaired, or replaced.

CAUTION

If system maintenance or repair will disable a sprinkler system, follow the manufacturer's instructions for shutting down the system as well as company procedures for informing building occupants and the supervisory service. A fire watch may be necessary to avoid a potential catastrophe. Protect yourself and your company from liability by obtaining a signed release before taking any action.

After maintenance or repairs have been completed, perform hydrostatic and pneumatic tests according to system requirements. Document all actions taken. Perform a visual inspection and ensure pressure readings meet the valve manufacturer's requirements. If the initial problem is not solved by the actions performed, troubleshoot the system again, analyzing the data for any possible causes that were overlooked.

Table 2 and *Table 3* show sample troubleshooting charts for differential and mechanical dry pipe valves. Always follow the manufacturer's instructions for the equipment in use.

Contractor's Material and Test Certificate for Aboveground Piping

PROCEDURE
Upon completion of work, inspection and tests shall be made by the contractor's representative and witnessed by the property owner or their authorized agent. All defects shall be corrected and system left in service before contractor's personnel finally leave the job.

A certificate shall be filled out and signed by both representatives. Copies shall be prepared for approving authorities, owners, and contractor. It is understood the owner's representative's signature in no way prejudices any claim against contractor for faulty material, poor workmanship, or failure to comply with approving authority's requirements or local ordinances.

AFSA
American Fire Sprinkler Association

Property name	Date

Property address

New Installation?	☐ Yes	☐ No
Modification? If yes, complete applicable portions of the form.	☐ Yes	☐ No

Provide a description of the scope of work on page 3.

Plans	Accepted by approving authorities (names)		
	Address		
	Installation conforms to accepted plans	☐ Yes	☐ No
	Equipment used is approved If no, explain deviations	☐ Yes	☐ No
Instructions	Has person in charge of fire equipment been instructed as to location of control valves and care and maintenance of this new equipment? If no, explain	☐ Yes	☐ No
	Have copies of the following been left on the premises?	☐ Yes	☐ No
	1. System components instructions	☐ Yes	☐ No
	2. Care and maintenance instructions	☐ Yes	☐ No
	3. NFPA 25	☐ Yes	☐ No
Location of system	Supplies buildings		

Sprinklers	Make	Model	Year of manufacture	Orifice size	Quantity	Temperature rating

Pipe and fittings	
	Type of pipe
	Type of fittings

Alarm valve or flow indicator	Alarm device			Maximum time to operate through test connection	
	Type	Make	Model	Minutes	Seconds

Dry pipe operating test	Dry valve			Q. O. D.		
	Make	Model	Serial no.	Make	Model	Serial no.

	Time to trip through test connection[a,b]		Water pressure	Air pressure	Trip point air pressure	Time water reached test outlet[a,b]		Alarm operated properly	
	Minutes	Seconds	psi	psi	psi	Minutes	Seconds	Yes	No
Without Q.O.D.									
With Q.O.D.									

If no, explain

[a] Measured from time inspector's test connection is opened
[b] *NFPA 13* only requires the 60-second limitation in specific sections

Figure 25A Example Contractor's Material and Test Certificate for Aboveground Piping (1 of 3).

Deluge and preaction valves							
Operation	☐ Pneumatic	☐ Electric	☐ Hydraulics				
Piping supervised	☐ Yes ☐ No	Detecting media supervised	☐ Yes ☐ No				
Does valve operate from the manual trip, remote, or both control stations?	☐ Yes ☐ No						
Is there an accessible facility in each circuit for testing? ☐ Yes ☐ No	If no, explain						

Make	Model	Does each circuit operate supervision loss alarm?		Does each circuit operate valve release?		Maximum time to operate release	
		Yes	No	Yes	No	Minutes	Seconds

Pressure-reducing valve test

Location and floor	Make and model	Setting	Static pressure		Residual pressure (flowing)		Flow rate
			Inlet (psi)	Outlet (psi)	Inlet (psi)	Outlet (psi)	Flow (gpm)

Backflow device forward flow test

Indicate means used for forward flow test of backflow device: ______________

When means to test device was opened, was system flow demand created? ☐ Yes ☐ No ☐ N/A

Test description

Hydrostatic: Hydrostatic tests shall be made at not less than 200 psi for 2 hours or 50 psi above static pressure in excess of 150 psi for 2 hours. Differential dry-pipe valve clappers shall be left open during the test to prevent damage. All aboveground piping leakage shall be stopped.

Pneumatic: Establish 40 psi air pressure and measure drop, which shall not exceed 1½ psi in 24 hours. Test pressure tanks at normal water level and air pressure and measure air pressure drop, which shall not exceed 1½ psi in 24 hours.

Tests

All piping hydrostatically tested at _____ psi for _____ hours		If no, state reason
Dry piping pneumatically tested	☐ Yes ☐ No	
Equipment operates properly	☐ Yes ☐ No	

Do you certify as the sprinkler contractor that additives and corrosive chemicals, sodium silicate or derivatives of sodium silicate, brine, or other corrosive chemicals were not used for testing systems or stopping leaks?

☐ Yes ☐ No

Drain test	Reading of gauge located near water supply test connection: _____ psi	Residual pressure with valve in test connection open wide: _____ psi

Underground mains and lead-in connections to system risers flushed before connection made to sprinkler piping

Verified by copy of the Contractor's Material and Test Certificate for Underground Piping.	☐ Yes ☐ No	Other	Explain
Flushed by installer of underground sprinkler piping	☐ Yes ☐ No		

If powder-driven fasteners are used in concrete, has representative sample testing been satisfactorily completed?	☐ Yes ☐ No	If no, explain

Blank testing gaskets

Number used	Locations	Number removed

Welding

Welding piping ☐ Yes ☐ No

If yes . . .

Do you certify as the sprinkler contractor that welding procedures used complied with the minimum requirements of *AWS B2.1, ASME Section IX Welding and Brazing Qualifications,* or other applicable qualification standard as required by the AHJ?	☐ Yes ☐ No
Do you certify that all welding was performed by welders or welding operators qualified in accordance with the minimum requirements of *AWS B2.1, ASME Section IX Welding and Brazing Qualifications,* or other applicable qualification standard as required by the AHJ?	☐ Yes ☐ No
Do you certify that the welding was conducted in compliance with a documented quality control procedure to ensure that (1) all discs are retrieved; (2) that openings in piping are smooth, that slag and other welding residue are removed; (3) the internal diameters of piping are not penetrated; (4) completed welds are free from cracks, incomplete fusion, surface porosity greater than 1⁄16 in. diameter, undercut deeper than the lesser of 25% of the wall thickness or 1⁄16 in.; and (5) completed circumferential butt weld reinforcement does not exceed 3⁄32 in.?	☐ Yes ☐ No

Figure 25B Example Contractor's Material and Test Certificate for Aboveground Piping (2 of 3).

<table>
<tr><td>Cutouts (disccs)</td><td colspan="2">Do you certify that you have a control feature to ensure that all cutouts (discs) are retrieved? ☐ Yes ☐ No</td></tr>
<tr><td>Hydraulic data nameplate</td><td>Nameplate provided
☐ Yes ☐ No</td><td>If no, explain</td></tr>
<tr><td colspan="3">Sprinkler contractor removed all caps and straps? ☐ Yes ☐ No</td></tr>
<tr><td>Remarks</td><td colspan="2">Date left in service with all control valves open</td></tr>
<tr><td rowspan="4">Signatures</td><td colspan="2">Name of sprinkler contractor</td></tr>
<tr><td colspan="2">Tests witnessed by</td></tr>
<tr><td colspan="2">The property owner or their authorized agent (signed) | Title | Date</td></tr>
<tr><td colspan="2">For sprinkler contractor (signed) | Title | Date</td></tr>
<tr><td colspan="3">Additional explanations and notes</td></tr>
</table>

Figure 25C Example Contractor's Material and Test Certificate for Aboveground Piping (3 of 3).

Table 2 Differential Dry Pipe Valve Troubleshooting Chart

Machine Component or Problem Area	Possible Cause	Corrective Action
Piping corrosion	Piping not installed at proper pitch	Pitch piping properly.
Valve trips when no sprinkler has activated	Inspection introduced cold water in hot riser room piping	Consider conditioning air in riser room.
	Temperature fluctuations in combination wet/dry system	Consider conditioning air in riser room.
	Loss of air pressure in the system	Check system for leaks and proper air supply. An air maintenance device should be installed on each system equipped with an automatic air supply. Consider adding a maintenance air compressor.
	Extreme pressure surge in water supply	Increase system air pressure per manufacturer's instructions. Note that increasing system pressure may increase dry pipe valve trip time.
Water constantly passing through drip check when valve is in set position	Water leaking over water seat into intermediate chamber	Inspect and clean water seat and clapper rubber. Replace clapper rubber if necessary. If water seat is pitted or damaged, replace base assembly.
	Alarm test valve in bypass connection not tightly closed	Verify that water is not getting past alarm test valve.
Air constantly passing through drip check when the valve is in set position	Air leaking over air seat into intermediate chamber	Inspect and clean air seat and clapper. Replace clapper rubber if necessary. If air seat is pitted or damaged, replace air plate assembly.
	Air leaking past rubber diaphragm	Inspect rubber diaphragm for deterioration. Replace if necessary.
Clapper will not latch	Incorrect resetting tool	Verify that resetting tool is smooth and sized correctly to provide required force at correct angle.
	Hook not sliding on resetting tool	File or grind resetting tool to provide smooth surface and proper clearance.
	Clapper rubber worn	Replace clapper rubber.
	Clapper damaged by high pressure	Replace valve member assembly.
Valve latches but will not remain set	Improper resetting procedure	Reset per manufacturer's instructions.
	Inadequate air supply	Adjust per manufacturer's instructions.
	Air pressure and priming water passing through intermediate chamber and out of drip check	Clean air seat and clapper rubber. Replace clapper rubber if necessary.

Table 3 Mechanical Dry Pipe Valve Troubleshooting Chart

Machine Component or Problem Area	Possible Cause	Corrective Action
Piping corrosion	Piping not installed at proper pitch	Pitch piping properly.
Actuator damaged	Excess pressure	Reset pressure per manufacturer's instructions. Install pressure relief kit where ambient temperatures are high.
		Check for water hammer caused by improper fire pump settings, underground construction work, or improper venting of trapped air in piping.
Mechanical sprinkler alarm not operating	Clogged screen in strainer of water motor	Remove plug from strainer thenremove and clean screen.
Air leakage out of drip check	Damage to airside of seal assembly, upper seat O-ring or mounting O-ring if supplied	Inspect seat, clapper, O-ring(s), and seal assembly for damage. Replace if necessary.
Water leakage out of drip check	Damage to seal assembly, seat, or lower seat O-ring	Inspect seat, clapper, O-ring, and seal assembly for damage. Replace if necessary.
Water leakage out of pushrod chamber vent hole	Leaking diaphragm	Inspect pushrod chamber cover to determine cause of damage and correct. Install new diaphragm.
Air leakage out of pushrod chamber vent hole	Defective O-ring assembled to the pushrod guide	Replace all O-rings and pushrod guide.

4.0.0 Section Review

1. A pneumatic test on a dry pipe sprinkler system runs for a period of _____.
   a. 2 hours
   b. 8 hours
   c. 16 hours
   d. 24 hours

2. Water leakage out of the pushrod chamber vent hole in a mechanical dry pipe valve is *most* likely caused by low system pressure.
   a. True
   b. False

Review Questions

1. Which of the following is *true* regarding dry pipe sprinkler systems?
   a. Dry pipe sprinkler systems are simpler than wet pipe sprinkler systems.
   b. Dry pipe sprinkler systems operate faster than wet pipe sprinkler systems.
   c. Dry pipe sprinkler systems are more prone to corrosion than wet pipe sprinkler systems.
   d. Dry pipe sprinkler systems may only be installed in conditioned spaces not subject to freezing.

2. If the city water pressure acting on the clapper of a differential dry pipe valve is 100 psi, and the clapper differential is 5:1, what is the trip pressure of the dry pipe valve?
   a. 20 psi
   b. 30 psi
   c. 40 psi
   d. 50 psi

3. System air pressure gauges designed for fire protection use typically read from 0 psi to ______.
   a. 80 psi with a retard to 250 psi
   b. 500 psi with a retard to 750 psi
   c. 1,000 psi
   d. 4,000 psi

4. Which of the following devices is required if a dry pipe sprinkler system with a 600-gallon piping system capacity does *not* meet the required water delivery time of 60 seconds?
   a. AMD
   b. PRV
   c. QOD
   d. AHU

5. The *most* common type of QOD is a(n)______.
   a. accelerator
   b. regulator
   c. reducer
   d. AMD

6. In refrigerated applications, main piping *must* be pitched *at least* ______.
   a. ⅛" per 10' of run
   b. ¼" per 10' of run
   c. ⅜" per 10' of run
   d. ½" per 10' of run

7. What is the change in elevation for a 50' branch line run in a parking garage?
   a. 2.0"
   b. 2.5"
   c. 3.0"
   d. 3.5"

8. In a dry pipe sprinkler system where the system main piping runs through a refrigerated area for a distance of 40', the change in elevation from one end of the pipe to the other is ______.
   a. 1"
   b. 2"
   c. 3"
   d. 4"

9. A drum drip is used as a(n) ______.
   a. auxiliary drain
   b. check valve
   c. main drain
   d. test valve

10. If the capacity of trapped piping sections exceeds five gallons, NFPA 13 requires a _____.
    a. drip check
    b. velocity check
    c. drum drip
    d. sump pump

11. The inspector's test connection on a dry pipe sprinkler system is normally located _____.
    a. in the most remote and highest branch line
    b. as close as possible to the riser
    c. outside of the building near the drain
    d. at the air supply

12. Per NFPA 13, all except very small air compressors *must* be equipped with a(n) _____.
    a. self-test device
    b. HEPA filter
    c. air receiver
    d. battery backup

13. The procedure used to test system operation under simulated fire conditions is a _____.
    a. pneumatic test
    b. hydrostatic test
    c. partial flow trip test
    d. full flow trip test

14. Piping corrosion in a dry pipe sprinkler system is *most* likely to be caused by _____.
    a. excess system pressure
    b. use of nitrogen in the system
    c. using a vacuum system
    d. pipe installed at the wrong pitch

15. A possible cause of a dry pipe valve tripping following an inspection is _____.
    a. cold underground water introduced into hot riser room piping
    b. low system pressure as a result of cold temperatures
    c. the inspector inadvertently left the test valve open
    d. system testing introduced backpressure into the piping

Trade Terms Introduced in This Module

Accelerator: A type of quick-opening device (QOD) used to speed up the trip time of a dry pipe valve.

Air maintenance device (AMD): A Listed assembly of valves, nipples, fittings, and actuators used to automatically control the air pressure in the piping of dry pipe sprinkler systems.

Air receiver: A chamber, compatible with an air compressor, that can store air under pressure that is higher in pressure than that in the dry pipe or preaction system piping.

Alarm test valve: A shutoff valve used to allow the testing of system alarms without activating the dry pipe valve.

Differential: The ratio between the surface area on the top of the clapper, where the supply air applies pressure and the bottom of the clapper, where the supply water applies pressure.

Differential dry pipe valve: A valve that is held in the closed position by the system air/nitrogen pressure exposed to the larger surface area on the air/nitrogen side of the clapper where it is at least five times that of the surface area on the water supply side.

Drip check: A spring-loaded check valve that is primarily used to evaluate the seating of the dry pipe valve clapper, as well as to retain atmospheric pressure in the intermediate chamber of a differential dry pipe valve.

Dry pipe sprinkler system: A sprinkler system employing automatic sprinklers that are attached to a piping system containing air or nitrogen under pressure, the release of which (as from the opening of a sprinkler) permits the water pressure to open a valve known as a *dry pipe valve*, and the water then flows into the piping system and out the opened sprinklers.

Intermediate chamber: A chamber in a differential dry pipe valve that water flows into when the clapper is unseated. A port in the intermediate chamber allows water to flow into the alarm line, where it activates the waterflow alarm pressure switch. In addition, the intermediate chamber is used to introduce pressure below the clapper when an accelerator is used.

Mechanical dry pipe valve: A valve that uses a series of mechanical devices such as levers, springs, diaphragms, and latches to hold the valve in the closed position using air/nitrogen pressure and without using the clapper surface areas to provide a differential between the air/nitrogen and water pressures.

Nitrogen generator: A device used to extract nitrogen from compressed air.

Pitch: The ratio of rise over run.

Pneumatic test: A test that introduces air pressure into the system at no less than 40 psi for 24 hours. The pressure drop cannot exceed $1\frac{1}{2}$ psi in the 24-hour period.

Quick-opening device (QOD): A device used in dry pipe sprinkler systems to reduce the trip time of the dry pipe valve.

Restricted orifice: The orifice through which most automatic sources of air are supplied to a dry pipe valve. It prevents system air from being replenished faster than the air that is being discharged through an activated sprinkler or test connection.

Trip pressure: The air pressure at which a dry pipe valve operates.

Additional Resources

This module presents thorough resources for task training. The following reference material is suggested for further study.

NFPA 13, *Standard for the Installation of Sprinkler Systems*, Latest Edition. Quincy, MA: National Fire Protection Association.

NFPA 25, *Standard for the Inspection, Testing, and Maintenance of Water-Based Fire Protection Systems*, Latest Edition. Quincy, MA: National Fire Protection Association.

Figure Credits

Cen-Cal Fire Systems, Inc., Module Opener, Figure 9

Reliable Automatic Sprinkler Co., Inc., Figures 1, 3–5, 24

Courtesy of Viking Group, Inc., Figures 2, 6, 10

ARGCO, Figures 7, 8, 15

Used with permission of Johnson Controls, Figures 11, 22

AGF Manufacturing, Figure 16

Courtesy of Victaulic Company, Figure 19

Courtesy of General Air Products, Figures 20–21

Courtesy of the Harris Products Group, Figure 23

Section Review Answer Key

Section 1.0.0

Answer	Section Reference	Objective
1. c	1.1.0	1a
2. b	1.2.1	1b
3. b	1.3.4	1c

Section 2.0.0

Answer	Section Reference	Objective
1. b	2.1.0	2a
2. b	2.2.1	2b

Section 3.0.0

Answer	Section Reference	Objective
1. d	3.1.2	3a
2. a	3.2.0	3b

Section 4.0.0

Answer	Section Reference	Objective
1. d	4.1.0	4a
2. b	4.2.0, *Table 3*	4b

User Update

NCCER makes every effort to keep its textbooks up-to-date and free of technical errors. We appreciate your help in this process. If you find an error, a typographical mistake, or an inaccuracy in NCCER's curricula, please submit a User Update form by visiting **https://www.nccer.org/olf**. You can also scan the QR code using the camera on your phone or mobile device to access the form.

Glossary

Above the finished floor (A.F.F.): A measurement taken from the surface of the installed flooring (e.g., tile), rather than the underlayment.

Accelerator: A type of quick-opening device (QOD) used to speed up the trip time of a dry pipe valve.

Air maintenance device (AMD): A Listed assembly of valves, nipples, fittings, and actuators used to automatically control the air pressure in the piping of dry pipe sprinkler systems.

Air receiver: A chamber, compatible with an air compressor, that can store air under pressure that is higher in pressure than that in the dry pipe or preaction system piping.

Alarm check valve: A device installed in a wet sprinkler system that triggers an audible alarm when water flow is detected.

Alarm test valve: A shutoff valve used to allow the testing of system alarms without activating the dry pipe valve.

All-thread rod (ATR): Rod that is fully threaded end to end and often cut to size and used to suspend hangers and fasteners.

Anchor: A device used to attach a hanger to a structure.

Building Information Modeling (BIM): An interactive building modeling and design methodology that uses special software to design and manage the construction and operation of systems and structures.

Bypass: Alarm check valve trim where pressures are equalized.

Clapper: A valve component that restricts water flow under normal pressure.

Cold soldering: A condition that occurs when sprinklers are positioned in close proximity to one another. It results when one sprinkler distributes water on another sprinkler, preventing or delaying the adjacent sprinkler's activation by cooling it below the fusing temperature.

Column: A round structural member used to support beams and girders where loadbearing walls do not exist.

Combustible: Refers to materials that can burn.

Control valve: A valve controlling flow to water-based fire protection systems.

Decimal equivalents: Numbers that represent fractions, with every decimal place indicating a multiple of 10.

Differential: The ratio between the surface area on the top of the clapper, where the supply air applies pressure and the bottom of the clapper, where the supply water applies pressure.

Differential dry pipe valve: A valve that is held in the closed position by the system air/nitrogen pressure exposed to the larger surface area on the air/nitrogen side of the clapper where it is at least five times that of the surface area on the water supply side.

Distort: To bend, warp, or damage a system component.

Double-pole (DP) switch: A switch that can control two circuits.

Double-throw (DT) switch: A switch that transmits a signal in two directions.

Drip check: A spring-loaded check valve that is primarily used to evaluate the seating of the dry pipe valve clapper, as well as to retain atmospheric pressure in the intermediate chamber of a differential dry pipe valve.

Dry pipe sprinkler system: A sprinkler system employing automatic sprinklers that are attached to a piping system containing air or nitrogen under pressure, the release of which (as from the opening of a sprinkler) permits the water pressure to open a valve known as a *dry pipe valve*, and the water then flows into the piping system and out the opened sprinklers.

Elevation view: Usually a two-dimensional view on a vertical plane showing internal and external dimensions of buildings. This view helps sprinkler fitters determine roof contours and floor levels. This is sometimes called a *sectional view.*

Endothermic: A property of a material that causes it to release moisture when heated.

Erode: To deteriorate the surfaces in a valve due to water flow.

Exponents: Numbers or symbols, such as the 2 in 10 ft^2, that indicate the number of times an expression is multiplied by itself.

Factor: An amount by which a value is multiplied to produce a specific result.

False alarm: A signal transmitted from a system that should not be in operation.

Firestopping: A means of cutting off the oxygen supply so that fire and smoke cannot readily move from one location to another.

Fusing (operating) temperature: The temperature at which a sprinkler activates.

General notes: A written list on a drawing that provides information such as applicable standards, types of pipe, and fittings.

Globe valve: A type of valve consisting of a movable plug or disc and a stationary seat.

Grade beam: The structural beam forming the perimeter of the building and used as the ultimate reference point for all construction dimensions.

Hanger: A device from which a pipe is suspended that is used to carry the weight of the pipe from above.

Hose header: A pipe that acts as a central connection point for one or more hoses.

Hydrostatic test: A wet pipe sprinkler system test that uses a test pump to increase the pressure of the system to 200 pounds per square inch (psi) or 50 psi above static pressure in excess of 150 psi for two hours to determine whether there are any leaks. If the system holds the pressure for the specified time period, it passes the test.

Hypotenuse: The longest side of a triangle that contains a right angle.

Intermediate chamber: A chamber in a differential dry pipe valve that water flows into when the clapper is unseated. A port in the intermediate chamber allows water to flow into the alarm line, where it activates the waterflow alarm pressure switch. In addition, the intermediate chamber is used to introduce pressure below the clapper when an accelerator is used.

Intumescent: A property of a material that causes it to expand when heated, acting as an insulator in the event of a fire.

Joist: A structural member that provides support.

Keyed notes: A system by which numbered notes are assigned to associated numbers on a drawing to provide detailed information without crowding the drawing itself.

Legend: A list of symbols and associated definitions used on a drawing.

Limited-combustible: A material with a structural base of noncombustible material; an example is gypsum wallboard.

Load rating: The usable weight capacity of a hanger or support.

Machined: A metal surface that has been held to precise tolerances during the manufacturing process.

Main drain: The drain used to empty the system for servicing, to check the alarm check valve, and to perform main drain tests.

Mathematical operator: A symbolic instruction to perform a specific type of calculation. The four basic mathematical operators are addition (+), subtraction (−), multiplication (×), and division (÷).

Mechanical dry pipe valve: A valve that uses a series of mechanical devices such as levers, springs, diaphragms, and latches to hold the valve in the closed position using air/nitrogen pressure and without using the clapper surface areas to provide a differential between the air/nitrogen and water pressures.

Nitrogen generator: A device used to extract nitrogen from compressed air.

Noncombustible: A material that, in the form in which it is used, and under the conditions anticipated, will not ignite, burn, support combustion, or release flammable vapors when subjected to fire or heat.

Obstructed construction: Beams, trusses, or other structural members that impede heat flow or water distribution and affect the ability of sprinklers to control or suppress a fire.

Pendent: Refers to a sprinkler installed so that the deflector is below the fusible element and designed to direct the water stream downward against the deflector.

Pi: A mathematical constant defined as the ratio of a circle's circumference to its diameter. Pi is equal to approximately 3.1416 or $^{22}/_{7}$ and is represented by the symbol π.

Pipe riser: A vertical pipe extending from an opening in a floor or ceiling.

Pitch: The ratio of rise over run.

Plan view: A two-dimensional layout or drawing of a building.

Pneumatic test: A test that introduces air pressure into the system at no less than 40 psi for 24 hours. The pressure drop cannot exceed 1½ psi in the 24-hour period.

Pressure: The amount of force applied to a given unit of surface area. Pressure is often expressed in *pounds per square inch (psi).*

Pressure-regulating device: A device designed for the purpose of reducing, regulating, controlling, or restricting water pressure.

Pressure switch: A device used to detect changes in pressure within a system and to transmit a signal when fluctuations occur.

Pressurized system: A piping system that is filled with pressure and has a pressure source.

Quick-opening device (QOD): A device used in dry pipe sprinkler systems to reduce the trip time of the dry pipe valve.

Reflected ceiling plan: An architectural term used to describe a plan view of a ceiling that typically shows the lights, fixtures, diffusers, and sprinklers within a given area.

Residual pressure: Pressure in a piping system under flow conditions.

Restraint: A device used to prevent pipe from shifting in any direction other than downward.

Restricted orifice: The orifice through which most automatic sources of air are supplied to a dry pipe valve. It prevents system air from being replenished faster than the air that is being discharged through an activated sprinkler or test connection.

Retard chamber: A device in variable-pressure systems that is commonly used with pressure switches to prevent false alarms from occurring.

Revision block: A box or table on a drawing that lists the date of every revision to the drawing.

Riser: The vertical supply pipe in a sprinkler system.

Scale: A ratio representing the relationship between drawing dimensions and actual dimensions. Sprinkler shop drawings are often drawn to a scale of ⅛" = 1'.

Sealed valves: Valves that are wired or strapped in the open position.

Section modulus: A measurement of the load capacity of a trapeze hanger or other support.

Seismic bracing: Bracing installed on a piping system to protect it from damage caused by seismic activity, such as an earthquake.

Seismicity: The tendency of a location to be subjected to seismic forces.

Sidewall: Refers to a sprinkler installed so that the deflector discharges most of the water away from the nearby wall in a pattern resembling one quarter of a sphere, with a small portion of the discharge directed at the wall behind the sprinkler.

Single-pole (SP) switch: A switch that can control only one circuit.

Single-throw (ST) switch: A switch that transmits a signal in one direction.

Site plan: A bird's-eye view of a building in small scale.

Small room: A compartment of light hazard occupancy classification having unobstructed construction and a floor area not exceeding 800 ft^2.

Sprig: A pipe that rises vertically and supplies a single sprinkler.

Sprinkler identification number (SIN): A unique manufacturer identification number for every sprinkler type, change in K-factor or shape, deflector characteristic, pressure rating, or thermal sensitivity.

Sprinkler lead-in: The pipe supplying water to the sprinkler system.

Sprinkler riser: The vertical supply pipe in a sprinkler system.

Static pressure: A measurement of the pressure of still water. A column of water that is 1' high will produce a static pressure of 0.433 psi.

Support: A structure or saddle on which a pipe rests that is designed to support the weight of the pipe from below.

Sway brace: An assembly attached to sprinkler system piping that is designed to resist earthquake (seismic) loads in two directions.

Symbol: A written or printed character or sign representing a drawing element.

Takeout: The distance from the center of a pipe to the bottom of the threaded rod in a hanger assembly.

Throttling: Using a valve to regulate flow through a system.

Title block: A boxed area on a drawing that shows the installation location, address, contractor, owner, occupant, and other identifying information as required by NFPA 13.

Trip pressure: The air pressure at which a dry pipe valve operates.

Turbine: A component part of a water motor gong that produces the sound.

Unobstructed construction: Construction where beams, trusses, or other members do not impede heat flow or water distribution in a manner that materially affects the ability of sprinklers to control or suppress a fire.

Upright: Refers to a sprinkler installed so that water distribution is directed upward against a deflector.

Water hammer: A damaging force that occurs when the flow of water is started or stopped, or when the direction of flow has changed; the pressure or shock wave within piping caused by a sudden change in water flow, such as when starting a pump or quickly closing a valve.

Water motor gong: An alarm that is used in wet pipe sprinkler systems.

www.ingramcontent.com/pod-product-compliance
Lightning Source LLC
Jackson TN
JSHW052155020626
103301JS00005B/10
9780137475711